Engineering Materials

This series provides topical information on innovative, structural and functional materials and composites with applications in optical, electrical, mechanical, civil, aeronautical, medical, bio- and nano-engineering. The individual volumes are complete, comprehensive monographs covering the structure, properties, manufacturing process and applications of these materials. This multidisciplinary series is devoted to professionals, students and all those interested in the latest developments in the Materials Science field, that look for a carefully selected collection of high quality review articles on their respective field of expertise.

Indexed at Compendex (2021) and Scopus (2022)

Shiban K. Koul · M. Praveen · Karthikeya G. S.

Microwave Absorbers

Materials Engineering

Springer

Shiban K. Koul
Centre for Applied Research in Electronics
Indian Institute of Technology Delhi
New Delhi, Delhi, India

Karthikeya G. S.
Advanced RF & Wireless Communication
Lab (ARiWu), Department of Electronics
and Communication Engineering
B.M.S. College of Engineering
Bengaluru, KA, India

M. Praveen
Department of Chemistry
PES Institute of Technology
and Management
Shivamogga, KA, India

ISSN 1612-1317 ISSN 1868-1212 (electronic)
Engineering Materials
ISBN 978-981-96-9595-9 ISBN 978-981-96-9596-6 (eBook)
https://doi.org/10.1007/978-981-96-9596-6

This Springer imprint is published by the registered company Springer Nature Singapore Pte Ltd.
The registered company address is: 152 Beach Road, #21-01/04 Gateway East, Singapore 189721, Singapore

To Baby Norah Koul

—Shiban K. Koul

To all struggling Ph.D. research scholars

—M. Praveen

To all the creative deeds of my paternal grandfather, Shri. Gulur Basavaraj

—Karthikeya G. S.

Preface

EMI shielding using nanotechnology leverages the unique properties of nanoscale materials like carbon nanotubes, graphene, and magnetic nanoparticles to create highly effective, lightweight, and flexible shields against electromagnetic interference. These nanomaterials, even at low loadings, form conductive or magnetic networks within polymer matrices, providing superior shielding performance by reflecting and absorbing electromagnetic waves across a broad frequency range. Unlike traditional heavy metal-based shields, nanocomposite shields are lightweight. They can be customized for specific absorption–reflection ratios, making them ideal for modern telecommunications, aerospace, automotive, and wearable electronics applications. Additionally, nanomaterial-based shields often enhance mechanical, thermal, and multifunctional properties, offering a versatile solution that meets the growing demand for efficient EMI shielding in next-generation electronic devices.

This book provides a comprehensive and up-to-date exploration of materials engineering tailored specifically for electromagnetic interference (EMI) shielding. It delves into the latest advancements in material design, highlighting the development of innovative composites and nanomaterials engineered to meet the growing demands for effective EMI shielding in modern technology. Through detailed descriptions of conductive polymers, nanocomposites, and other advanced materials, the book showcases how researchers are pushing the boundaries of traditional shielding solutions, replacing heavy metals with lightweight, flexible, and high-performance alternatives. It also discusses the complex mechanisms by which these materials reflect, absorb, or dissipate electromagnetic waves across broad frequency ranges, making them suitable for aerospace, automotive, telecommunications, and consumer electronics applications. Ultimately, this book aims to serve as both a reference and an inspiration for researchers and engineers dedicated to advancing EMI shielding materials for the rapidly evolving landscape of electronic and wireless technologies.

As mentioned, this book delves into the world of materials engineering, exploring a wide range of nanomaterials and their transformative potential to shape the future of technology. Examining cutting-edge materials such as carbon nanotubes, graphene, metal oxides, and magnetic nanoparticles reveals how these nanoscale innovations pave the way for revolutionary advancements across industries. The book highlights how these materials, with their unique electrical, thermal, and mechanical properties, are engineered to solve critical challenges in fields like EMI shielding, energy storage, biomedicine, and environmental protection. Through detailed discussions on synthesis methods, material properties, and real-world applications, this book not only offers insights into current research but also envisions the possibilities nanotechnology holds for creating more efficient, durable, and sustainable solutions for the future.

The book's structure is designed to provide a systematic and in-depth journey through the field of electromagnetic (EM) shielding materials, from foundational concepts to advanced applications and future directions. Chapter 1 introduces Electromagnetic Absorbers, outlining the principles and importance of materials that can effectively absorb EM waves to prevent interference. Chapter 2, Synthesis of Nanomaterials Used as EM Absorbers, dives into the methods for creating nanoscale materials with tailored properties for optimal absorption. Chapter 3, Microwave Characterization, discusses the techniques used to evaluate materials' effectiveness across microwave frequencies, while Chap. 4, Material Characterization, focuses on analyzing structural, thermal, and electrical properties crucial for EM performance.

Building on this foundation, Chap. 5, Inorganic Materials for Electromagnetic Shielding, explores metals, ceramics, and other inorganic compounds used for EM protection. Chapter 6, Carbonaceous Nanofillers in Polymer Matrix, examines how carbon-based nanomaterials like graphene and carbon nanotubes are integrated into polymer matrices to create lightweight, flexible, and efficient shielding composites. In Chap. 7, Polymers for Electromagnetic Shielding, various polymers' roles are explored, highlighting their advantages for lightweight and versatile shielding applications. Chapter 8, PCB Based Absorbers, addresses printed circuit board (PCB) materials, underscoring their importance in electronic devices where localized EM shielding is critical. Finally, Chap. 9, Future Scopes and Research Gaps, focuses on the current limitations and identifies promising areas for future exploration, aiming to inspire continued research and development in this rapidly evolving field.

New Delhi, India — Shiban K. Koul
Shivamogga, India — M. Praveen
Bengaluru, India — Karthikeya G. S.

Competing Interests The authors have no competing interests to declare that are relevant to the content of this manuscript.

Contents

About the Authors

Shiban K. Koul received his Bachelor of Technology degree in electrical engineering from the Regional Engineering College, Srinagar, in 1977, Master of Technology degree in microwave engineering in 1979, and Ph.D. degree in the same field in 1983 from the Indian Institute of Technology Delhi, India. He is currently Honorary Professor at the Indian Institute of Technology, Delhi. His research interests include RF MEMS, high-frequency wireless communication, microwave engineering, microwave passive and active circuits, device modeling, millimetre and sub-millimetre wave IC design, body area networks, flexible and wearable electronics, medical applications of sub-terahertz waves, EMI/EMC suppression techniques, and reconfigurable microwave circuits including miniaturized antennas.

Shiban has completed 38 major sponsored projects, 52 consultancy projects, and 61 technology development projects. He has authored/co-authored 635 research papers, 24 state-of-the-art books, 4 book chapters, and 2 e-books. He holds 26 patents, 6 copyrights, and one trademark. He has guided 31 Ph.D. theses and more than 120 master's theses.

Shiban is a Life Fellow of IEEE and a Fellow of INAE and IETE. He is the Chief Editor of *IETE Journal of Research* and has earlier served as Associate Editor of the *International Journal of Microwave and Wireless Technologies*, Cambridge University Press. From 2012 to 2014, he served as Distinguished Microwave Lecturer (DML) for IEEE MTT-S. Previously, he served as a Speaker Bureau Lecturer for IEEE MTT-S. In addition to serving as an AdCom member of the IEEE MTT Society from 2010 to 2018, he served as a member of the society's Awards, Nominations and Appointments, and Education Committees. Before this, he served for over a decade as regional coordinator for R-10. He has recently been selected as the Distinguished Lecturer (DL) of the IEEE EMC Society for two years (2024–2025).

Shiban is recipient of numerous awards including Indian National Science Academy (INSA) Young Scientist Award (1986); Top Invention Award (1991) of the National Research Development Council (NRDC) for his contributions to the indigenous development of ferrite phase shifter technology; VASVIK Award (1994) for the development of Ka-band components and phase shifters; Ram Lal Wadhwa Gold Medal (1995) from IETE; Academic Excellence Award (1998) from Indian Government for his pioneering contributions to phase control module development for Rajendra Radar; Shri Om Prakash Bhasin Award (2009) in the field of Electronics and Information Technology; VASVIK Award (2012) for the contributions made to the area of Information, Communication Technology (ICT); Teaching Excellence Award (2012) from IIT Delhi; Award for donations made to the growth of smart materials technology (2012) by Institute of Smart Materials and Systems (ISSS) Bengaluru; M. N. Saha Memorial Award (2013) from IETE, IEEE MTT Society Distinguished Educator Award (2014); Lifetime Achievement Award-2023 from the IETE; India and Lifetime Achievement Award-2024 from Broadcast Society of India.

M. Praveen is a passionate teacher, researcher and application scientist with expertise in industrial chemistry and nanotechnology. He earned his undergraduate degree in Industrial Chemistry from Sahyadri Science College, Shimoga, in 2007, followed by a Master's in Industrial Chemistry (2010) and an M.Tech in Nanoscience and Technology (2012) from Kuvempu University, Karnataka. In 2023, he completed his Ph.D. in Chemistry from Visvesvaraya Technological University (VTU), Belgaum, focusing on polymer composites and electromagnetic interference (EMI) shielding materials.

Dr. Praveen has contributed significantly to both academia and applied research. He served as an Assistant Professor at engineering colleges in Bangalore from 2012 to 2017. He then worked at The Energy and Resources Institute (TERI), Bangalore, contributing to research funded by agencies like the Naval Research Board (NRB), the Department of Science and Technology (DST), and the Council of Scientific and Industrial Research (CSIR). His research focused on nanocomposites for dental applications and flame retardants for polymer composites.

From 2021 to 2024, Dr. Praveen was a Research Associate at MS Ramaiah Institute of Technology, CAMT, centre of excellence lab where he contributed to advanced materials research. He also worked with Ad-Nano Technology Pvt. Ltd. and Atsbia Innovation Pvt. Ltd. in Shimoga, supporting nanomaterials development and commercialization.

Currently, Dr. Praveen is an Assistant Professor in the Department of Chemistry at PES Institute of Technology and Management (PESIT&M), Karnataka. With over 30 + publications, his research interests include nanomaterials, graphene, carbon nanotubes, polymer hybrids, and advanced analytical techniques, reflecting his commitment to innovation in materials science.

Karthikeya G. S. received his undergraduate degree in electronics and communication engineering, with distinction, in the year 2010 from Visvesvaraya Technological University, Belgaum. He received his master's degree, with distinction, in Microwave Engineering from the University of Kerala in 2012. He worked as an assistant professor in affiliated colleges of Visvesvaraya Technological University from 2013 to 2016, where he established the Antenna Architects' lab. He joined the Centre for Applied Research in Electronics (CARE), IIT Delhi, as a Ph.D. student in 2017 and defended his thesis in 2019. He worked as a Project Scientist at the Foundation for Innovation and Technology Transfer (FITT) in association with Synergy Microwave Corporation, USA from 2019 to 2021. From 2021 to 2022, he served as an Assistant Professor at the Centre for Antennas and Radio Frequency Systems (CARFS), Ramaiah Institute of Technology. Since 2023, he has been an Assistant Professor at the Advanced RF and

Wireless Communication Lab (ARiWu), Department of Electronics and Communication, B.M.S College of Engineering, Bengaluru. He has authored or co-authored more than 90 articles in peer-reviewed journals and international conference proceedings. He holds two US and four Indian patents. He has also filed 11 Indian patents. He has co-authored four state-of-the-art books and a book chapter. He also holds a trademark on 5G antenna module design. One of his articles has featured on the cover page of Microwave Journal, an industry-leading magazine. His research interests include multi-octave antenna systems for portable devices, novel material design for antenna applications and shielding material designs. He serves as a research consultant to Space Pulse Techno-solutions Pvt Ltd. He is a senior member of the IEEE-Antenna Propagation Society and a life member of the Antenna Test and Measurement Society. He has been a reviewer of several IEEE and John Wiley journals. He received the Dr. N. Vasudevan Nair Memorial Award (2011), Research Excellence Travel Award (2019) and Dr. Desmond Sim Award for Best Paper in Antenna Design for Mobile Devices (2025).

Abbreviations

0D	Zero Dimensional
2D	Two Dimensional
3D	Three Dimensional
4D	Four Dimensional
AAC	Active Antenna Components
ABS	Acrylonitrile Butadiene Styrene
AC	Alternating Current
AFM	Atomic Force Microscopy
AFSS	Active Frequency Selective Surface
AgNPs	Silver Nanoparticles
Al_2O_3	Aluminum Oxide
APTES	3-Aminopropyltriethoxysilane
ATR	Attenuated Total Reflectance
$BaFe_{12}O_{19}$	Barium Ferrite
BFO	Bismuth Ferrite
BMN	Bismuth Magnesium Niobate
BPs	Bucky Papers
BTE	Butylated Toluene Ether
BTO	Barium Titanate
CB	Carbon Black
CdS	Cadmium Sulfide
CEM	Computational Electromagnetics
CFRP	Carbon Fiber Reinforced Polymer
CMF	Cellulose Microfibril
CMS	Carboxymethyl Starch
CN	Cellulose Nanofibers
CNB	Cellulose Nanobeads
CNF	Carbon Nanofibers
CNT	Carbon Nanotube
$CoFe_2O_4$	Cobalt Ferrite
CQDs	Carbon Quantum Dots

CRP	Carbon reinforced polymer
CS	Chitosan
CTAB	Cetyltrimethylammonium Bromide
CuO	Copper(II) Oxide
CVD	Chemical Vapor Deposition
DA	Dielectric Absorber
DC	Direct Current
DI H_2O	Deionized Water
DP	Degree of Polymerization
DSC	Differential Scanning Calorimetry
DSP	Digital Signal Processing
DUT	Device Under Test
DWCNT	DoubleWalled Carbon Nanotube
EBC	Ethyl Butyl Cellulose
EDS	Energy Dispersive X-Ray Spectroscopy
EHF	Extremely High Frequency
ELF	Extremely Low Frequency
EM	Electromagnetic
EMA	Ethylene Methyl Acrylate
EMC	Electromagnetic Compatibility
EMI	Electromagnetic Interference
EN	Epoxy Nanocomposites
EPS	Expanded Polystyrene
ES	Electro Spun
ESD	Electrostatic Discharge
ETFE	Ethylene Tetrafluoroethylene
FCC	Federal Communications Commission
FDM	Fused Deposition Modeling
Fe_2O_3	Iron(III) Oxide
FEG-SEM	Field Emission Gun Scanning Electron Microscope
FR-4	Flame Retardant
FSS	Frequency Selective Surface
FTIR	Fourier Transform Infrared Spectroscopy
FWHM	Full Width at Half Maximum
GF	Glass Fiber
g-Fe_2O_3	Gamma Iron Oxide
GNP	Graphene Nanoparticles
GNRs	Graphene Nanoribbons
GO	Graphene Oxide
GQDs	Graphene Quantum Dots
HF	High Frequency
HFSS	High-Frequency Structure Simulator
HRL	High-Resolution Lithography
HRTEM	High-Resolution Transmission Electron Microscopy
HT	Hydrothermal

IBD	Ion Beam Deposition
IC	Integrated Circuit
ICP	Inductively Coupled Plasma
IoT	Internet of Things
iPP	Isotactic Polypropylene
IR	Infrared
ITO	Indium Tin Oxide
KBr	Potassium Bromide
LAS	Laser Ablation Synthesis
LC	Liquid Crystal
LDPE	Low-Density Polyethylene
MATLAB	Matrix Laboratory
MC	Methylcellulose
MEMS	Micro Electro Mechanical Systems
MFI	Melt Flow Index
$MgAl_2O_4$	Magnesium Aluminate
MgO	Magnesium Oxide
MI	Magnetic Induction
MIP	Molecularly Imprinted Polymer
MLD	Molecular Layer Deposition
MMW	Millimetre Wave
$MnZnFe_2O_4$	Manganese Zinc Ferrite
MOF	Metal-Organic Framework
MoS_2	Molybdenum Disulfide
MR	Magnetic Resonance
MSA	Microstrip Antenna
MUT	Material Under Test
MWCNTF	Functionalized Multi-Walled Carbon Nanotubes
MWCNTs	Multi-Walled Carbon Nanotubes
MXene	A type of 2D material, typically metal carbide/nitride
NC	Nanocomposites
NCNC	Nitrogen-Doped Carbon Nanocages
NCS	Nanocellulose Structure
NEMS	Nanoelectromechanical Systems
NFC	Near-Field Communication
NiO	Nickel Oxide
NIR	Near-Infrared
$NiZnFe_2O_4$	Nickel Zinc Ferrite
NMP	N-Methyl-2-Pyrrolidone
NMR	Nuclear Magnetic Resonance
NPs	Nanoparticles
NS	Nanospheres
OC	Open Circuit
OM	Optical Microscopy
PAC	Polyamide Composite

PAN	Polyacrylonitrile
PANi	Polyaniline
PBT	Polybutylene Terephthalate
PC	Polycarbonate
PCB	Printed Circuit Board
PCC	Portland Cement Composite
PCE	Power Conversion Efficiency
PDMS	Polydimethylsiloxane
PE	Polyethylene
PEDOT: PSS	Poly(3,4-ethylene dioxythiophene) Polystyrene Sulfonate
PEEK	Polyether Ether Ketone
PEG	Polyethylene Glycol
PEI	Polyetherimide
PI	Polyimide
PL	Photoluminescence
PLC	Programmable Logic Controller
PLGA	Poly (lactic-co-glycolic acid)
PM	Particulate Matter
PMC	Polymer Matrix Composite
PMMA	Polymethyl Methacrylate
PMT	Photomultiplier Tube
PNA	Precision Network Analyzer
PO	Propylene Oxide
PP	Polypropylene
PPy	Polypyrrole
PS	Polystyrene
PSMA	Polystyrene Maleic Anhydride
PSS	Polystyrene Sulfonate
PT	Plasma Treatment
PTFE	Polytetrafluoroethylene
PVAc	Polyvinyl Acetate
PVD	Physical Vapor Deposition
PVDF	Polyvinylidene Fluoride
PXRD	Powder X-Ray Diffraction
QY	Quantum Yield
RCS	Radar Cross-Section
RF	Radio Frequency
RFID	Radio Frequency Identification
rGO	Reduced Graphene Oxide
RIE	Reactive Ion Etching
RL	Reflection Loss
Rx	Receiver
SAC	Selective Area Coating
SAED	Selected Area Electron Diffraction
SAN	Styrene Acrylonitrile

SAR	Specific Absorption Rate
SAW	Surface Acoustic Wave
SAXS	Small Angle X-Ray Scattering
SCS	Solution Combustion Synthesis
SE	Shielding Effectiveness
SE_A	Shielding Effectiveness due to Absorption
SEM	Scanning Electron Microscope
SE_{MR}	Shielding Effectiveness due to Multiple Reflections
SE_R	Shielding Effectiveness due to Reflection
SE_T	Total Shielding Effectiveness
Si_3N_4	Silicon Nitride
SiC	Silicon Carbide
SINR	Signal to Interference Plus Noise Ratio
SiO_2	Silicon Dioxide
SLM	Selective Laser Melting
SM	Shape Memory
SMA	Sub Miniature Version A
SMPS	Switch Mode Power Supply
SNR	Signal-to-Noise Ratio
S-parameters	Scattering Parameters
SPR	Surface Plasmon Resonance
SQUID	Superconducting Quantum Interference Device
$SrFe_{12}O_{19}$	Strontium Ferrite
SS	Stainless Steel
STEM	Scanning Transmission Electron Microscopy
SWCNT	Single-Walled Carbon Nanotube
$\tan \delta_e$	Dielectric Loss Tangent
TCF	Transparent Conductive Film
TCP	Tricalcium Phosphate
TEM	Transmission Electron Microscopy
TGA	Thermogravimetric Analysis
THF	Tetrahydrofuran
TiC	Titanium Carbide
TiN	Titanium Nitride
TiO_2	Titanium Dioxide
TPU	Thermoplastic Polyurethane
TRP	Total Radiated Power
UHF	Ultra High Frequency
UV	Ultraviolet–Visible Spectroscopy
VHF	Very High Frequency
VLSI	Very-Large-Scale Integration
VNA	Vector Network Analyzer
VO_2	Vanadium Dioxide
VOCs	Volatile Organic Compounds
VSM	Vibrating Sample Magnetometer

WC	Tungsten Carbide
WLAN	Wireless Local Area Network
WO_3	Tungsten Trioxide
XRD	X-Ray Diffraction
XRF	X-Ray Fluorescence
YAG	Yttrium Aluminum Garnet
ZnO	Zinc Oxide
$ZrFe_2O_4$	Zirconium Ferrite
ZrO_2	Zirconium Dioxide
α-Fe_2O_3	Alpha Phase Iron(III) Oxide

Chapter 1
Introduction to Electromagnetic Absorbers

Abstract This chapter discusses the importance of Electromagnetic Radiation (EMR) and its effects on biological organisms, followed by a discussion on Electromagnetic Interference (EMI). Popular techniques to mitigate EMI are explored in this chapter. Conventional metal shielding is one of the most straightforward solutions for EMI mitigation. However, metallic shields suffer from drawbacks like lack of flexibility, processability, difficulty in realizing conformal geometries, and lack of immunity to electrostatic charge buildup. On the other hand, microwave absorbers designed with polymer nanocomposites exhibit characteristics such as ease of manufacturing, relatively lower manufacturing costs, low weight, and corrosion resistance. Hence, polymer composite-based absorbers are investigated in this chapter. The design characteristics of microwave-absorbing materials are also explored.

This chapter presents a brief glimpse of microwave absorbers with appropriate examples. Nanomaterials are among the most promising candidates for microwave absorption. This chapter provides an overview of these nanomaterial designs for the X-band. Polymers are also an essential class of lightweight, flexible absorber media. A glimpse of polymer-based designs for high-frequency absorption are included in this chapter. Nanocomposites act as a performance booster to the base polymer for EMI shielding solutions for aerospace and defense applications. Polymers blended with nanocomposites are also discussed in this chapter.

1.1 Introduction

Electromagnetic radiation is one of the most used resources for mankind's benefit. The frequency of radiation dictates the nature and behavior of the respective radiation [1]. The electromagnetic spectrum starts at a shallow frequency band (1 to 300 Hz) and extends up to very high-frequency band (10^{18} Hz) [2, 3]. Electromagnetic spectra could be loosely classified into two broad categories: ionizing and non-ionizing radiation [4]. Figure 1.1 illustrates the electromagnetic spectrum with examples of

S. K. Koul et al., *Microwave Absorbers*, Engineering Materials,
https://doi.org/10.1007/978-981-96-9596-6_1

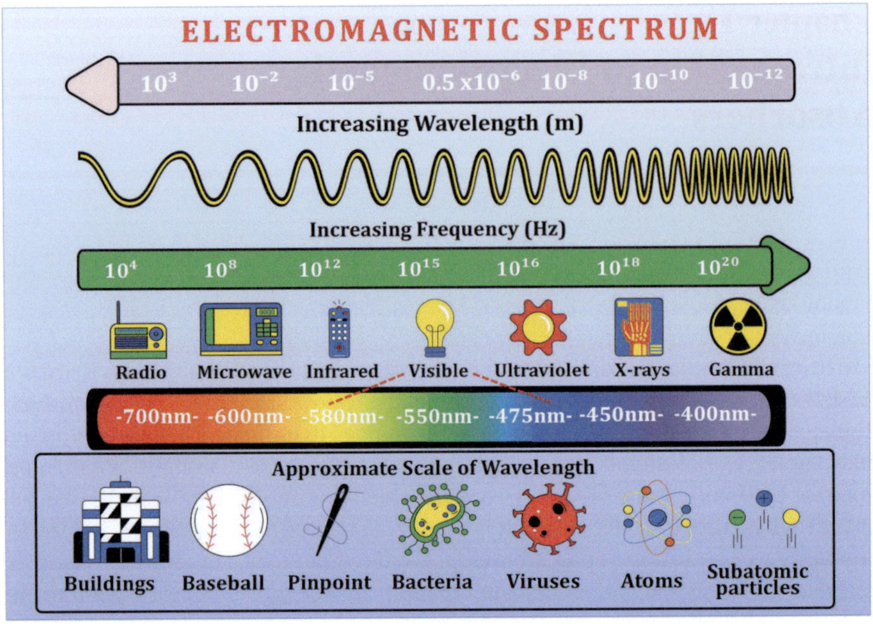

Fig. 1.1 Electromagnetic spectrum illustrating ionizing and non-ionizing radiation

ionizing and non-ionizing radiation classes [5, 6]. Ionizing radiation has enough energy to remove tightly bound electrons from atoms, creating ions. Examples of ionizing radiation include X-rays, Gamma rays, and high-energy Ultraviolet (UV) rays. Prolonged exposure to ionizing radiation can be harmful and might increase the risk of cancer [5, 6].

Non-ionizing radiation, on the other hand, does not have enough energy to ionize atoms or molecules. Examples of non-ionizing radiation include radio waves, microwaves, and visible light. While non-ionizing radiation is frequently encountered in workplaces and homes, prolonged exposure to certain non-ionizing radiation can also harm living organisms [7, 8].

1.2 Biological Effects of Electromagnetic Radiation

Electromagnetic (EM) radiations, a diverse spectrum of energy waves, can interact with living organisms in various ways due to their distinct properties. High-frequency EM waves, such as X-rays and Gamma rays, possess sufficient energy to ionize atoms and molecules, disrupting the delicate balance of biological structures like Deoxyribonucleic Acid (DNA) and potentially leading to mutations, genetic damage, and cancer. Ionizing radiation is, in general, harmful to living organisms, as it can cause cellular damage, tissue damage, and even death at high doses.

EM waves, like visible light, radio, and microwaves, are classified as non-ionizing radiation. These waves do not have enough energy to cause ionization and are considered less harmful to living organisms. However, excessive exposure to certain non-ionizing EM waves can still cause biological effects. For example, ultraviolet (UV) light is a non-ionizing EM wave that can damage living tissue, leading to sunburn, skin aging, and an increased risk of skin cancer [9].

In addition to their potentially harmful effects, EM waves have beneficial applications in various fields. Visible light is essential for photosynthesis in plants, while radio waves and microwaves are used in communication, navigation, and remote sensing technologies. In medicine, X-rays and gamma rays are used for diagnostic imaging and cancer treatment.

The biological aspects of EM radiations are crucial for ensuring the safe use of these waves in various applications while minimizing potential harm to living organisms. Research in this area focuses on understanding the mechanisms of interaction between EM radiation and biological systems, evaluating the risks associated with different types of radiation, and developing strategies to mitigate those risks. This knowledge is essential for protecting human health and preserving the environment.

With the advancements in numerous wireless technologies, the effects of EMR on biological organisms have become increasingly relevant. Figure 1.2 depicts the daily commercial and household radiation sources with corresponding operational frequencies.

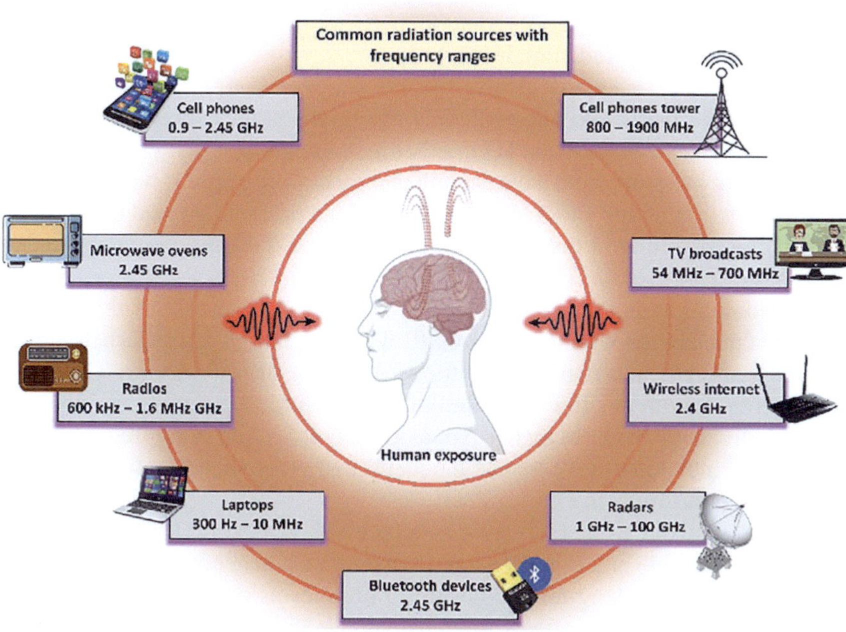

Fig. 1.2 Various types of possible EM radiation on an average human [10]

The ever-increasing exponential growth of many electronic systems results in electromagnetic interference (EMI) across numerous bands within the electromagnetic spectrum. Electromagnetic (EM) radiation can cause various health issues in living organisms, especially when exposed to high-frequency ionizing radiation. Moreover, it can cause the ionization of water molecules within living cells, forming reactive oxygen species (ROS) that can cause oxidative stress and cellular damage [11]. At high doses, ionizing radiation can lead to acute radiation syndrome, characterized by symptoms such as nausea, vomiting, and, ultimately, death [12].

Lower frequency non-ionizing EM waves, such as radio and microwaves, are generally considered less harmful to living organisms [13]. However, excessive exposure to certain non-ionizing EM waves, like radiofrequency radiation from mobile phones and Wireless Fidelity (Wi-Fi), has been debated regarding its potential impact on human health. Some studies suggest possible links to increased cancer risk [13], while others dispute these findings [14].

Electromagnetic waves can interact with human cells in different ways. One of the most common interactions is the transfer of thermal energy from radiation to the tissue exposed. This principle is the basis for medical applications of electromagnetic waves, such as microwave radiation, for cancer treatment [15]. However, electromagnetic waves can also interact with human cells in non-thermal ways, such as by inducing changes in the permeability of the cell membrane or by altering the expression of specific genes. There is ongoing research in this area to better understand the mechanisms by which electromagnetic waves interact with human cells and the potential health effects of these interactions. Additionally, there are regulations to limit exposure to electromagnetic radiation, particularly in industries where workers may be exposed to high radiation levels [16].

Electromagnetic waves, specifically ionizing radiation, can potentially cause DNA mutations. This is particularly true when living organisms are exposed to high radiation levels for an extended period [17]. Ionizing radiation, like X-rays, Gamma rays, and high-energy Cosmic rays, have enough energy to ionize atoms, which can break the bonds between atoms and release electrons. This process can lead to damage to cellular DNA, resulting in the formation of DNA strand breaks, base damage, and other types of DNA lesions [18].

When ionizing radiation damages DNA, it can lead to mutations that alter the sequence of nucleotides within the DNA [19]. These mutations can disturb the normal functioning of cells and may cause various diseases, such as cancer. The risk of cancer increases with the dose and duration of exposure to ionizing radiation, with younger individuals being at higher risk as their cells divide more frequently than those of older individuals [20]. Cell components experience deterioration due to EM irradiation, as illustrated in Fig. 1.3 [21].

DNA mutation due to the EM radiation effect is shown in Fig. 1.4 [22]. DNA will experience nucleotide breakages leading to disturbance in genomic sequence.

Apart from cancer, long-term exposure to high levels of ionizing radiation can cause other harmful health effects, such as damage to the central nervous system, cataracts, and reduced fertility [23]. Therefore, it is crucial to implementing appropriate shielding measures when working with ionizing radiation and adhering to

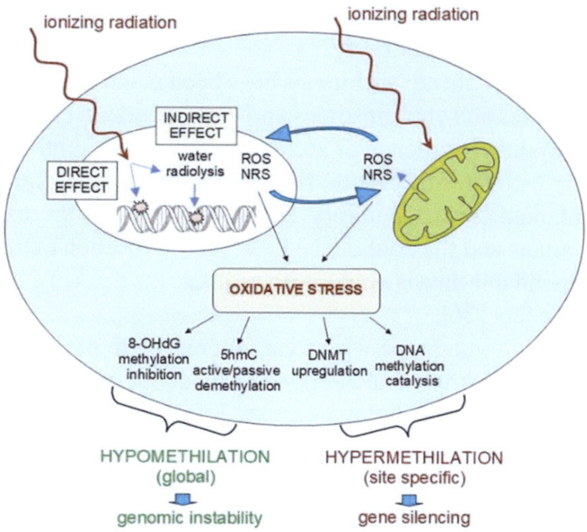

Fig. 1.3 DNA mutation due to EM effect [21]

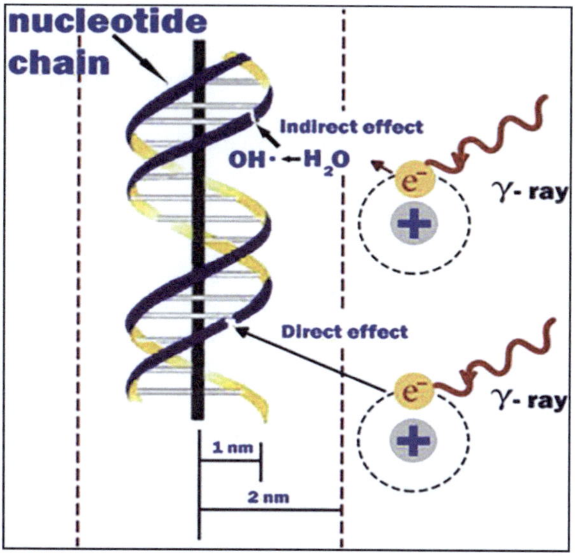

Fig. 1.4 Direct and indirect actions of radiation [22]

exposure guidelines established by regulatory agencies to minimize the risk of DNA mutations and other health issues.

Studies have shown that exposure to EM radiation can have carcinogenic effects on various organs such as bone marrow, breast, thyroid gland, lung, stomach, colon, ovary, and others [24]. However, it is worth noting that these effects have been observed at high-intensity doses and high dose rates. In contrast, studies of populations exposed to low-level radiation, such as those living in regions with elevated natural background radiation, have not provided conclusive evidence of an associated increase in cancer risk [25].

Significant advancements have been made in comprehending the mutation process on genes and chromosomes and their repulsion on genetic disorders. However, there is no direct evidence of increased human heritable effects from radiation exposure. The estimates of genetic risks in humans are primarily based on experimental data obtained from laboratory animals. However, the extrapolation of such results to humans and the confidence to be placed on such extrapolation remain uncertain. No dependable data is available to estimate the risks of complex, multifactorial hereditary disorders [26].

Electromagnetic waves can interact with the brain in different ways. One of the most well-known techniques is magnetic resonance imaging (MRI), which uses strong magnetic fields to create brain images. However, ongoing research is also into how electromagnetic waves from sources such as cell phones and Wi-Fi may affect the brain [27]. Some studies have suggested that long-term exposure to electromagnetic waves from these sources may negatively affect cognitive functions, such as memory and attention [28]. Other studies have indicated that electromagnetic waves may affect the brain‘s electrical activity and alter the release of certain neurotransmitters [29, 30]. However, the evidence on the effects of electromagnetic waves on the brain are not entirely conclusive. It is essential to taking precautions to limit exposure to electromagnetic radiation, particularly for individuals who may be more sensitive to its effects [31, 32].

Electromagnetic waves have the potential to interfere with pacemakers and other cardiac-regulating medical devices. As these devices depend on precise electrical signals to maintain proper heart function, electromagnetic interference may disrupt their performance. However, modern pacemakers are designed with shielding to minimize the risk of interference from electromagnetic waves. Additionally, there are guidelines to limit exposure to electromagnetic radiation for individuals with pacemakers, such as avoiding using specific electronic devices or maintaining a safe distance from sources of electromagnetic radiation. Individuals with pacemakers need to follow these guidelines to minimize the risk of interference and ensure the device’s proper functioning. In Fig. 1.5, an angiogram of a patient having an artificial pacemaker with EM radiation exposure leading to inappropriate ventricular tachycardia is shown.

There is ongoing research to explore the potential link between exposure to electromagnetic waves and breast cancer. Some studies have suggested that long-term exposure to high levels of electromagnetic radiation may increase the risk of breast cancer, particularly in cases where the exposure occurs at a young age. It is essential to take precautions to limit exposure to electromagnetic radiation. Regular breast cancer screening is also essential for early detection and treatment, regardless of any potential link to electromagnetic waves [24].

Evidence suggests that exposure to electromagnetic radiation can cause tension and stress in some individuals, particularly in long-term exposure or exposure to high radiation levels. Some individuals may experience headaches, fatigue, and irritability from exposure to electromagnetic waves, contributing to tension and stress. However,

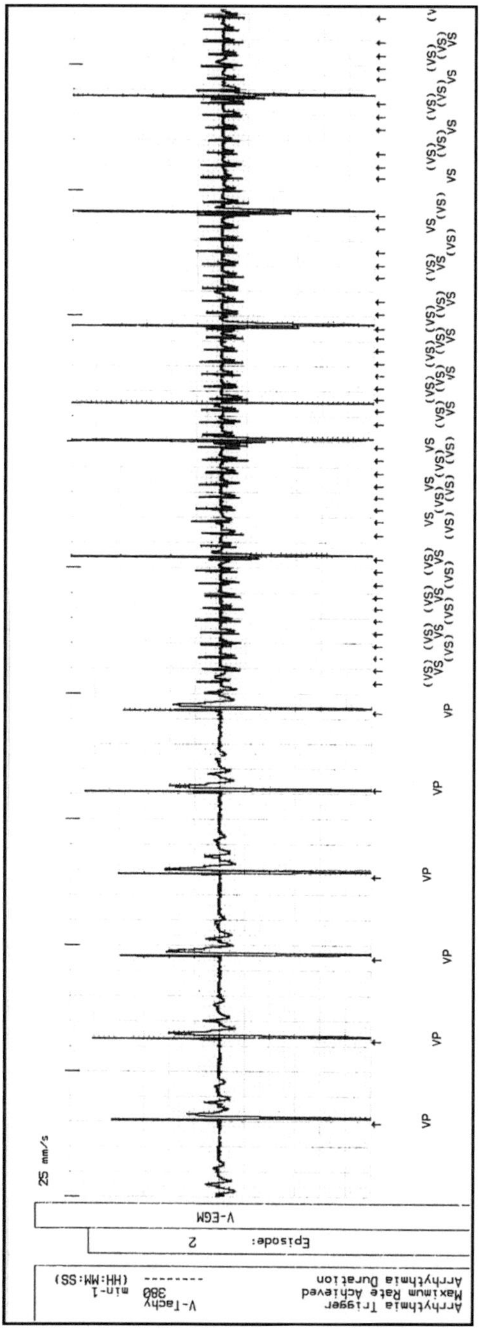

Fig. 1.5 Inappropriate ventricular tachycardia detection [33]

the evidence on the effects of electromagnetic waves on tension and anxiety is not entirely conclusive. It's essential to taking precautions to limit exposure to electromagnetic radiation and seek medical attention if symptoms of tension or stress persist [34].

There is some evidence to suggest that exposure to electromagnetic waves can have adverse effects on reproductive organs. Long-term exposure to high levels of radiation can potentially cause damage to reproductive cells, which might lead to fertility issues or an increased risk of birth defects [35]. Additionally, exposure to electromagnetic waves can cause physiological stress, disrupting hormonal balance and affecting reproductive function. It's essential to limiting exposure to electromagnetic radiation, particularly for individuals more sensitive to its effects, such as pregnant women [36].

There is some evidence to suggest that electromagnetic waves can have adverse effects on honeybees and birds [37]. For example, studies have found that exposure to electromagnetic radiation from sources such as power lines and cell phone towers can disrupt the navigation systems of honeybees and birds, which rely on Earth's magnetic field for navigation [37]. Additionally, exposure to electromagnetic waves can cause physiological stress in honeybees and birds, weakening their immune systems and making them more susceptible to diseases. However, limiting electromagnetic radiation, particularly in sensitive habitats, is essential. Figure 1.6 shows the effects of elevated electromagnetic radiation (RF-EMF) on honeybee colony decrement.

Exposure to electromagnetic waves can potentially cause habitat loss, particularly in areas where large structures such as power lines and cell phone towers are built. These structures can disrupt animal migration patterns and breeding behaviors, leading to population decline or habitat loss. Additionally, exposure to electromagnetic radiation can negatively affect plant growth, indirectly affecting wildlife that relies on these plants for food and shelter.

It is essential to taking precautions when building large structures in areas important for wildlife and considering the potential effects of electromagnetic radiation

Fig. 1.6 Experimental colony setup showing placement of mobile phones and power density meter [38]

Fig. 1.7 X-ray diffractometer with lead-coated glass door

on the environment [30, 39]. As mentioned above, materials with specific absorption properties can be used as absorbers to mitigate the potential health risks of EM radiation. These absorbers effectively attenuate EM waves by converting them into heat or other forms of energy. Common absorbing materials include carbon-based materials and composite materials.

Absorbers are widely used in various applications to protect human health and the environment from the harmful effects of EM radiation. For example, EM absorbers are employed in shielding devices [40].

In the telecommunications industry, absorbers are used in cell phone cases and base stations to reduce exposure to radiofrequency radiation. Furthermore, absorbers are utilized in radar systems to minimize interference with other electronic devices and to protect against the potential hazards of radar-induced heating [41]. Figure 1.7 shows the X-ray diffractometer with lead-coated glass to protect the operator from harmful exposure to the X-rays.

1.3 Need for EMI Shielding

The increasing need for electromagnetic interference (EMI) shields has become essential for industries like airlines, nuclear reactors, transformers, defense, control systems, etc. Polymeric EMI shielding materials are gaining popularity due to their ease of processing, less weight, immunity to corrosion, and affordable fabrication methods [42]. Typically, conductive fillers like metallic powders, metal flakes, metal nanowires, carbon black, graphite [43], graphene [44], and carbon nanotubes [45] are mixed into the polymeric matrix to create an electrically absorptive composite for EMI shielding applications. Apart from polymeric EMI shields, there are other materials like biodegradable EMI shields, EMI shields based on textile materials, cement-based EMI shields, EMI shields that can be used at high temperatures, EMI shields based on metallic alloys, ceramic EMI shields, etc. that are being researched and developed by scientists and researchers in this field [46, 47]. Figure 1.8a shows the image of copper metal nanoparticles, and Fig. 1.8b is the SEM image of the copper metal nanoparticles under an electron microscope, which could be used as EMI shields.

Some common uses of microwave absorbers include

- **Reducing electromagnetic interference (EMI)**: Microwave absorbers can reduce the level of EMI generated by electronic devices. They are often used in electronic enclosures and other devices to reduce the amount of EMI emitted [46].
- **Improving antenna performance**: Microwave absorbers can improve antenna performance by reducing reflections and standing waves. They are often used in antenna design to improve the efficiency and bandwidth of the antenna [49]. Figure 1.9 shows a circular antenna integrated with an EM absorber. The absorber improves antenna performance by reducing backscatter, and improving signal transmission and reception.

- **Reducing radar signature**: Microwave absorbers can reduce the radar signature of military vehicles and structures. They are often used in stealth technology to reduce the detectability of radar signals [47]. Figure 1.10a shows an image of the metamaterial coating on the structure; the simulated and measured reflection reduction under normal incidence for 90° dihedral meta-surface is shown in Fig. 1.10b. These types of absorbers are covered in Chap. 7.

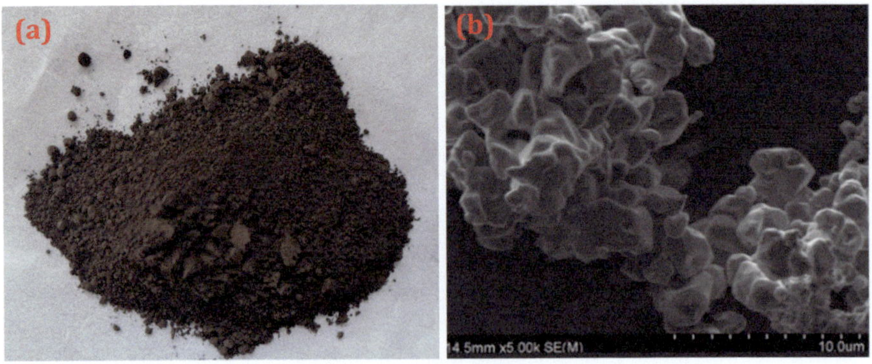

Fig. 1.8 **a** Metal nanoparticles powder; **b** SEM images of Cu nanoparticles [48]

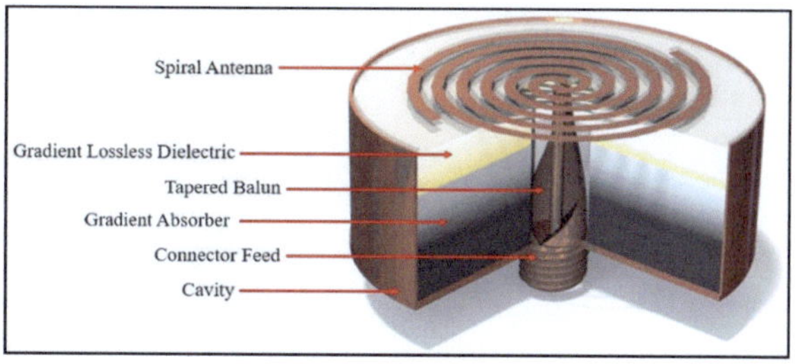

Fig. 1.9 Illustration of the proposed additively manufactured absorber-loaded cavity-backed spiral antenna [50]

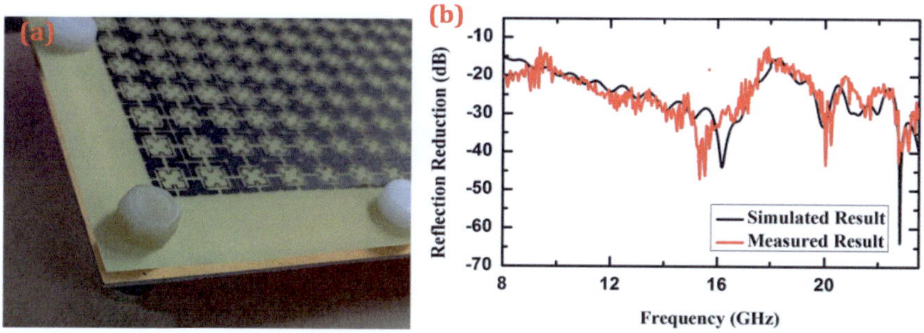

Fig. 1.10 **a** Photograph of the fabricated sample with metamaterial coating; **b** the simulated and measured reflection reduction [51]

- **Medical applications**: Microwave absorbers can be used in medical applications such as hyperthermia therapy, which selectively heats cancer cells using microwave radiation [52].

Overall, microwave absorbers play an essential role in various applications where controlling or eliminating unwanted electromagnetic radiation is necessary.

1.4 Popular Techniques to Mitigate EMI

EMI originates due to the poor design of wireless systems. The transmitters within the wireless communication systems do not strictly radiate the intended or designed bands. The harmonics of the carrier frequency are radiated unintentionally in most wireless transmitters. These unintended signal parts would create undesired effects on the wireless devices in close electrical proximity. These waves might also affect human tissue depending on the radiating source.

Various shielding techniques have been developed to mitigate the impact of Electromagnetic (EM) radiation on living organisms and sensitive equipment. Some

Fig. 1.11 Polymer composite cap for EMI shielding in X- and Ku-bands

of these methods include shielding materials, gaskets, and scattering techniques. Figure 1.11 shows a polymer composite enclosure sample operating in the X- and Ku-bands used for EMI shielding.

1.4.1 Shielding Materials

These are materials designed to absorb, reflect, or block EM radiation. Common shielding materials include metals, such as Copper and Aluminum, which reflect EM waves, and ferromagnetic materials, which can deflect or absorb EM energy due to their magnetic properties. Specialized shielding materials, like metal mesh or composites, can also optimize performance and reduce weight.

1.4.2 Gaskets

Shielding gaskets are seals made from conductive or magnetic materials that prevent EM radiation from leaking through gaps or joints in enclosures, equipment, or structures. Metal gaskets, such as those made from Copper, Aluminum, and Stainless Steel, are often used to create an effective barrier against EM radiation. Shielding gaskets can be designed in various shapes and sizes to fit specific applications and provide a reliable seal. A nanocomposite rubber-based gasket generally used for filling the gaps of shielded room doors and windows is shown in Fig. 1.12.

Fig. 1.12 EMI shielding gasket with conducting carbon

1.4.3 Scattering Techniques

These involve redirecting or dispersing EM radiation to reduce its intensity at a specific orientation. Reflective surfaces, such as mirrors or coated surfaces, can redirect radiation, while diffusive materials and surfaces can scatter EM waves to dissipate their energy over a wider area. This approach can minimize radiation exposure at critical points or sensitive areas while allowing it to be safely redirected or dissipated [53].

In addition to these techniques, combining multiple shielding techniques, such as using a metal gasket with a scattering surface, can improve performance and provide a more effective solution. Furthermore, incorporating innovative materials with tunable EM properties, such as metamaterials or liquid crystals, could enable adaptive shielding systems that automatically adjust to changing EM environments [54].

1.5 Classification of EMI Shielding Materials

Electromagnetic interference (EMI) shielding materials can be classified based on various criteria, such as composition, form, and shielding mechanism. Here's a classification based on these factors [55].

Fig. 1.13 Polyaniline powder, a conducting polymer

1.5.1 Composition

- **Metallic materials**: These materials are typically metals or metal alloys with high electrical conductivity, such as Copper, Aluminum, Nickel, and Stainless Steel. They provide excellent shielding effectiveness due to their ability to reflect electromagnetic waves.
- **Conductive polymers**: Polymers infused with conductive fillers like carbon nanotubes, graphene, or metal particles offer flexibility and lightweight properties while providing effective shielding. Figure 1.13 shows an example of a conducting polymer.
- **Metalized films and foils**: Thin films or foils coated with a layer of metal, such as Aluminum or Copper, are often used as tapes or sheets that can be applied to surfaces or enclosures to provide shielding.
- **Form**: Solid materials in sheets or panels, such as metal sheets, conductive fabrics, and composite materials are often used to construct enclosures or shields around electronic devices or rooms.
- **Tapes and gaskets**: Thin, flexible materials like conductive tapes or gaskets that can be applied between mating surfaces to seal against EMI.
- **Coatings and paints**: Liquid or spray-on coatings containing conductive particles or metallic flakes that can be applied onto surfaces to provide EMI shielding.

1.5.2 Mechanism of Shielding

- **Reflection**: Materials that reflect electromagnetic waves away from sensitive components, such as metallic enclosures or conductive coatings.
- **Absorption**: Materials that absorb electromagnetic energy and convert it into heat, such as ferrites or carbon-loaded polymers. Figure 1.14 shows the resistive film used for electronic applications such as pressure sensors.

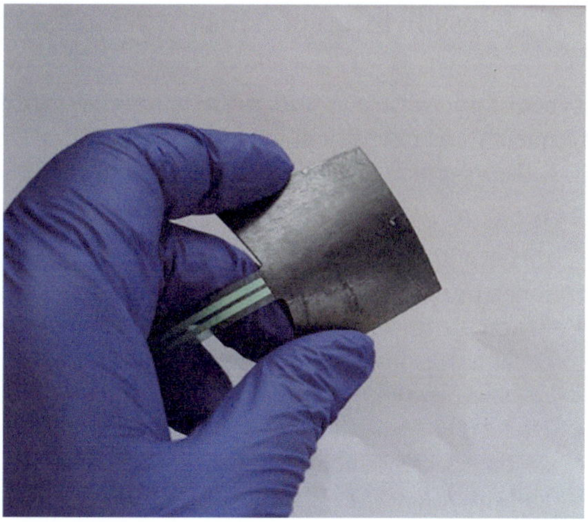

Fig. 1.14 Photograph of sample resistive film

1.5.3 Frequency Range

- **Low-frequency shielding materials**: These materials effectively block electromagnetic interference in the lower frequency range, typically below 1 MHz. They may include magnetic alloys or conductive foams.
- **High-frequency shielding materials**: These are designed to block electromagnetic interference in the higher frequency range, including radio frequencies (RF) and microwave frequencies. Examples include metal meshes, conductive fabrics, and metalized films.

1.5.4 Application-Specific

- Aerospace: Materials specifically designed to meet the stringent requirements of aerospace applications, such as lightweight composites with high shielding effectiveness.
- Automotive: Materials for automotive electronics, such as conductive coatings for automotive enclosures or cables.
- Consumer electronics: EMI shielding materials for consumer electronic devices, including smartphones, tablets, and laptops, often focus on lightweight and aesthetic considerations.

By understanding these classifications, engineers can select the most appropriate EMI shielding materials for their specific application, considering factors such as frequency of operation, form factor, shielding effectiveness, and cost.

1.6 Classification of Microwave Absorbers

Types of microwave absorbers are tailored to specific requirements, considering the frequency and intensity of the microwaves they need to absorb. Common varieties of absorbers include resistive absorbers, magnetic absorbers, and dielectric absorbers:

- **Resistive absorbers**: Typically comprising layers of a conductor, such as carbon, mixed with a binder, these absorbers are relatively thin but boast a broad frequency range [56].
- **Magnetic absorbers**: Made of ferrite or other magnetic materials, these absorbers excel at absorbing lower frequency microwaves and are generally thicker than resistive absorbers [57].
- **Dielectric absorbers**: Constructed from specialized rubber or materials, these absorbers can effectively absorb microwaves depending on the specific material and structure [58].

Microwave absorbers are engineered to possess impedance characteristics closely resembling free space. This enables them to effectively absorb incoming microwave energy, converting it into heat while reducing reflections to a minimum. There is a growing need for microwave absorbers in various applications, including telecommunications, radar systems, and electronic devices [59].

One common type of microwave absorber is made up of magnetic materials, such as Iron or Ferrites, which can absorb electromagnetic radiation by converting it into heat energy [60–62]. Another type of microwave absorber comprises carbon-based materials, such as carbon black or carbon nanotubes, which can absorb electromagnetic radiation through electrical and magnetic interactions [63]. Other microwave absorbers, such as polymers or ceramics, have also been studied for their shielding effectiveness in the microwave band. These materials have the potential to be used in a variety of applications, particularly in the aerospace and defense industries. Ongoing research efforts are focused on improving the performance of microwave absorbers and exploring new materials and synthesis methods for microwave-absorbing applications. Figure 1.15 shows the pyramidal EM absorbers typically used inside the anechoic chambers.

Microwave absorbers can also be classified as bio-derived and biodegradable materials. Bio-derived and biodegradable materials have been studied for their potential use in EMI shielding applications. These materials are typically made of renewable resources and can be designed to degrade naturally, making them environmentally friendly. One common bio-derived material used in EMI shielding applications is cellulose-based composites, which can be engineered to have specific electrical and mechanical properties. For example, a Graphene Oxide/cellulose aerogels nanocomposite material has been shown to have good EMI shielding effectiveness, and a favorable EMI shielding capacity (SE_{Total} = 58.4 dB) was achieved. This work expands the application scopes of both cellulose and GO-based materials [64].

Another study investigated using bio-derived composite material of soy protein isolate and graphene oxide for EMI shielding applications. The study found that

Fig. 1.15 Broadband pyramidal microwave absorbers are used in anechoic chambers

adding Graphene Oxide to the Soy protein isolate matrix improved the EMI shielding effectiveness of the material. Bio-derived materials such as Chitosan and Lignin have also been studied for their EMI shielding properties. These materials have the potential to be used in a variety of EMI shielding applications, particularly in the packaging and biomedical industries. Ongoing research efforts are focused on improving the performance of bio-derived and biodegradable materials for EMI shielding applications [65–67]. Figure 1.16 shows nanocellulose powder for EMI shielding application.

The utilization of epoxidized vegetable oil as a new biodegradable polymer has gained attention due to its low cost and environment-friendly nature. Biopolymers offer a renewable, environment-friendly, and economical alternative to traditional polymers, and their flexibility in processing parameters has led to their use in

Fig. 1.16 Nanocellulose powder for EMI shielding applications

Fig. 1.17 Starch-based polymer composite sheet

various applications. Conductive materials derived from biopolymers have applications in robotics, medical imaging, sensitive membranes, actuators, visual displays, electronic wiring and shielding, and battery components [68].

Bio-based composites have become the material of choice in every field of human life, ranging from daily packaging materials to automobile and aerospace applications. Many biopolymers exhibit piezoelectricity due to the internal rotation of polar atomic groups associated with asymmetric carbon atoms, which makes them suitable for various applications. Several naturally derived and biodegradable polymeric materials have recently been used as efficient EMI shields [42, 69]. A starch-based polymer nanocomposite sheet is shown in Fig. 1.17.

1.6.1 Thermoplastic Composite Materials

Electrically conductive composite materials that rely on a thermoplastic matrix are essential functional materials utilized in the electronics industry for purposes such as EMI shielding, electrostatic discharge (ESD) protection sensors, and more. Incorporating carbon-based conductive fillers into insulating thermoplastic polymers can enhance these composites' conductive, EMI, and thermal properties. Currently, the focus on developing new and efficient EMI shielding materials is mainly directed toward polymeric composites due to the drawbacks of conventional metallic EMI shields, including high cost, difficulty in processing, weight, and the possibility of corrosion [70].

Recent research proposes the use of ZrO_2-coated Graphene Oxide (GO)/Styrene-Ethylene-Butylene-Styrene (SEBS)-based nanocomposites for EMI shielding purposes, as reported by Zhao et al. [71]. An Electromagnetic Interference-Shielding Effectiveness (EMI-SE) of approximately 37.9 dB was observed for SEBS/ZrO_2-coated GO nanocomposite in the X-band region. This increase in EMI-SE was attributed to the large available surface area, moderate electrical conductivity property of GO, and high permittivity of ZrO_2, leading to the formation of good conductive networks in the SEBS matrix. This conductive network interacts with the EM waves and improves EMI shielding. Moreover, Ye et al. [72] employed a nano-assembly freeze extraction route. To strengthen conductive graphene/nano-Cu/crosslinking polyurethane (CPU), foam is useful. It was observed that an EMI-SE > 55 dB in the frequency range of 1–9 GHz was demonstrated on the interface, suggesting that thermoplastic composites have great potential to be used as lightweight, highly conductive, anti-penetrative, and stable EMI shielding materials [73].

1.6.2 Thermosetting Polymers for EMI Shielding

Thermosetting polymers have been studied for their potential use in EMI shielding applications. These materials are typically cured under high temperatures and pressure to form a rigid, cross-linked structure, resulting in a composite material with unique electrical and mechanical properties favorable for EMI shielding. One common type of thermosetting polymer used in EMI shielding applications is epoxy, which has good electrical insulation properties and can be easily modified to have high electrical conductivity. For example, an epoxy composite material of multi-walled carbon nanotubes (MWCNTs) has been shown to have excellent EMI shielding effectiveness [44, 45, 60, 74].

Another study investigated using an epoxy composite material of Silver nanowires for EMI shielding applications. The study found that adding Silver nanowires to the epoxy matrix improved the EMI shielding effectiveness of the material. Other thermosetting polymers, such as phenolic resin, polyurethane, and polyester, have been studied for their EMI shielding properties. These materials have the potential to be used in a variety of EMI shielding applications, particularly in the aerospace and automotive industries. Ongoing research efforts are focused on improving the performance of thermosetting polymers for EMI shielding applications, as well as exploring new materials and synthesis methods [75]. Scanning electron microscope (SEM) images of silver nanowires are shown in Fig. 1.18.

Thermosetting polymers are widely used in civilian and defense applications due to their advantages, such as easy processing, good mechanical properties, and significantly lower coefficients of thermal expansion. These materials can also withstand temperatures up to ~ 350–400 °C, making them highly suitable for EMI shielding applications. Epoxy matrices are commonly used in the aerospace industry. However, some thermoset drawbacks include high dipole activity and hydrogen bonding, which may lead to high $\tan\delta$ values. These drawbacks can be overcome by using special classes of polyester resins. Recent thermosets, including Polyimides and Cyanate Esters, can withstand temperatures close to 400 °C, making them attractive for high-temperature applications [77].

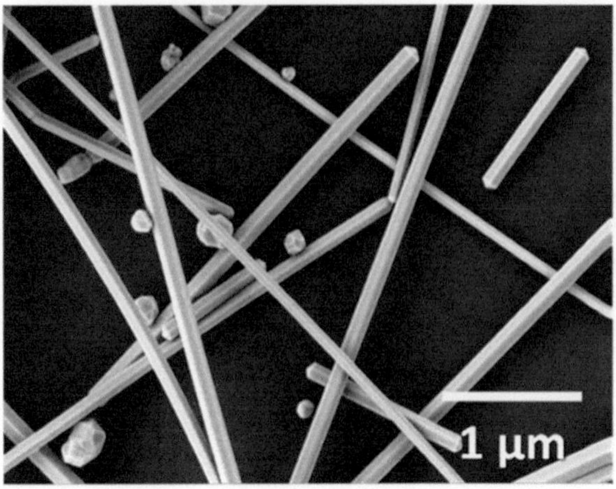

Fig. 1.18 SEM image of silver nanowires [76]

- **Metal-embedded matrices**: Metal-embedded matrices have been widely studied for their potential use in EMI shielding. These materials typically comprise a polymer or ceramic matrix embedded with metal particles or fibers. The metal particles or fibers provide a conductive pathway for electromagnetic waves, resulting in a composite material with high EMI shielding effectiveness. For example, a metal-embedded epoxy composite material made of nickel-coated carbon fibers has been shown to have excellent EMI shielding effectiveness. Another study investigated using a metal-embedded ceramic composite material of Silicon carbide and Aluminum for EMI shielding applications [78]. The study found that adding Aluminum to the ceramic matrix improved the EMI shielding effectiveness of the material. Other metal-embedded matrices, such as those made up of copper or silver, have been studied for their EMI shielding properties. These materials have the potential to be used in a variety of EMI shielding applications, particularly in the electronics industry. Ongoing research efforts are focused on improving the performance of metal-embedded matrices for EMI shielding applications and exploring new materials and synthesis methods [79, 80].

1.6.3 Metal-Based EMI Shields

These are still widely used despite drawbacks such as high density, being prone to corrosion, and high cost. This is because metals such as copper, iron, silver, and nickel offer superior conductivity and mechanical strength compared to polymeric and other materials commonly used in EMI shielding applications. Thin metal foils of copper, silver, etc. are widely embedded into polymeric matrices with suitable grounding in electrical appliances. Alternatively, metal particles can also be embedded into

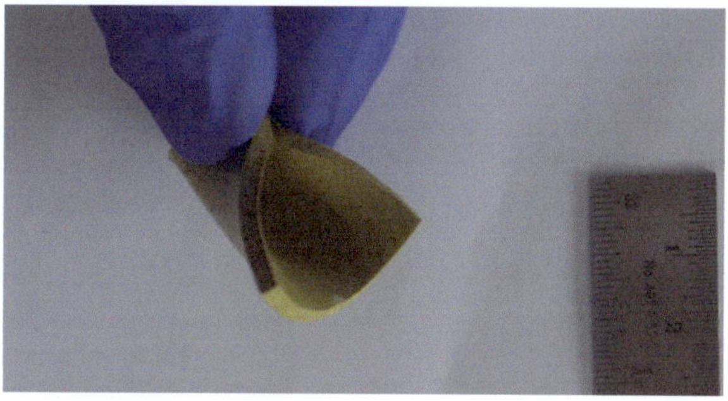

Fig. 1.19 Image of elastomer-based radiation shielding material

polymeric or other matrices. The EMI-SE of metal-based materials depends on the metal used, the thickness of the foil or particles embedded, etc. [81].

1.6.4 Elastomer-Based EMI Shields

Rubber-based materials, or elastomers, are suitable candidates for creating efficient EMI shields. With the onset of nanotechnology, several fillers have been available that can significantly improve the characteristics of rubber matrix at minimal loading, making it appropriate for various uses, including EMI shielding. The benefit of elastomer-based EMI shields is that they are the best materials for EMI shielding applications due to the characteristics of nanofillers, ease of manufacturing, and rubber‘s elasticity. It should be emphasized that most elastomers are transparent to electromagnetic waves because they are made of dielectric materials. Conductive materials such as metals, metallic nanoparticles, and carbon nanomaterials (graphene, CNT, carbon fiber, etc.) are added to make elastomers more conducive and effective EMI shielding materials. The type of conductive filler (carbon-based or noncarbon-based) being employed is the primary determinant of research on elastomer-based EMI shields [82, 83]. An image of elastomer-based radiation shielding materials is shown in Fig. 1.19.

1.6.5 EMI Shields Based on Polymeric Blends

Researchers frequently blend polymers to fine-tune the qualities of the final product because using a single polymer may not produce the required results in some specific applications. This process is known as polymer blending. Adding two distinct types of polymers in various quantities will disclose two different morphologies. The final properties of the polymer blends are determined by the additive’s dispersion in one of the phases or at the interphase. The localization of the additive or filler must occur in one of the polymer‘s phases to produce customized qualities for a polymer mix. It

Fig. 1.20 Flexible polymer composite

is possible to achieve high conductivity for polymer blends at very low percolation thresholds [84].

EMI shields based on polymeric blends can effectively protect electronic devices from EMI. These shields combine conductive fillers with a polymer matrix, such as metal particles or carbon nanotubes. The resulting mixture is then molded into the desired shape and smeared onto the surface of the electronic device. Polymeric blends offer several advantages over traditional metal shields, including lower weight, flexibility, and ease of processing [42, 85]. The flexible polymer composite sheet is shown in Fig. 1.20.

1.6.6 Nanomaterials with Potential EMI Shielding Properties

Nanomaterials show unexpectedly better qualities than their bulk equivalents, including electrical, magnetic, and optical characteristics. Because of these features, various nanomaterials have been added to polymeric, ceramic, and other matrices—such as metal, metal oxide, and carbon nanomaterials—to produce materials with improved qualities and uses. Effective EMI shields have been made using a number of these nanomaterials as conductive fillers. The size, morphology, form, and other characteristics of the nanomaterials utilized as the filler affect how these systems shield electromagnetic interference [73, 86, 87].

1.6.7 Carbon-Containing Materials

Carbon black has been widely used as the main reinforcement or filler in the rubber and plastic industries. The graphitic nature of carbon black gives it semiconductor properties, making it an ideal choice for antistatic and conductive applications when added to rubbers or plastics. Carbon nanotubes (CNTs) have been researched extensively due to their unique physical and chemical properties since their discovery by Ijiima [88]. CNT-based composites have been utilized in various applications in the past decade, owing to the benefits of being lightweight, corrosion- and chemical-resistant, and having high electrical conductivity, aspect ratio, and mechanical strength. Carbon nanomaterials such as carbon fibers, graphene, and quantum dots have also been studied for their potential as fillers for EMI shielding applications [70, 89–91].

1.6.8 Hybrid Polymer Nanocomposites with EMI Shielding Applications

These materials typically consist of a polymer matrix, such as polyurethane or epoxy, and filler material, such as metal nanoparticles or carbon nanotubes. Combining these materials can produce a composite material with unique electrical and mechanical properties favorable for EMI shielding. For example, a hybrid polymer nanocomposite made up of polycarbonate and silver nanoparticles has been shown to have excellent EMI shielding effectiveness, good thermal stability, and mechanical properties [92].

Another hybrid polymer nanocomposite comprises epoxy and carbon nanotubes (CNTs), which have high electrical conductivity and can be dispersed evenly throughout the polymer matrix to create a composite material with high EMI shielding effectiveness. Other hybrid polymer nanocomposites, such as polymer/carbon black/magnetic particle composites, have also been studied for their EMI shielding properties. Ongoing research efforts are focused on improving the performance of hybrid polymer nanocomposites for EMI shielding applications and exploring new materials and synthesis methods [93].

According to researchers, the critical requirements for EMI shielding materials for next-generation electronic devices include good structural strength and a broad bandwidth operation. To meet these needs, a combination of carbon with other organic or inorganic nanofillers must be encapsulated into different matrices. As a result, there has been a growing interest in EMI shielding of hybrid filler-reinforced composites over the past few years [94].

1.6.9 Conducting Polymer Composites

They have been extensively studied for their potential use in EMI shielding. These materials are lightweight, flexible, and can be engineered to have specific electrical and mechanical properties. One common type of conducting polymer composite comprises polyaniline (PANI) and a polymer matrix, such as polyethene [45] or polypropylene. PANI is an intrinsically conducting polymer that can be easily synthesized and has suitable EMI shielding properties. The composite material exhibits high electrical conductivity and EMI shielding effectiveness when incorporated into a polymer matrix. Another conducting polymer composite comprises carbon nanotubes (CNTs) and a polymer matrix [45]. CNTs have excellent electrical conductivity and can be dispersed evenly throughout a polymer matrix to create a composite material with high EMI shielding effectiveness. Other conducting polymers, such as Polypyrrole (PPy) and Polythiophene (PT), have also been studied for their EMI shielding properties. These materials have the potential to be used in a variety of EMI shielding applications, particularly in the electronics industry [95]. Figure 1.21a is a polymer composite containing MWCNT, and Fig. 1.21b shows the SEM image of MWCNT.

Conducting polymers [95], such as Polyaniline and Polypyrrole, are characterized by their conjugated double bonds on their backbone, which result in excellent electrical and optical properties. These properties make them useful in various applications, including batteries, light-emitting diodes, solar cells, and EMI shielding materials. However, these polymers have drawbacks, such as poor solubility, complex processing, and low stability. To overcome these issues, conducting polymers can be modified through composites, hybrids, co-polymers, blends, and other methods to obtain the desired properties [42, 95].

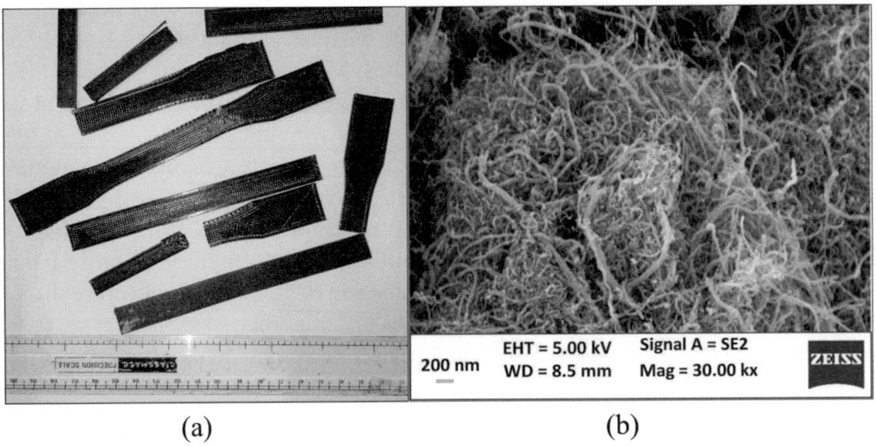

Fig. 1.21 **a** Polymer composite containing MWCNT for EMI shielding applications; **b** SEM image of MWCNT

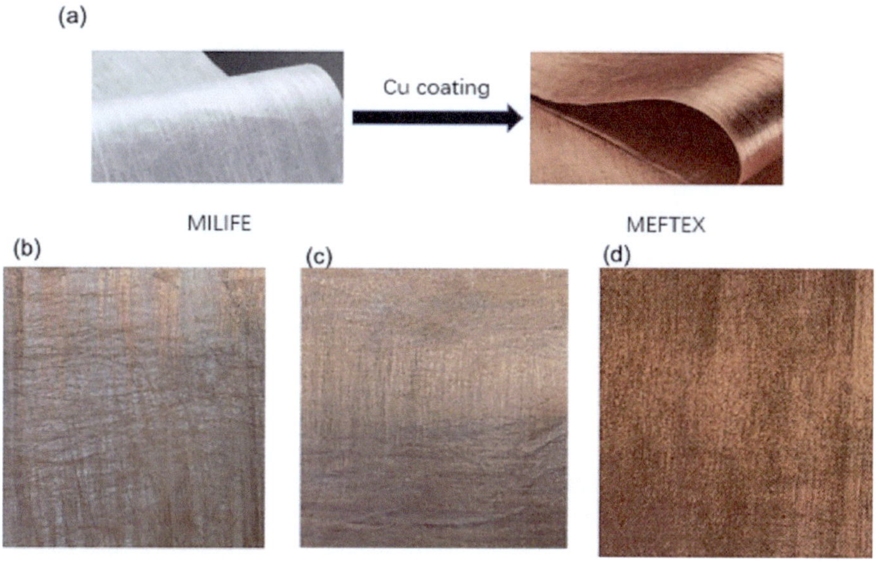

Fig. 1.22 Copper-coated MEFTEX fabric cloth [96]

1.6.10 Textile-Based Materials

Fabrics with EMI shielding capabilities can efficiently shield EM radiation from mobile phones. The fabric structure, metal coverage, and metal content can be manipulated to yield fabrics with varying EMI shielding effectiveness [95]. Figure 1.22 shows an image of an EMI shielding textile called MEFTEX.

1.6.11 Porous Materials for EMI Shielding

Porous materials have also been explored for their potential use in EMI shielding. Porous materials have a high surface area and can be engineered to have specific pore sizes and structures, making them favorable for EMI shielding applications. Metal-organic frameworks (MOFs) are a type of porous material that has been studied for their EMI shielding properties. Metal-organic frameworks (MOFs) comprise metal ions or clusters linked by organic ligands, resulting in a highly porous structure [97]. Qingganget et al. [98] reported that MOFs can be used as a coating material to enhance the EMI shielding effectiveness of Polyurethane (PU) foam. Other porous materials, such as carbon-based materials, aerogels, and zeolites, have also been studied for their EMI shielding properties. These materials can be fabricated into various shapes and sizes, making them suitable for multiple EMI shielding applications [99].

Fig. 1.23 Yttria-stabilized zirconium ceramics nanoparticle powder

1.6.12 Ceramic Materials for EMI Shielding

Ceramic materials are another class of materials explored for their potential use in EMI shielding. Ceramic materials have high thermal and mechanical stability, making them suitable for demanding applications. Additionally, they can be engineered to have specific electrical properties, such as high dielectric constant or low loss tangent, which are essential for EMI shielding. Alumina, for example, has been studied for its EMI shielding properties due to its high dielectric constant and low loss tangent [100].

In another study, a composite material was prepared by incorporating conductive particles into a ceramic matrix, resulting in high electrical conductivity and EMI shielding effectiveness [101]. Ceramic materials such as Ferrites, Titanium Carbide, and Silicon Carbide have also been studied for their EMI shielding properties. These materials can be used in various EMI shielding applications, particularly in high-temperature environments or where high mechanical strength is required. Ceramic material is used to coat metal surfaces and work in shilling material on high-temperature surfaces, as shown in Fig. 1.23.

1.6.13 Concrete for Radiation Shielding

Nanomaterials have been extensively studied for their potential use in concrete as radiation shielding materials. By incorporating nanomaterials into concrete, it is possible to significantly increase the material's resistance to radiation while maintaining its structural integrity. One of the most promising nanomaterials for this application is Graphene Oxide (GO). Graphene Oxide is a form of graphene that has been oxidized, which makes it more soluble in water and allows it to be easily incorporated into concrete. Studies have shown that adding Graphene Oxide to concrete can significantly reduce the radiation that passes through the material [102]. Other nanomaterials studied for radiation shielding concrete include carbon nanotubes, metal nanoparticles, and nano-clay. Each of these materials have unique properties that make it suitable for use in concrete. Overall, using nanomaterials in concrete for radiation shielding can significantly improve the safety of nuclear facilities and other applications that require radiation shielding.

1.6.14 Metal Alloys and Composites

Metal alloys and composites have been widely studied for their potential use in EMI shielding. These materials have excellent electrical conductivity and can be engineered for specific properties important for EMI shielding, such as high magnetic permeability or low electrical resistivity. In one study, a metal matrix composite material was prepared by incorporating magnetic particles into a metal matrix, resulting in high magnetic permeability and effective EMI shielding [78]. Another study investigated using carbon fiber-reinforced polymer (CFRP) composites as a lightweight and flexible EMI shielding material. The study found that the CFRP composites exhibited good EMI shielding effectiveness in a wide frequency range [54]. Other metal alloys, such as Aluminum and Copper, have also been studied for their EMI shielding properties. These materials have the potential to be used in a variety of EMI shielding applications, particularly in the aerospace and automotive industries.

Extensive research on multifunctional, electrically conductive metallic alloys and their composites has led to the need for novel and broad-range EMI shielding materials specifically designed for high-power electronics. Conductive alloys and composites of metals like Aluminum, Iron, Copper, etc. have great potential for commercial EMI shielding applications due to their high conductivity, environmental stability, and accessible synthesis methods. These materials can be engineered to have specific properties important for EMI shielding, such as high magnetic permeability or low electrical resistivity. Moreover, they can be designed to be lightweight and flexible, making them suitable for various applications. An image of polymer composite containing MWCNT: Graphene: Ferrite nanoparticle in LDPE polymer matrix is shown in Fig. 1.24.

1.6.15 High-Temperature Shielding Materials

High-temperature shielding materials are essential for various applications, including aerospace, automotive, and nuclear industries, where exposure to high temperatures and radiation is expected. These materials must withstand extreme temperatures without any compromise in EMI shielding. One common type of high-temperature shielding material is ceramic-based composites, which can be engineered to have specific thermal and electrical properties. For example, a ceramic-based composite material of Silicon Carbide fibers has been shown to have excellent EMI shielding

Fig. 1.24 Conductive polymer composite made from MWCNT and graphene in LDPE polymer matrix

effectiveness at high temperatures. Another type of high-temperature shielding material is metal-based composites, such as Nickel-based alloys, with high melting points and good thermal stability. These materials can be used in high-temperature environments to provide effective EMI shielding. Ongoing research efforts are focused on developing new high-temperature shielding materials and improving the performance of existing materials for use in extreme environments [103]. Metal alloy-based gasket for shielding the windows and spaces between edges is shown in Fig. 1.25.

Fig. 1.25 Alloy-based ventilation gasket for filtering EM waves

1.7 Characteristics of EMI Shielding Materials

Here are some critical characteristics of EMI shielding materials:

- **Conductivity**: EMI shielding materials are typically conductive or have conductive properties. This conductivity allows them to absorb or reflect electromagnetic waves effectively, preventing them from interfering with sensitive electronic components.
- **High Permeability and Permittivity**: Materials with high permeability and permittivity are more effective at shielding against electromagnetic radiation. Permeability refers to a material's ability to concentrate magnetic flux, while permittivity refers to its ability to concentrate electric flux.
- **Frequency Range**: EMI shielding materials vary in effectiveness across different frequencies. Some are better suited for protecting against low-frequency electromagnetic radiation, while others excel at blocking high-frequency signals such as radio waves and microwaves.
- **Thickness**: The thickness of the shielding material can impact its effectiveness. Thicker materials may provide better shielding but can also be heavier and less flexible, which may not be suitable for all applications.
- **Flexibility**: Flexibility is essential, especially in applications where the shielding material needs to conform to irregular shapes or fit into tight spaces. Flexible materials can be more versatile and easier to work with during installation.
- **Durability**: EMI shielding materials should be durable and resistant to environmental factors such as temperature extremes, moisture, corrosion, and mechanical stress. This ensures that the shielding remains effective, even in harsh operating conditions.
- **Compatibility**: The shielding material should be compatible with the electronic components and materials intended to protect. It should not interfere with the device's performance. Whether it is shielding or causing any unwanted interactions is important to know.
- **Cost**: Cost is a significant factor in choosing EMI shielding materials, especially for large-scale applications. Materials should provide adequate shielding performance at a reasonable cost to be economically viable.
- **Ease of Installation**: Easy installation and integration into existing systems can reduce downtime and labor costs associated with implementing EMI shielding solutions. Some materials, such as tapes, films, and coatings, are convenient for more straightforward applications.
- **Regulatory Compliance**: EMI shielding materials may need to meet certain regulatory standards and certifications, especially in industries with strict electromagnetic compatibility (EMC) requirements, such as aerospace, defense, and medical devices.

By considering these characteristics, engineers and designers can select the most suitable EMI shielding materials for their specific application, balancing factors such as performance, cost, and ease of implementation [104].

1.8 Drawbacks of EMI Shielding Materials

There are still a lot of drawbacks to materials that absorb microwaves. Composites combine the benefits of each building block unit with the shortcomings of each other; they have emerged as a potential future source of microwave absorbers. Consequently, composite materials will be crucial to eliminating this bottleneck and producing high-performance materials for microwave absorption. Materials that are thin, lightweight, and broadband with high shielding effectiveness have attracted much attention because of their practical and efficient roles in lowering electromagnetic interference pollution. The choice of material should be based on the application at hand and the operating frequency [105]. The development and application of advanced composite materials as microwave absorbers have faced several challenges or bottlenecks, which have hindered their widespread use. Some of these challenges include

- **Fabrication complexity**: Producing composite materials with tailored properties often requires intricate processes that could be costly and time-consuming. Developing new composite materials with optimized microwave absorption properties requires extensive research, experimentation, and the latest manufacturing techniques.
- **Material cost**: The raw materials used to produce advanced composites can be high, particularly those containing rare earth elements or other expensive components. This can make it challenging to produce cost-effective microwave-absorbing materials at a large scale.
- **Environmental impact**: Electromagnetic (EM) absorbing materials can have positive and negative environmental impacts. On the one hand, they can be used to reduce the amount of EM radiation emitted from electronic devices, which can help mitigate potential health risks associated with prolonged exposure to EM radiation. However, the production and disposal of these materials can have negative environmental impacts. The output of EM-absorbing materials can require large amounts of energy, contributing to greenhouse gas emissions and other forms of pollution. Additionally, the disposal of these materials can be problematic if they are not recycled or disposed of properly. Therefore, it's essential to considering the environmental impact of EM-absorbing materials and choosing materials with minimal negative impact while still providing effective EM absorption.
- **Limited absorption frequency**: When it comes to electromagnetic (EM) absorbing materials, it's essential to considering their limited absorption frequency. This refers to the frequencies over which the material can effectively absorb EM radiation. If the absorption frequency is too narrow, the material may not be able to provide adequate protection from EM radiation at other frequencies. Therefore, choosing a material with an absorption frequency that matches the intended application is crucial.
- **Quick oxidation**: One potential issue with electromagnetic (EM) absorbing materials is their susceptibility to quick oxidation, which can reduce their effectiveness over time. Oxidation occurs when the material reacts with oxygen in the air, which

can cause it to break down and lose its EM-absorbing properties. This can be a significant problem for materials exposed to the elements or used in harsh environments, such as those used in military or aerospace applications. To address this issue, manufacturers may use coatings or other treatments to protect the material from oxidation. Additionally, it's essential to monitoring the condition of EM-absorbing materials over time and replacing them as needed to ensure continued effectiveness. Overall, while quick oxidation is a potential concern with EM-absorbing materials, some strategies can be employed to mitigate this issue and provide reliable performance over time.

- **Undesirable skin effect**: The skin effect is a phenomenon that occurs when electromagnetic waves propagate through a conductor. In this effect, the wave travels along the conductor's surface rather than penetrating the interior. This can lead to an undesirable skin effect, where the EM-absorbing material only absorbs radiation at the surface rather than throughout its volume. This can limit the effectiveness of the material, as it may not be able to absorb all the radiation present. To minimize the skin effect, manufacturers may use materials with specific conductive properties or alter the geometry of the material to increase its surface area and improve absorption. Additionally, it's essential to choosing a material with an absorption frequency that matches the intended application, as this can help ensure that the material can absorb radiation effectively. Overall, while the skin effect can be a concern with EM-absorbing materials, some strategies can be employed to mitigate this issue and improve absorption.
- **Low heat stability**: Low heat stability is a potential issue with electromagnetic (EM) absorbing materials that can limit their effectiveness in high-temperature environments. When exposed to high temperatures, these materials can degrade or melt, reducing their ability to absorb EM radiation. This can be a significant problem in applications such as aerospace or military, where materials may be exposed to high temperatures during operation. To address this issue, manufacturers may use materials with higher heat stability or develop coatings or other treatments to protect the material from high temperatures. Additionally, it's crucial to ensuring that the material is tested in conditions that simulate the intended operating environment to verify its effectiveness under these conditions. While low heat stability can be a concern with EM-absorbing materials, some strategies can be employed to mitigate this issue and ensure reliable performance in high-temperature environments.

1.9 Conclusion

Material engineering is central to advancing microwave absorbers, which are essential components in attenuating electromagnetic interference (EMI) and enhancing the performance of electronic systems. With careful selection and optimization of materials, engineers can effectively customize absorber properties to dampen electromagnetic waves across various frequencies, including microwaves. The field has witnessed the development of advanced microwave absorber materials, ranging from traditional ferrite-based composites to modern metamaterials and nanostructured coatings. These materials offer improved performance, lightweight characteristics, and increased tunability, meeting the evolving needs of industries such as telecommunications, aerospace, automotive, and consumer electronics. Furthermore, the interdisciplinary nature of material engineering fosters collaboration with fields like Electromagnetics, Physics, and Chemistry. This collaboration drives the innovation of fabrication methods, characterization techniques, and computational models to optimize microwave absorber efficiency. With the growing demand for high-frequency electronic devices, the significance of materials engineering in microwave absorbers is poised to expand further. By harnessing emerging materials, fabrication processes, and simulation tools, researchers and engineers can pioneer advancements in EMI mitigation, realizing more efficient, reliable, and compact electronic systems in the years ahead.

References

1. Griffiths, D.J.: Introduction to electrodynamics. Cambridge University Press (2023)
2. Zabarmawi, Y.: Assessment of DNA damage in healthcare workers occupationally exposed to low doses of ionizing radiation in the comet assay: meta-analysis. J. Rad. Res. Appl. Sci. **17**(2), 100864 (2024)
3. Hannachi, E., Sayyed, M.I., Slimani, Y., Elsafi, M.: Gamma radiation shielding efficiency of some different ceramic composites: a comparative study. Radiat. Phys. Chem. **220**, 111695 (2024)
4. Calvente, I., Núñez, M.I.: Is the sustainability of exposure to non-ionizing electromagnetic radiation possible? Med. Clin. **162**(8), 387–393 (2024)
5. Hou, X., Feng, X.-R., Jiang, K., Zheng, Y.-C., Liu, J.-T., Wang, M.: Recent progress in smart electromagnetic interference shielding materials. J. Mater. Sci. Technol. **186**, 256–271 (2024)
6. Ossenkopp, K.-P., Barbeito, R.: Bird orientation and the geomagnetic field: a review. Neurosci. Biobehav. Rev. **2**(4), 255–270 (1978)
7. Chu, K.Y., Khodamoradi, K., Blachman-Braun, R., Dullea, A., Bidhan, J., Campbell, K., Zizzo, J., Israeli, J., Kim, M., Petrella, F., Ibrahim, E., Ramasamy, R.: Effect of radiofrequency electromagnetic radiation emitted by modern cellphones on sperm motility and viability: an in vitro study. Eur. Urol. Focus **9**(1), 69–74 (2023)
8. Tyagi, A., Prasad, A.K., Bhatia, D.: Effects of excessive use of mobile phone technology in India on human health during COVID-19 lockdown. Technol. Soc. **67**, 101762 (2021)
9. Tang, X., Yang, T., Yu, D., Xiong, H., Zhang, S.: Current insights and future perspectives of ultraviolet radiation (UV) exposure: friends and foes to the skin and beyond the skin. Environ. Int. **185**, 108535 (2024)
10. Mumtaz, S., Rana, J.N., Choi, E.H., Han, I.: Microwave radiation and the brain: mechanisms, current status, and future prospects. **23**(16), 9288 (2022)
11. Nilsson, R., Liu, N.-A.: Nuclear DNA damages generated by reactive oxygen molecules (ROS) under oxidative stress and their relevance to human cancers, including ionizing radiation-induced neoplasia part I: physical, chemical and molecular biology aspects. Rad. Med. Protect. **1**(3), 140–152 (2020)
12. Mohan, S., Chopra, V.: Chapter 18—Biological effects of radiation. In: Dhoble, S., Chopra, V., Nayar, V., Kitis, G., Poelman, D., Swart, H. (eds.) Radiation Dosimetry Phosphors, pp. 485–508. Woodhead Publishing (2022)

13. Ben Ishai, P., Davis, D., Taylor, H., Birnbaum, L.: Problems in evaluating the health impacts of radio frequency radiation. Environ. Res. **243**, 115038 (2024)
14. Ataei, M., Abdollahi, M.: Cell phones. In: Wexler, P. (ed.) Encyclopedia of Toxicology (Fourth Edition), pp. 675–683. Academic Press, Oxford (2024)
15. Kerr, J.F.R., Winterford, C.M., Harmon, B.V.: Apoptosis. Its significance in cancer and cancer therapy. **73**(8), 2013–2026 (1994)
16. Chitranshi, R., Mehrotra, R.K., Pancoli, P., Ghaziabad, A.K.G.: Analysis of cell tower radiation, RF safety and practical realization of compliance distance. Int. J. Scien. Res. Publ. **4**(4), 1–14 (2014)
17. Car, C., Quevarec, L., Gilles, A., Réale, D., Bonzom, J.-M.: Evolutionary approach for pollution study: the case of ionizing radiation. Environ. Poll., 123692 (2024)
18. Caldecott, K.W.: Causes and consequences of DNA single-strand breaks. Trends Biochem. Sci. **49**(1), 68–78 (2024)
19. Perry, C.C., Le, N.T., Allard, M.M., Milligan, J.R.: DNA condensation by cysteine containing tetra-arginine ligands as a model for ionizing radiation damage to chromatin. Radiat. Phys. Chem. **213**, 111216 (2023)
20. Narasimhamurthy, R.K., Mumbrekar, K.D., Satish Rao, B.S.: Effects of low dose ionizing radiation on the brain—a functional, cellular, and molecular perspective. Toxicology **465**, 153030 (2022)
21. Belli, M., Tabocchini, M.A.: Ionizing radiation-induced epigenetic modifications and their relevance to radiation protection. **21**(17), 5993 (2020)
22. Desouky, O., Ding, N., Zhou, G.: Targeted and non-targeted effects of ionizing radiation. J. Rad. Res. Appl. Sci. **8**(2), 247–254 (2015)
23. Yadav, H., Rai, U., Singh, R.: Radiofrequency radiation: a possible threat to male fertility. Reprod. Toxicol. **100**, 90–100 (2021)
24. Kavithaa, K., Paulpandi, M., Ponraj, T., Murugan, K., Sumathi, S.: Induction of intrinsic apoptotic pathway in human breast cancer (MCF-7) cells through facile biosynthesized zinc oxide nanorods. Karbala Int. J. Modern Sci. **2**(1), 46–55 (2016)
25. Ford-Glanton, B.S., Brisbon, J.L.: Male reproductive toxicants: electromagnetic radiation and heat. In: Reference module in biomedical sciences. Elsevier (2024)
26. Jayaraju, N., Pramod Kumar, M., Sreenivasulu, G., Lakshmi Prasad, T., Lakshmanna, B., Nagalaksmi, K., Madakka, M.: Mobile phone and base stations radiation and its effects on human health and environment: a review. Sustain. Technol. Entrepreneur. **2**(2), 100031 (2023)
27. Ramirez-Vazquez, R., Escobar, I., Vandenbosch, G.A.E., Vargas, F., Caceres-Monllor, D.A., Arribas, E.: Measurement studies of personal exposure to radiofrequency electromagnetic fields: a systematic review. Environ. Res. **218**, 114979 (2023)
28. Hashemi, S.A., Ghaffarkhah, A., Hosseini, E., Bahrani, S., Najmi, P., Omidifar, N., Mousavi, S.M., Amini, M., Ghaedi, M., Ramakrishna, S., Arjmand, M.: Recent progress on hybrid fibrous electromagnetic shields: key protectors of living species against electromagnetic radiation. Matter **5**(11), 3807–3868 (2022)
29. Ohayon, M.M., Stolc, V., Freund, F.T., Milesi, C., Sullivan, S.S.: The potential for impact of man-made super low and extremely low frequency electromagnetic fields on sleep. Sleep Med. Rev. **47**, 28–38 (2019)
30. Liu, L., Huang, B., Lu, Y., Zhao, Y., Tang, X., Shi, Y.: Interactions between electromagnetic radiation and biological systems. iScience **27**(3), 109201 (2024)
31. Lorenz, E., Congdon, C.C.: Radioactivity biologic effects of ionizing radiations. **5**(1), 323–338 (1954)
32. Cerutti, P.A.: Effects of ionizing radiation on mammalian cells. **61**, 51–59 (1974)
33. Tiikkaja, M., Aro, A.L., Alanko, T., Lindholm, H., Sistonen, H., Hartikainen, J.E.K., Toivonen, L., Juutilainen, J., Hietanen, M.: Electromagnetic interference with cardiac pacemakers and implantable cardioverter-defibrillators from low-frequency electromagnetic fields in vivo. EP Europace **15**(3), 388–394 (2012)
34. Martel, J., Chang, S.-H., Chevalier, G., Ojcius, D.M., Young, J.D.: Influence of electromagnetic fields on the circadian rhythm: implications for human health and disease. Biomed. J. **46**(1), 48–59 (2023)

35. Ray, K., Choudhuri, R.: Chapter 22—Effects of radiation on the reproductive system. In: Gupta, R.C. (ed.) Reproductive and Developmental Toxicology, pp. 291–299. Academic Press, San Diego (2011)
36. Kesari, K.K., Agarwal, A., Henkel, R.: Radiations and male fertility. Reprod. Biol. Endocrinol. **16**(1), 118 (2018)
37. Treder, M., Müller, M., Fellner, L., Traynor, K., Rosenkranz, P.: Defined exposure of honey bee colonies to simulated radiofrequency electromagnetic fields (RF-EMF): negative effects on the homing ability, but not on brood development or longevity. Sci. Total. Environ. **896**, 165211 (2023)
38. Sharma, V.P., Kumar, N.R.J.C.S.: Changes in honeybee behaviour and biology under the influence of cellphone radiations. **98**, 1376–1378 (2010)
39. Balmori, A.: Radio-tracking systems emit pulsed waves that could affect the health and alter the orientation of animals. J. Nat. Conserv. **77**, 126520 (2024)
40. Hashemi, S.A., Mousavi, S.M., Faghihi, R., Arjmand, M., Sina, S., Amani, A.M.: Lead oxide-decorated graphene oxide/epoxy composite towards X-Ray radiation shielding. Radiat. Phys. Chem. **146**, 77–85 (2018)
41. Dehghan, N., Taeb, S.: Adverse health effects of occupational exposure to radiofrequency radiation in airport surveillance radar operators. **17**(1), 7–11 (2013)
42. Kumari, S., Dalal, J., Kumar, A., Pal, R., Chahal, R., Ohlan, A.: Enhanced microwave absorption properties of conducting polymer@graphene composite to counteract electromagnetic radiation pollution: green EMI shielding††. Electronic supplementary information (ESI) available. RSC Adv. **14**(1), 662–676 (2024). https://doi.org/10.1039/d3ra07245b
43. Chung, D.D.L.: A perspective on electromagnetic interference shielding materials comprising exfoliated graphite. Carbon **216**, 118569 (2024)
44. Manjappa, P., Rajan, H.K., Mahesh, M.G., Sadananda, K.G., Channegowda, M., Shivashankar, G.K., Mutt, N.B.: effective attenuation of electromagnetic waves by synergetic effect of α-Fe_2O_3 and MWCNT/graphene in LDPE-based composites for EMI applications. **15**(24), 9006 (2022)
45. Praveen, M., Karthikeya, G.S., Krishna, R.H., Mamatha, G.M., Manjunatha, C., Khosla, A., Nagabhushana, B.M.: The role of magnetic nano $CoFe_2O_4$ and conductive MWCNT/graphene in LDPE-based composites for electromagnetic interference shielding in X-band. Diam. Relat. Mater. **130**, 109501 (2022)
46. Kondawar, S.B., Modak, P.R.: Chapter 2—Theory of EMI shielding. In: Joseph, K., Wilson, R., George, G. (eds.) Materials for Potential EMI Shielding Applications, pp. 9–25. Elsevier (2020)
47. Przybył, W., Januszko, A., Radek, N., Szczepaniak, M., Bogdanowicz, K.A., Plebankiewicz, I., Szczodrowska, B., Mazurczuk, R.: Microwave absorption properties of carbonyl iron-based paint coatings for military applications. Defence Technol. **22**, 1–9 (2023)
48. Cebula, F., Wojtaszek, K., Shabbir, H., Kula, A., Pędzich, Z., Kornaus, K., Małecki, S., Kołczyk-Siedlecka, K., Socha, R.P., Escribà-Gelonch, M., Rogóż, K., Goodsite, M., Hessel, V., Wojnicki, M.: Circular solid state reduction process of fine copper powder synthesis with life cycle assessment for photovoltaics application. Sustain. Mater. Technol. **33**, e00486 (2022)
49. Karki, A., Cincotti, G., Chen, S., Stanishev, V., Darakchieva, V., Wang, C., Fahlman, M., Jonsson, M.P.: Electrical tuning of plasmonic conducting polymer nanoantennas. **34**(13), 2107172 (2022)
50. Gupta, E., Bonner, C., Muhammed, F., McParland, K., Mirotznik, M.: Design and fabrication of multi-material broadband electromagnetic absorbers for use in cavity-backed antennas. Heliyon **9**(3), e14164 (2023)
51. Tiwari, P., Pathak, S.K., Siju, V.: Development of resistive-ink based planar and conformal metasurfaces for RCS reduction. IEEE Access **10**, 61472–61483 (2022)
52. Mendecki, J., Friedenthal, E., Botstein, C., Sterzer, F., Paglione, R., Nowogrodzki, M., Beck, E.: Microwave-induced hyperthermia in cancer treatment: apparatus and preliminary results. Int. J. Rad. Oncol.*Biol.*Phys. **4**(11), 1095–1103 (1978)

53. Bhardwaj, A., Puthoor, N.M., Nair, G.G.: Tunable directional scattering from high-refractive-index particles dispersed in an anisotropic medium. J. Phys. Chem. C **124**(34), 18698–18706 (2020)
54. Wanasinghe, D., Aslani, F.: A review on recent advancement of electromagnetic interference shielding novel metallic materials and processes. Compos. Part B: Eng. **176**, 107207 (2019)
55. Abbasi, H., Antunes, M., Velasco, J.I.: Recent advances in carbon-based polymer nanocomposites for electromagnetic interference shielding. Prog. Mater. Sci. **103**, 319–373 (2019)
56. Qin, G., Dong, Y., Zhang, P., Liu, J., Li, C., Huang, W., Hong, Z., Shen, C., Jing, X.: Broadband absorber coupled with anti-reflection inverted pyramid type all-dielectric microstructure. Opt. Lasers Eng. **160**, 107288 (2023)
57. Govindasamy, T., Mathew, N.K., Asapu, V.K., Subramanian, V., Subramanian, B.: Modulating the structural and magnetic properties of Fe_3O_4 NPs for high-performance supercapattery and EMI shielding applications. J. Energy Storage **79**, 110243 (2024)
58. Cai, B., Wu, L., Zhu, X., Cheng, Z., Cheng, Y.: Ultra-broadband and wide-angle plasmonic light absorber based on all-dielectric gallium arsenide (GaAs) metasurface in visible and near-infrared region. Resul. Phys. **58**, 107509 (2024)
59. Sahoo, P., Saini, L., Dixit, A.: Microwave-absorbing materials for stealth application: a holistic overview. Oxford Open Mater. Sci. **3**(1) (2022)
60. Basa Reddy, S., Praveen, M., Basavarajappa, T., Bapu Tandale, V., Pachipulusu, R., Hegde, S.: EMI shielding performance of electroless coated iron nanoparticles on graphite in low-density polyethylene composite for X-band applications. J. Indian Chem. Soc. **100**(4), 100962 (2023)
61. Dai, Y., Xu, Y., Jiang, D., Bai, L., Li, Z., Huo, P., Liu, C., Liu, Y.: Electromagnetic interference shielding of flexible carboxymethyl cellulose/MWCNT@Fe_3O_4 composite film with ultralow reflection loss. Int. J. Biol. Macromol. **257**, 128604 (2024)
62. Dang-Bao, T., Tran, T.-A., Lam, H.-H.: First-row transition-metal hybrid oxide nanoparticles (MFe_2O_4) as UV-fenton-like catalysts for Rhodamine B treatment: a comparative study. Mater. Today: Proc. (2023)
63. Elmahaishi, M.F., Azis, R.A.S., Ismail, I., Muhammad, F.D.: A review on electromagnetic microwave absorption properties: their materials and performance. J. Mater. Res. Technol. **20**, 2188–2220 (2022)
64. Wan, C., Li, J.: Graphene oxide/cellulose aerogels nanocomposite: preparation, pyrolysis, and application for electromagnetic interference shielding. Carbohyd. Polym. **150**, 172–179 (2016)
65. Wilson, R., George, G., Joseph, K.: Chapter 1—An introduction to materials for potential EMI shielding applications: status and future. In: Joseph, K., Wilson, R., George, G. (eds.) Materials for Potential EMI Shielding Applications, pp. 1–8. Elsevier (2020)
66. Gokce, E.C., Calisir, M.D., Selcuk, S., Gungor, M., Acma, M.E.: Electromagnetic interference shielding using biomass-derived carbon materials. Mater. Chem. Phys. **317**, 129165 (2024)
67. Nath, K., Bhattacharyya, S.K., Das, N.C.: Chapter 10—Biodegradable polymeric materials for EMI shielding. In: Joseph, K., Wilson, R., George, G. (eds.) Materials for Potential EMI Shielding Applications, pp. 165–178. Elsevier (2020)
68. Green, R.A., Goding, J.A.: 11—Biosynthetic conductive polymer composites for tissue-engineering biomedical devices. In: Poole-Warren, L., Martens, P., Green, R. (eds.) Biosynthetic Polymers for Medical Applications, pp. 277–298. Woodhead Publishing (2016)
69. Ahmadi, Y., Iqbal, S., Ahmad, S.: Chapter 4—Significance of naturally derived materials for potential EMI shields. In: Joseph, K., Wilson, R., George, G. (eds.) Materials for Potential EMI Shielding Applications, pp. 63–72. Elsevier (2020)
70. Ghosh, S.K., Nath, K., Nath Chowdhury, S., Paul, S., Ghosh, T., Katheria, A., Das, P., Das, N.C.: Combination effect of functionalized high aspect ratio carbonaceous nanofillers and carbon black on electrical, thermal conductivity, dielectric and EMI shielding behavior of co-continuous thermoplastic elastomeric blend composite films. Chem. Eng. J. Adv. **15**, 100505 (2023)
71. Zhao, S., Sun, P.P., Shao, X., Liu, X., Zhao, Z., Li, L., Xin, Z.: ZrO_2 functionalized graphene oxide/SEBS-based nanocomposites for efficient electromagnetic interference shielding applications. **25**(S1), E130–E136 (2019)

72. Ye, X., Hu, J., Li, B., Hong, M., Zhang, Y.: Graphene loaded with nano-Cu as a highly efficient foam interface material with excellent properties of thermal-electronic conduction, anti-permeation and electromagnetic interference shielding. Chem. Eng. J. **361**, 1110–1120 (2019)
73. Khan, T., Khalid, M.Y., Andrew, J.J., Ali, M.A., Zheng, L., Umer, R.: Co-cured GNP films with liquid thermoplastic/glass fiber composites for superior EMI shielding and impact properties for space applications. Compos. Commun. **44**, 101767 (2023)
74. Kumar, S., Walia, R., Kumar, A., Verma, V.: Hybrid structure of MWCNT/ferrite and GO incorporated composites for microwave shielding properties and their practical applications. RSC Adv. **11**(17), 9775–9787 (2021)
75. Zhang, L., Bi, S., Wang, X., Li, J.: Chapter 7—Thermosetting nanocomposite foams for electromagnetic interference shielding. In: Thomas, S., Paoloni, C., Pai, A.R. (eds.) Porous Nanocomposites for Electromagnetic Interference Shielding, pp. 161–178: Woodhead Publishing (2024)
76. Bardet, L., Papanastasiou, D.T., Crivello, C., Akbari, M., Resende, J., Sekkat, A., Sanchez-Velasquez, C., Rapenne, L., Jiménez, C., Muñoz-Rojas, D., Denneulin, A., Bellet, D.: Silver nanowire networks: ways to enhance their physical properties and stability. **11**(11), 2785 (2021)
77. Deeraj, B.D.S., Mathew, M.S., Parameswaranpillai, J., Joseph, K.: Chapter 6—EMI shielding materials based on thermosetting polymers. In: Joseph, K., Wilson, R., George, G. (eds.) Materials for Potential EMI Shielding Applications, pp. 101–110. Elsevier (2020)
78. Pandey, R., Tekumalla, S., Gupta, M.: Chapter 21—EMI shielding of metals, alloys, and composites. In: Joseph, K., Wilson, R., George, G. (eds.) Materials for Potential EMI Shielding Applications, pp. 341–355. Elsevier (2020)
79. Sun, Y.-X., Zhao, J., Li, X.-Z., Jiang, H., Cai, Y.-J., Yang, X., Liu, Y., Wei, N.-J., Li, Y.-B., Wu, Y.-G., Yang, Z.-H., Gai, J.-G.: Metal-grade laminated nanofiber films with outstanding EMI shielding performances and high-temperature resistance. Colloids Surf. A **680**, 132701 (2024)
80. Rayar, A., Naveen, C.S., Onkarappa, H.S., Betageri, V.S., Prasanna, G.D.: EMI shielding applications of PANI-Ferrite nanocomposite materials: a review. Synth. Met. **295**, 117338 (2023)
81. Li, J., Zhang, W., Yang, L., Yin, S.: Conductive fabrics based on carbon nanotube/Ti_3C_2Tx MXene/polyaniline/liquid metal quaternary composites with improved performance of EMI shielding and joule heating. Compos. Commun. **38**, 101476 (2023)
82. Saritha, A., Thomas, B., George, G., Wilson, R., Joseph, K.: Chapter 8—Elastomer-based materials for EMI shielding applications. In: Joseph, K., Wilson, R., George, G. (eds.) Materials for Potential EMI Shielding Applications, pp. 121–143. Elsevier (2020)
83. Shetty, H.D., Ashok Reddy, G.V., Ramasamy, V., Kaliprasad, C.S., Daruka Prasad, B., Yogananda, H.S., Naik, R., Prasad, V., Koyyada, G., Anil Kumar, Y.: Electrical conductivity and electromagnetic interference shielding effectiveness of elastomer composites: comparative study with various filler systems. Inorg. Chem. Commun. **151**, 110578 (2023)
84. Choudhury, S.N., Das, P., Bhawal, P., Pal, A., Banerji, P., Das, N.C.: Double percolation behavior through the preferential distribution of conductive black in polymer blends to boost electrical properties and EMI shielding effectiveness. Mater. Today Commun. **35**, 106109 (2023)
85. Calheiros Souto, L.F., Soares, B.G.: Electromagnetic wave absorption, EMI shielding effectiveness and electrical properties of ethylene—vinyl Acetate (EVA)/Polyaniline (PAni) blends prepared by in situ polymerization. Syn. Metals. **298**, 117441 (2023)
86. Altin, Y., Bedeloglu, A.: Chapter 11—Nanomaterials with potential EMI shielding properties. In: Joseph, K., Wilson, R., George, G. (eds.) Materials for Potential EMI Shielding Applications, pp. 179–199. Elsevier (2020)
87. Dj, B.G., Wessley, G.J.J.: A study on EMI shielding in aircraft: introduction, methods and significance of using electrospun nanocomposites. J. Space Saf. Eng. **11**(1), 150–160 (2024)
88. Iijima, S.: Helical microtubules of graphitic carbon. Nature **354**(6348), 56–58 (1991)

89. Chakraborty, T., Saha, S., Gupta, K., Dutta, S., Mahapatra, A.S., Mondal, D., Pradhan, A.K., Chakraborty, C., Das, S., Sutradhar, S.: An effective microwave absorber using multi-layer design of carbon allotrope based Co2Z hexaferrite-Polymer nanocomposite film for EMI shielding applications. Chem. Eng. J. **487**, 150323 (2024)
90. Raji, S., Sharma, G.K., Aranya, B.R., Prabhakaran, K.: Carbon composite foams from the wasted banana leaf for EMI shielding and thermal insulation. Carbon **213**, 118259 (2023)
91. Wang, Y.-H., Li, C.-C.: Highly effective EMI shielding composites for 5G Ka-band frequencies. Appl. Mater. Today **36**, 102041 (2024)
92. Kausar, A.: Chapter 14—Hybrid polymeric nanocomposites with EMI shielding applications. In: Joseph, K., Wilson, R., George, G. (eds.) Materials for Potential EMI Shielding Applications, pp. 227–236. Elsevier (2020)
93. Ahmed, I., Khan, A.N., Jan, R., Gul, I.H.: Structure—properties relationships of graphene and spinel nickel ferrites based poly(vinylidene fluoride) hybrid polymer nanocomposites for improved dielectric and EMI shielding characteristics. Mater. Res. Bull. **148**, 111687 (2022)
94. Uma Varun, V., Rajesh Kumar, B., Etika, K.C.: Hybrid polymer nanocomposites as EMI shielding materials in the X-band. Mater. Today: Proc. **28**, 796–798 (2020)
95. Rybicki, T., Karbownik, I.: Chapter 17—EMI shielding textile materials based on conducting polymers. In: Joseph, K., Wilson, R., George, G. (eds.) Materials for Potential EMI Shielding Applications, pp. 267–285. Elsevier (2020)
96. Hu, S., Wang, D., Periyasamy, A.P., Kremenakova, D., Militky, J., Tunak, M.: Ultrathin multilayer textile structure with enhanced EMI shielding and air-permeable properties. **13**(23), 4176 (2021)
97. Liang, X., Ji, G.: Chapter 15—Metal organic framework (MOF)-anchored polymeric nanocomposite foams for electromagnetic interference shielding. In: Thomas, S., Paoloni, C., Pai, A.R. (eds.) Porous Nanocomposites for Electromagnetic Interference Shielding, pp. 337–362. Woodhead Publishing (2024)
98. Peng, Q., Gao, C., Song, C., Liu, Z., Fatehi, P., Wang, S., Kong, F.: Wood-derived porous carbon foams filled with Ti_3C_2TxMXene/CoFe-MOF for electromagnetic shielding with flame retardant, heat insulation and excellent cycle stability. J. Ind. Eng. Chem. **133**, 333–344 (2024)
99. Wu, Q., Zeng, Z., Xiao, L.: From 2D graphene and MXene nanolayers to 3D biomimetic porous composite aerogels for electromagnetic interference shielding. Compos. A Appl. Sci. Manuf. **177**, 107939 (2024)
100. Kumari, P., Tripathi, P., Singh, S.P., Kumar, D.: Chapter 19—Electromagnetic shielding using ceramic materials. In: Joseph, K., Wilson, R., George, G. (eds.) Materials for Potential EMI Shielding Applications, pp. 315–331. Elsevier (2020)
101. Tan, Y., Luo, H., Zhou, X., Peng, S., Zhang, H.: Dependences of microstructure on electromagnetic interference shielding properties of nano-layered Ti_3AlC_2 ceramics. Sci. Rep. **8**(1), 7935 (2018)
102. Han, B., Zhang, L., Ou, J.: Radiation shielding concrete. In: Han, B., Zhang, L., Ou, J. (eds.) Smart and Multifunctional Concrete Toward Sustainable Infrastructures, pp. 329–337. Springer Singapore, Singapore (2017)
103. Gogoi, J.P., & Shabir, A.: Chapter 23—High-temperature electromagnetic interference shielding materials. In: Joseph, K., Wilson, R., George, G. (eds.) Materials for Potential EMI Shielding Applications, pp. 379–390. Elsevier (2020)
104. Shu, J.-C., Yang, X.-Y., Zhang, X.-R., Huang, X.-Y., Cao, M.-S., Li, L., Yang, H.-J., Cao, W.-Q.: Tailoring MOF-based materials to tune electromagnetic property for great microwave absorbers and devices. Carbon **162**, 157–171 (2020)
105. Qoidah, S.N., Ulfawanti Intan Subadra, S.T., Taufiq, A., Mufti, N., Sunaryono, S., Hidayat, N., Handoko, E., Alaydrus, M., Amrillah, T.: Fe_3O_4/MWCNT/TiO_2 nanocomposites as excellent microwave absorber material. J. Alloys Compd. **970**, 172590 (2024)

Chapter 2
Synthesis of Nanomaterials Used as EM Absorbers

Abstract This chapter explores the synthesis of nanomaterials used as electromagnetic (EM) absorbers. Such materials have unique properties that allow them to selectively absorb and attenuate EM waves, making them useful in various applications. The chapter covers various methods for synthesizing nanomaterials, including their advantages and limitations. It also discusses the key factors that affect the EM absorption properties of nanomaterials, such as size, shape, composition, crystal structure, and surface morphology. The chapter concludes by highlighting the multidisciplinary nature of this field and the potential for continued developments in nanotechnology to produce new and improved EM absorbers.

2.1 Introduction

Naturally occurring materials often do not offer efficient Electromagnetic Interference (EMI) shielding because their properties and structures are not optimized. In contrast, materials specifically designed for EMI shielding are usually engineered at the nanoscale level to have unique properties, such as high conductivity and high loss tangent, that allow them to attenuate and absorb electromagnetic radiation effectively. Additionally, natural materials may not offer the same durability, reliability, and reproducibility required for applications that demand consistent EMI shielding performance [1].

Creating nanomaterials for electromagnetic (EM) absorbers involves synthesizing materials with unique properties that allow them to absorb and attenuate EM waves [2]. These materials have many applications, including stealth technology, electromagnetic interference (EMI) shielding, and energy harvesting [3]. Various methods for synthesizing nanomaterials include chemical vapor deposition (CVD), sol–gel, hydrothermal synthesis, and ball milling [4]. The choice of method depends on the desired properties of the final product. The size and shape of the nanoparticles are critical factors in synthesizing nanomaterials for EM absorption. Nanoparticles ranging from 1 to 100 nm have been found to exhibit unique properties that make them effective EM absorbers. The geometry of the nanoparticles can also affect their

S. K. Koul et al., *Microwave Absorbers*, Engineering Materials,
https://doi.org/10.1007/978-981-96-9596-6_2

EM absorption properties, with irregularly shaped particles being more effective absorbers than regularly shaped ones.

Other factors that can influence the EM absorption properties of nanomaterials include their composition, crystal structure, and surface morphology [5]. Nanoscale materials have unique properties due to increased surface area-to-volume ratio. This can lead to changes in reactivity, surface chemistry, and other surface-dependent phenomena. Quantum effects become prominent in nanomaterials, affecting their electronic, optical, and mechanical properties. Nanomaterials can deliver high thermal conductivity but may be more brittle due to defects and impurities. With the strategic choice of these factors, customization of the EM characteristics is possible [6]. The synthesis of nanomaterials for EM absorption is a complex and interdisciplinary field that requires expertise in material science, chemistry, and physics [7]. With the continuous advancement of nanotechnology, developing new and improved nanomaterials with enhanced EM absorption properties is expected to continue [8]. Figure 2.1 shows an image of a copper sulfate solution and Cu nanoparticles prepared from a chemical reduction process. This image of nanoparticle synthesis uses copper sulfate solution in 0.01 M concentration. It is added with a reducing agent of 0.01 M concentration of lithium borohydride as a reducing agent in deionized water with CTAB as a capping agent. In the colloidal solution of copper nanoparticles, the initial blue color turns red-brown, indicating that the formation of Cu^{2+} state copper was reduced to Cu^{0+} state and precipitated as metallic copper nanoparticles, which is an instant preparation technique for an immense researcher.

Nanoscience refers to the study of phenomena and manipulation of materials on a nanoscale, where properties differ significantly from those at a larger scale. This cross-disciplinary field encompasses chemistry, physics, mechanics, medicine, biology, computing, engineering, mathematics, and materials science. Furthermore, nanotechnology utilizes nanomaterials and their properties to create new structures, systems, and devices. This process involves designing, producing, characterizing, and

Fig. 2.1 Colloidal solution of copper nanoparticles

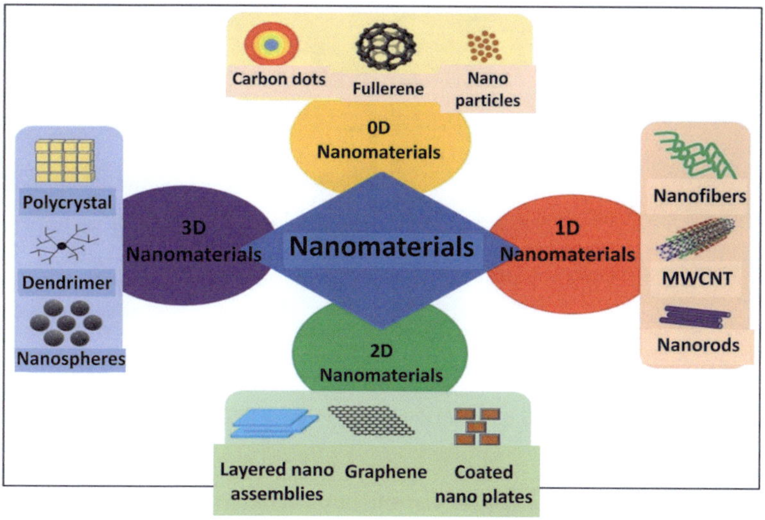

Fig. 2.2 Classification of nanomaterials according to their confined dimensionality [9]

applying materials at the nanometre scale. In Fig. 2.2, nanomaterials are classified according to their confined dimensionality.

Methods from nanoscience are applied to produce economically viable products. Nanomaterials exhibit unique properties, including mechanical, thermal, electrical, physical stability, magnetic [10], catalytic [11], diffusive, and optical properties attributed to their size. Various forms of nanostructures have been developed in different fields over the years for various applications. These include nanosheets, nanorods [12], nanowires [13], nanocomposites [14], nanocages, magnetic nano chains [15], nano fabrics [16], nanofibers [17], nanoflowers [18], nanofoams, nanoholes, nano meshes, nanopillars [19], nano pin films [20], nanoplatelets [21], nanoribbons [22], nanorings [23], nano shells [24], nanotips [25], nanostructured films [26], quantum dots [27], and lipid nanostructures [28]. In Fig. 2.3, an image of cadmium sulfate quantum (CdS) dots with the morphology of the flower looks prepared from the hydrothermal method [29] is shown.

The realm of quantum materials has expanded to include nanomaterials in two dimensions (2D). In 2D quantum nanomaterials [30], electrons can move in two sizes, while the remaining dimension is confined to the atomic scale. This results in a range of unique quantum properties in low dimensions, such as valleytronics physics [31, 32], topological insulator [33], integer and fractional quantum Hall effect [34, 35], high-temperature super conduction [36], ferroelectricity [37], and ferromagnetism [38]. 2D materials are essential for studying significant quantum phenomena [39]. The progress of 2D quantum materials depends on their synthesis and design, which are influenced by various chemical engineering techniques [35].

There are two classifications for the synthesis of nanomaterials. The first classification is based on the approach followed to realize nanomaterials, which could be either

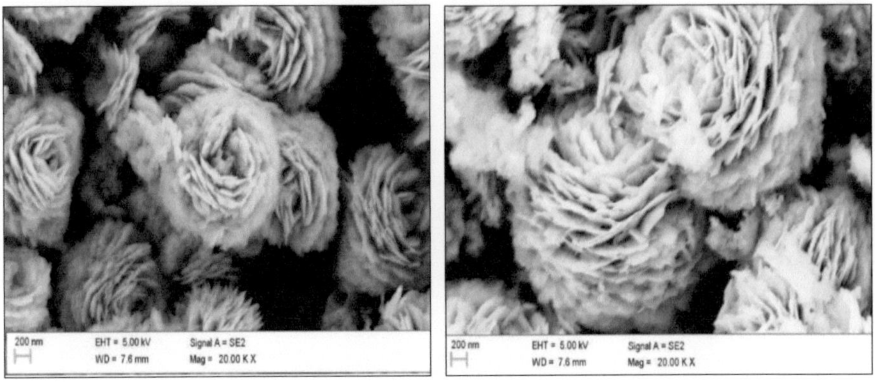

Fig. 2.3 SEM images of the CdS quantum dots having flower-like morphology [29]

top-down or bottom-up. The second classification is based on the process involved in synthesizing nanomaterials. This classification includes physical methods, chemical methods, hybrid and green methods.

There are two primary techniques for manufacturing nanoparticles, "Top-down" and "Bottom-up," which depend on the experimental procedure and protocols. The top-down approach involves breaking down larger molecules into smaller particles while maintaining their original purity and property, particularly the bulk material's physical characteristics and chemical composition. However, this method results in structural defects and crystallographic damage, making it slow and unsuitable for large-scale production. Despite its drawbacks, this method offers better control over nanoparticles. For example, ZnO nanoparticles are synthesized in spherical balls, spears, rods, fibers, platelets, polygonal crystals, flowers, needles, tubes, clusters, core shells, and quantum dots today by various global researchers in variable nano sizes to micro sizes.

On the other hand, the bottom-up method involves building structures from the bottom up, starting with atoms, molecules, and even nanoparticles to create complex nanoparticles. This approach is less expensive than the top-down method, more scalable, and generally yields a more uniform product. The methods that come under top-down are ball milling [40] and laser ablation [41]. The bottom-up approach includes sol–gel, coprecipitation, and solution combustion [42]. Figure 2.4 shows an image of the top-down and bottom-up methods.

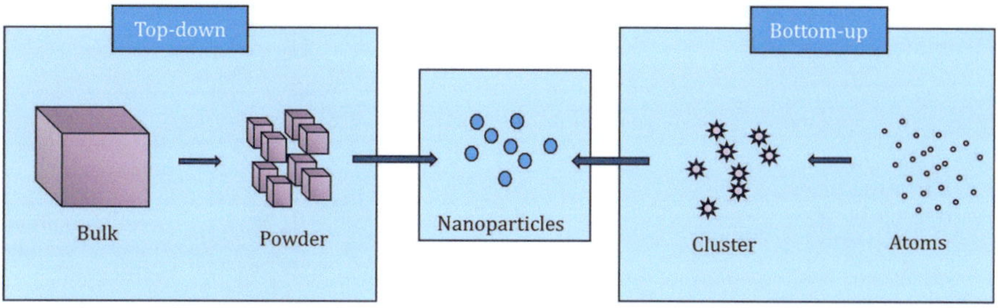

- Ball milling, Thermal evaporation, Laser ablation, Sputtering, Erosion, Slicing
- Thermal evaporation, Hydrothermal, Co-ppt, Chemical vapour deposition, Sol-gel, Solution combustion method

Fig. 2.4 Top-down and bottom-down methods

2.2 Methods to Synthesize Nanomaterials

2.2.1 Physical Method of Synthesis

The physical method of nanomaterial preparation involves creating new materials or substances using physical means such as heating, cooling, grinding, and mixing [4]. This method is widely used in nanotechnology to generate nanoparticles, nanowires, and thin films with specific properties. Some standard physical synthesis techniques used in nanomaterials preparation include ball milling, vapor deposition (VD), plasma techniques, and lithography. These methods have proved effective in creating materials with desired properties such as high surface area, improved conductivity, and enhanced mechanical strength [43]. The physical process of nanomaterial preparation has opened new avenues for research and development in various fields, including electronics, medicine, and energy [44].

Physical methods of nanomaterials synthesis are of great importance in nanotechnology. These methods enable the production of materials with unique properties that are not achievable through traditional methods. By using physical means such as heating, cooling, and mixing, it is possible to manipulate the size, shape, and composition of nanoparticles, nanowires, and thin films. This allows for the creation of materials with tailored properties suitable for specific applications in electronics, energy storage, and biomedical devices [45].

Additionally, physical synthesis methods are often more cost-effective and scalable than chemical synthesis methods, making them a popular choice for the industrial production of nanomaterials. Overall, physical methods of nanomaterial synthesis play a crucial role in advancing the field of nanotechnology and contributing to the development of innovative technologies with practical applications. Several types of physical methods are used to synthesize nanomaterials. These methods include ball milling [46], vapor deposition [47], plasma technology [48], laser ablation [49], and lithography [50].

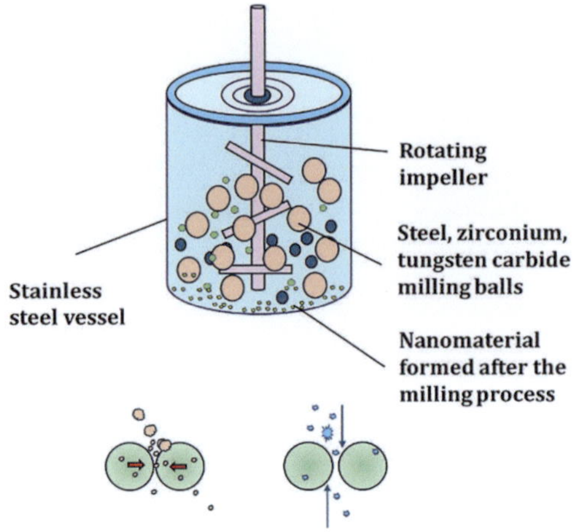

Fig. 2.5 Schematics of the ball milling process

2.2.1.1 Ball Milling

Ball milling involves using high-energy ball mills to produce nanoparticles by mechanical attrition. Ball milling effectively produces nanomaterials for various applications, including electromagnetic interference (EMI) shielding. Using high-energy ball mills, the mechanical attrition process can create nanoparticles with unique properties that make them ideal for EMI shielding applications. The resulting nanoparticles can be engineered to have specific sizes, shapes, and compositions, significantly influencing their EMI shielding effectiveness. Additionally, ball milling is a scalable process, which means that large quantities of nanoparticles can be produced relatively quickly, making it a cost-effective method for mass production. Ball milling is a promising technique for producing nanomaterials used for EMI shielding due to its ability to create highly customized nanoparticles with enhanced properties. Figure 2.5 shows a schematic image of a cross-section image of the ball milling process.

2.2.1.2 Physical Vapor Deposition (PVD)

This involves condensing material from a vapor phase onto a substrate to form thin films or coatings. In this method, a source material is heated to a high temperature until it vaporizes. The vapor then condenses onto the substrate, forming a thin film. The process can be carried out under vacuum or in the presence of a carrier gas. The thickness of the film can be controlled by adjusting the deposition time and the deposition rate. The method can deposit many materials, including metals, semiconductors, and insulators, and is widely used in the semiconductor, optics, and electronics industries. This technique will coat metals in a thin layer on another substrate. This method is appropriate for creating sensors, solar cells, and PCB boards [51]. Figure 2.6 shows RF and DC sputtering (Physical Vapor Deposition) instruments.

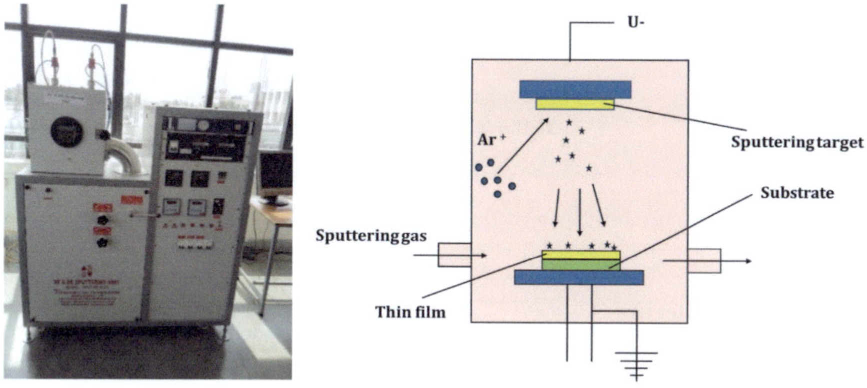

(a) (b)

Fig. 2.6 **a** RF and DC sputtering (physical vapor deposition) instrument; **b** schematics of working principle

2.2.1.3 Plasma Technology

This involves using plasma to deposit or etch materials at the nanoscale. Plasma technology can prepare Electromagnetic Interference (EMI) shielding materials by depositing thin films of conductive materials such as copper, silver, and gold onto the surface of the substrate material. The substrate may be non-conducting plates like flame retardant-4 (FR-4), ceramic plates, glass, or quartz, depending on the application, such as photovoltaic, sensor, antenna, and conductivity measurement. These thin films can effectively shield against electromagnetic radiation. Additionally, plasma technology can etch patterns into these conductive films, creating complex EMI shielding structures tailored to specific applications. Overall, the precise control of plasma technology makes it a valuable tool for preparing EMI shielding materials with optimal performance [48].

Thermal plasma torches, or arc plasma torches, use a high-voltage electric arc between electrodes to generate hot gases that maintain a temperature of 8000–16,000 °K through heat exchange and conductive and convective operations [52]. These torches come in two forms for nanoparticle synthesis: transferred and non-transferred. Each type has a cathode and a gas-flow nozzle cooled with water. A nozzle is placed at the end of the torch to create a plasma jet, and the jet's temperature and velocity can be customized for nanomaterial production. The high temperature and gas velocity make these torches valuable for nanoparticle synthesis [53]. Non-transferred arc plasma differs from transferred arc plasma because the anode is outside the torch, generating less erosion and more spherical nanoparticles. This technique is widely used in medicine, waste treatment technology, and microbiology [54]. Figure 2.7 illustrates the carbon nanotube synthesis system by arc discharge.

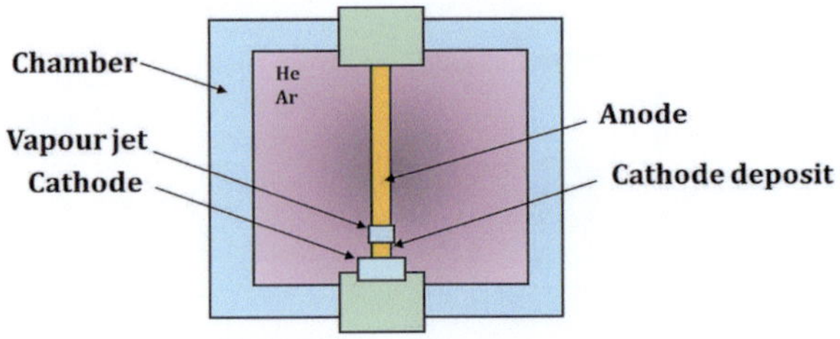

Fig. 2.7 Carbon nanotube synthesis process using an arc discharge process

2.2.1.4 Ion Sputtering

This technique also known as ion beam deposition (IBD) involves using an ion source to sputter the targeted material (metal or dielectric) to generate free ions or clusters that evaporate and deposit on the substrate.

2.2.1.5 Plasma Arc Discharge

This technique produces low-dimensional carbon materials (nanodiamonds, graphite) between graphite electrodes in a low-pressure inert gas environment in the presence of a catalyst.

2.2.1.6 Laser Ablation

This involves using a laser to vaporize a material, condensing it to form nanoparticles. The nanoparticles produced through laser ablation can be used to create electromagnetic interference (EMI) shielding materials. The advantage of using nanoparticles produced through laser ablation is that they are highly pure and uniform in size, which makes them more effective in providing EMI shielding [55]. Figure 2.8 shows the pulsed laser ablation using the liquid technique.

2.2.1.7 Lithography

It involves using light to pattern and selectively remove material to create nanostructures. Lithography is an exact and versatile technique that can prepare various nanomaterials with different shapes, sizes, and properties [56, 57]. By carefully controlling the light beam's intensity, wavelength, and pattern, lithography can selectively remove material from a substrate or deposit a thin layer onto a substrate. Usually, visible light and UV light are used as light sources. This can be used to create complex 2D or 3D nanostructures, such as nanowires, nanotubes, and nanodots with high precision and reproducibility. Moreover, lithography can be combined with other techniques, such as etching, deposition, and self-assembly, to create more complex

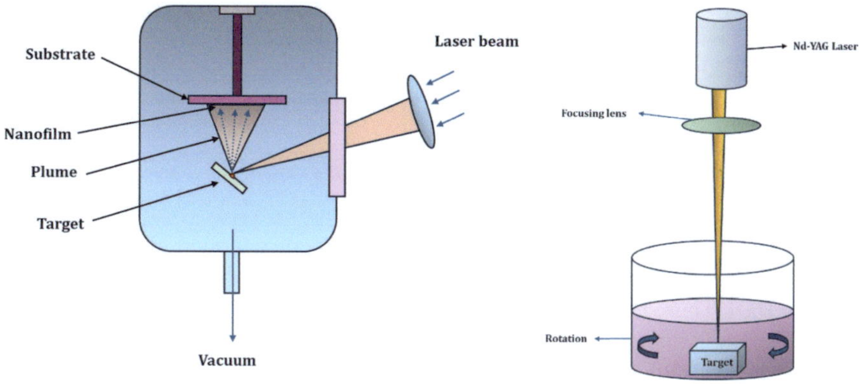

Fig. 2.8 Pulsed laser ablation in liquid technique

nanostructures and functional devices for various electronics, optics, sensors, and biomedicine applications. In Fig. 2.9, the nano-lithography technique is pictorially shown [58]. These physical methods offer different advantages and limitations regarding cost, scalability, and the properties of the resulting nanomaterials.

The advantage of physical methods for synthesizing nanomaterials is the ability to produce high-quality, pure nanoparticles with well-defined sizes, shapes, and structures. These methods are also generally more reproducible than chemical methods and can be easily scaled up for large-scale production. The disadvantages of physical methods are that they can be more expensive and time-consuming than chemical methods. They may also require specialized equipment and expertise. In addition to this, some physical techniques can result in high energy consumption, leading to

Fig. 2.9 Nano-lithography technique [58]

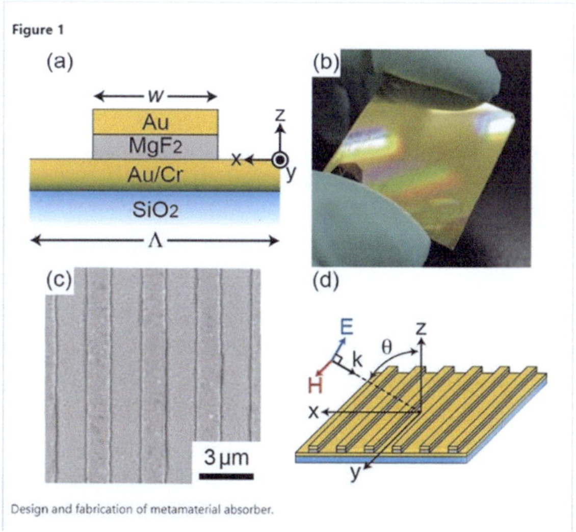

environmental concerns. Overall, the choice of synthesis method for nanomaterials depends on the specific needs of the application and the resources available [59].

2.2.2 Chemical Methods of Synthesis

These methods involve using chemical reactions to create nanoparticles with specific properties. These methods typically include precipitation, sol–gel, and hydrothermal synthesis [60]. These techniques are widely used in producing various nanomaterials, including metals, metal oxides, and quantum dots [61]. One of the advantages of chemical methods is that they can create nanoparticles in large quantities with a high degree of control over their size, shape, and composition. However, they also require careful attention to safety and environmental considerations, as some chemicals could be hazardous.

These methods are widely used in industry to produce various nanomaterials. For example, hydrothermal synthesis is commonly used to make metal oxides such as Titanium Dioxide and Zinc Oxide, which find applications in multiple industries, including cosmetics, electronics, and energy. Sol–gel synthesis is used to produce silica nanoparticles, which have applications in catalysis, coatings, and drug delivery. Precipitation synthesis is used to produce metal nanoparticles, which are used in various applications such as sensing, imaging, and catalysis. These techniques offer several benefits, including affordability, ease of synthesis, and high output. However, chemical methods can be detrimental to living organisms despite their ability to control particle size distribution accurately [21]. Factors such as reaction time, component concentration, and temperature influence the particle's shape and size. In certain situations, the growth conditions can be sensitive to process variables and may only allow a little room for process optimizations.

Chemical methods are crucial for synthesizing nanomaterials due to their high yield, reproducibility, and ability to produce materials with controlled size, shape, composition, and surface properties [62]. Various chemical methods have been explored for synthesizing nanomaterials, including solution combustion, sol–gel, hydrothermal synthesis, precipitation, chemical vapor deposition (CVD), sono chemical, mechanical, electrochemical, and cytochemical methods.

These methods allow for producing a wide range of nanomaterials, including nanoparticles, nanorods, nanowires, and nanotubes, with unique properties relevant to specific applications. Chemical methods of nanomaterial synthesis also offer the potential for incorporating dopants, heterostructures, and surface functionalization, which can enhance or modify the material's properties [63]. Overall, chemical nanomaterial synthesis methods are essential in developing nanotechnology and creating new and improved nanomaterials with tailored properties for various applications.

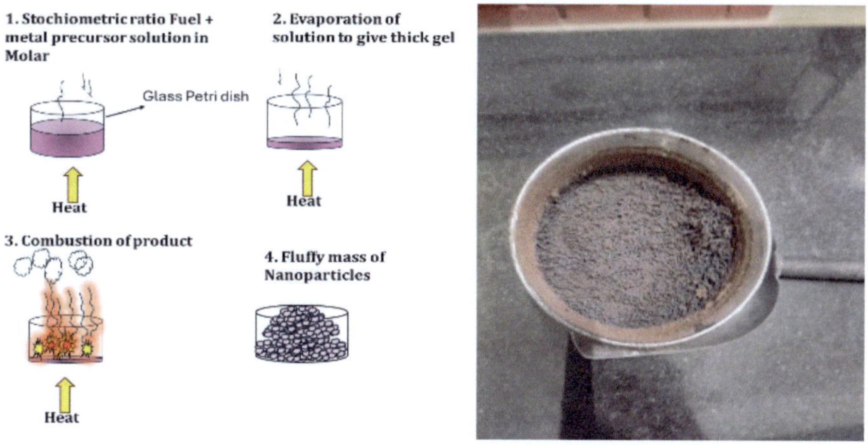

Fig. 2.10 The diagram provides an overview of the solution combustion process and the $ZrFe_2O_4$ nanoparticles prepared after the SCS process

2.2.2.1 Solution Combustion Synthesis Method (SCS)

This method involves the nitrate salts of the precursor and hydrocarbon fuel being taken in a 1:1 fuel to oxidizer's stoichiometric ratio. This is dissolved in minimum amount of water in a 100 ml Petri dish and transferred to a preheated muffle furnace. The temperature setting for the muffle furnace usually ranges from 300 to 1000 °C, depending on the desired product. The heating rate of the muffle furnace can vary, but it is typically in the range 1–5 °C per minute. This allows for a combustion reaction; the product will condense as a thick liquid after evaporation of water and then catch fire and burn itself in the fire with the liberation of the enormous number of gases; as a result, a fluffy mass of the material is obtained which is subjected for calcination; the method is suitable for producing all the metal oxide and ceramic oxide nanoparticles in large quantity and quick time [64]. Some of the common materials synthesized in this method include ZnO, Fe_2O_3, NiO, CuO, MgO, TiO_2, ZrO_2, $BaFe_{12}O_{19}$, $SrFe_{12}O_{19}$, g-Fe_2O_3, and $CoFe_2O_4$. The process of solution combustion synthesis with prepared nanoparticles in a ceramic dish is illustrated in Fig. 2.10 [65].

2.2.2.2 Sol–Gel Synthesis

This method involves the hydrolysis and condensation of metal alkoxides in a solution to form a gel-like material, which can be further processed to produce nanoparticles. Specific salts, oxides, hydroxides, complexes, alkoxides, acrylates, and amines are precursors for synthesizing nanomaterials if they are soluble in proper solvents. Several parameters influence the hydrolysis and condensation reactions (sol–gel

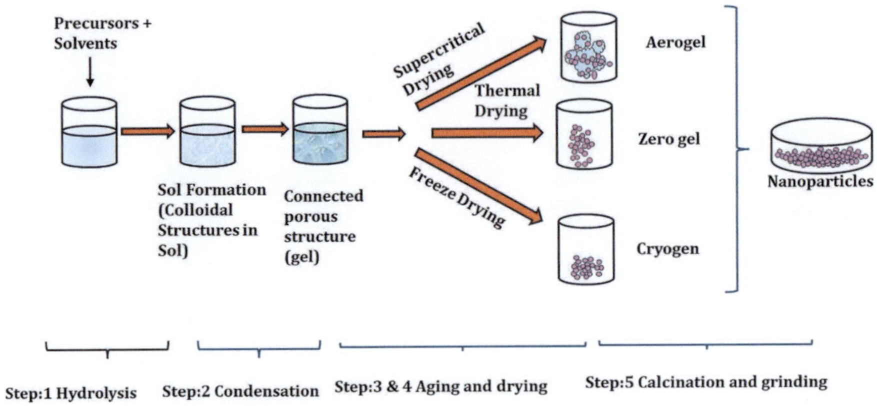

Fig. 2.11 Schematic of different stages of the sol–gel process

process), including the activity of the metal alkoxide, the water/alkoxide ratio, solution pH, temperature, nature of the solvent, and additives used. It is important to note that these technical specifications can vary depending on the specific materials and desired outcome and should be carefully controlled and monitored to ensure consistent and reproducible results [66].

Figure 2.11 shows a schematic of different stages of the sol–gel process.

2.2.2.3 Chemical Vapor Deposition (CVD)

This method differs from PVD since PVD offers a thin film with a few nanometres of thickness that can be coated on any dielectric substrate. Meanwhile, CVD is used to prepare carbonaceous material. This method involves exposing a high-temperature substance to a volatile precursor. The reactant is divided into products that uniformly spread on the surface, undergo a chemical reaction, nucleate, and grow to create the desired material. The chamber must be capable of withstanding the process temperature. It must also be compatible with reactive gases and commonly utilized materials, including quartz, stainless steel, and ceramics. The process can take place in a range of environments, from high vacuum (10^{-3} Torr) to low pressure (a few Torr) to atmospheric pressure (760 Torr). The temperature of the reaction is essential in controlling the deposition rate and film properties. The temperature can range from a few hundred degrees Celsius to over a thousand degrees Celsius. The precursor can be a single source (that decomposes to form the desired film) or involve multiple sources that react to form the film. Heated lines are typically used to prevent precursor condensation before reaching the reaction zone. Precise control of precursor gas-flow rates is crucial in achieving the desired film stoichiometry and thickness. The substrate is heated to promote desired chemical reactions and film adhesion. Plasma is usually used in most CVD processes to improve deposition rate, film quality, or control surface properties. The substrate must be correctly positioned within the

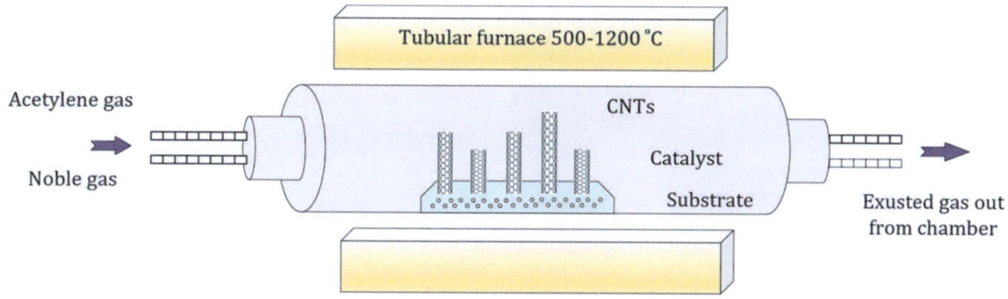

Fig. 2.12 The schematic of the CVD technique

reaction zone to ensure uniform film deposition. A proper exhaust system is required to remove unreacted gases and by-products from the chamber. For process control, ellipsometry or mass spectrometry can monitor film growth in real time [67]. The schematic of the CVD technique is shown in Fig. 2.12.

2.2.2.4 Wet Chemical Synthesis

It is a general category of methods that uses liquid as a reaction medium, in which some precursors are used and allowed to react with the reactant with optimal pH, pressure, and temperature. The prepared nanomaterials are separated from the reaction mixture to separate with other unreacted chemicals and surfactants, precursors, and by-products using filtration, centrifugation, etc., followed by evaporation or heat treatment to remove the solvent. This method involves reduction of metal salts in a solution to produce nanoparticles. The size and shape of the nanoparticles can be controlled by varying the reaction conditions. This includes coprecipitation, sol–gel, reduction, and hydrothermal or solvothermal processes [68].

2.2.2.5 Hydrothermal and Solvothermal Methods

These methods use high-pressure and high-temperature conditions to produce nanoparticles in a solution. The method is beneficial for making nanoparticles with high crystallinity. The hydrothermal method utilizes water as a reaction solvent and medium, and the solvothermal method utilizes organic solvents like ethanol, benzene, chloroform, etc. The procedure consists of two stages: nucleation and crystal growth using direct crystallization from a solution. An autoclave with a Teflon® or alloy lining is used to perform the hydrothermal technique and placed in a pressure container. The temperature inside the autoclave can be raised above the boiling point by gradually increasing the vapor pressure of the container. It is like a pressure cooker used in the typical household, and sometimes it can blast like a bomb due to improper threading and closing. In the chamber, considerable pressure builds up due to the phase transformation of solvent from liquid to gas or liquid plasma, which leads to a

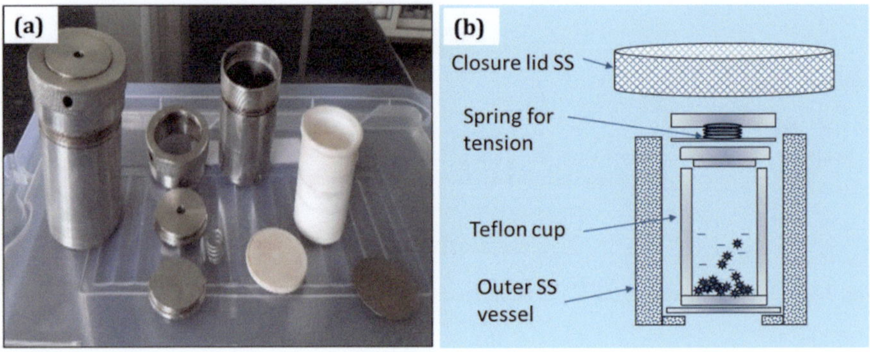

Fig. 2.13 **a** Photographic image of the autoclave, and **b** is the cross-sectional view of the autoclave

burst similar to a small bomb. The Chemistry lab nanotechnologists should carefully use this method with proper guidance [6]. Figure 2.13a is a photographic image of the autoclave, and Fig. 2.13b is the cross-sectional view of the autoclave.

2.2.2.6 Electrochemical Synthesis

This method uses an electric current to reduce metal ions in a solution to produce nanoparticles. This method is beneficial for making nanoparticles with controlled size and shape. The electrochemical method utilizes the interconversion of electrical energy and chemical energy. This principle is used for nanomaterial synthesis, coating on the surface of electrodes, or reducing the precursor by supplying electrical charges in a controlled manner by controlling the supply of voltage and current to the electrochemical reaction chamber. Since the electric current can control chemical reactions, this technique can precisely control the nanomaterial synthesis and coating. Usually, an instrument such as an electrochemical workstation is used for this process. The electrochemical workstation consists of three electrode systems: reference, counter, and working electrodes. In a typical experiment, potassium ferricyanide can be placed in an electrolyte of the desired concentration. The pH, temperature, electrolyte agitation, current density, and the distance between the electrodes are the variables in this process [69, 70]. In Fig. 2.14, the electrochemical synthesis process is shown.

2.2.2.7 Microwave-Assisted Synthesis

This method involves using microwave radiation to heat a reaction mixture, accelerating the rate of chemical reactions and producing nanoparticles with controlled size and shape. Microwave radiation is utilized as an energy source to heat the reactants. The energy and time duration are varied according to the requirement. About 5 min of microwave irradiation will provide sufficient energy to heat the reaction mixture,

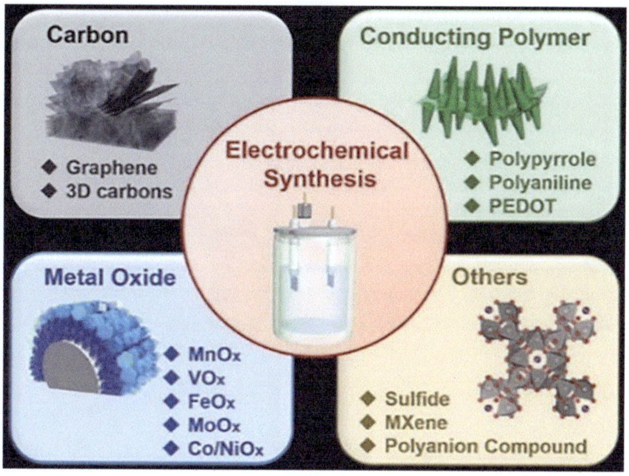

Fig. 2.14 The outcome of the electrochemical synthesis process

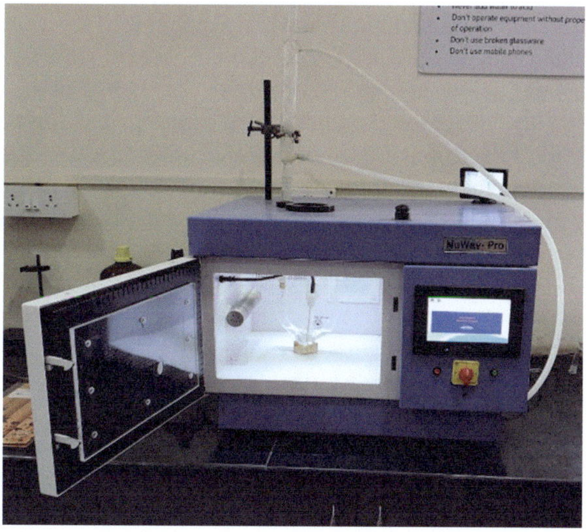

Fig. 2.15 Microwave reactor used to synthesize nanomaterials

equal to 60 min of conventional heating. The time of irradiation, agitation of the reaction mixture, and temperature-like parameters can be adjusted. Some reactions where volatile liquids are used for a long time are connected with water-cooled or ice-cooled condensers. In Fig. 2.15, the microwave-assisted process of nanoparticle synthesis is shown.

2.2.2.8 Sono Chemical Method

The sono chemical method, also known as ultrasonic irradiation, is a process that harnesses high-frequency sound energy to induce chemical reactions in a liquid

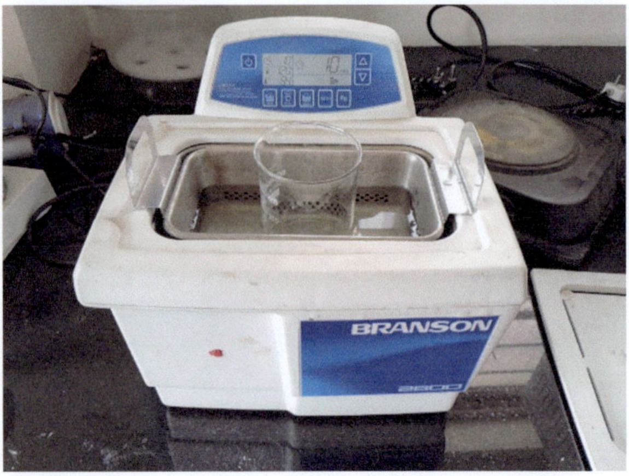

Fig. 2.16 Sono chemical process, an ultrasonic bath

medium. This technique is widely recognized for its ability to enhance reaction rates and yields unique product outcomes. Sound waves propagate through a liquid and create cavitation bubbles, tiny voids, or "holes" in the liquid. These bubbles collapse rapidly, generating intense local heating and high pressures. This phenomenon leads to forming free radicals and other reactive species, which can significantly accelerate chemical reactions. As a result, sonochemistry has found applications in various fields, such as organic synthesis, nanomaterials production, and environmental remediation. In Fig 2.16, the sono chemical process, an ultrasonic bath, is shown.

2.2.2.9 Cryogenic Synthesis

This method includes crystallizing different solutions at very low temperatures (cryogenic) and removing the dissolvent by freezing, drying, or spraying. The reaction occurs at low temperatures, or cryogenic cooling and separation methods can separate the synthesized product [71]. This method is preferred for specific applications.

2.2.2.10 Aerosol-Based Process

This method comprises three integral parts that work in cycles to yield the desired nanomaterials. First, a chamber atomizes a liquid by combining it with a high-energy carrier gas, creating a fine aerosol. Second, this aerosol is transferred into a tubular reactor, subjected to temperatures ranging 800–1000 °C to facilitate the necessary chemical reactions. Finally, a particle collector is employed to gather the particles resulting from the chemical reactions within the tubular reactor.

Each method has advantages and limitations, and the choice of method depends on the desired properties of the final product. Chemical methods of nanomaterial synthesis involve using chemical reactions to produce nanomaterials. These methods have several advantages and disadvantages, which are listed below.

Advantages

- **Precise control over size and shape**: Chemical methods provide precise control over the size and shape of nanomaterials, which is critical for achieving specific properties and applications.
- **High purity**: Chemical methods can produce high-purity nanomaterials, essential for ensuring consistent and reliable performance [72].
- **Scalability**: Chemical methods can be easily scaled up to produce large nanomaterials, making them suitable for industrial applications.
- **Various materials**: Chemical methods can synthesize many materials, including metals, metal oxides, and organic compounds [73].

Disadvantages

- **Complexity**: Chemical methods can be complex and require a high degree of Chemistry expertise, making them challenging to implement.
- **High cost**: Chemical methods can be expensive, especially those requiring specialized equipment or reagents.
- **Environmental impact**: Chemical methods can have a significant environmental impact, especially if they involve toxic or hazardous materials.
- **Reproducibility**: Chemical methods can be sensitive to minute variations in reaction conditions, making it challenging to reproduce results consistently.

Overall, chemical nanomaterial synthesis methods offer precise control over size and shape, high purity, scalability, and the ability to synthesize a wide range of materials. However, they can be complex, expensive, and challenging to reproduce consistently.

2.2.3 Green Synthesis Method

This is also known as sustainable synthesis. The green method can synthesize nanomaterials employing environmentally friendly techniques and reagents. This method reduces or eliminates hazardous chemicals and solvents and minimizes waste generation [74]. Green synthesis involves using renewable starting materials, such as plant extracts [75], enzymes [76], and bacteria [77], as well as non-toxic solvents, such as water or ethanol. The reaction conditions are mild and energy-efficient, reducing the synthesis process's energy consumption and carbon footprint. Figure 2.17 shows the advantages of green chemistry.

Green synthesis involves the use of environmentally friendly methods to produce nanomaterials. This approach utilizes various natural products such as

- Plant extracts: Natural substances extracted from plants with various uses.
- Bacteria: Microscopic organisms with diverse roles in nature.

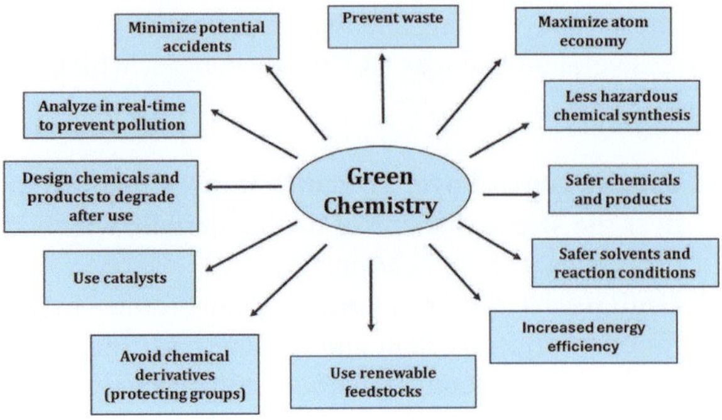

Fig. 2.17 Advantages of green chemistry

- Fungi: Diverse group of organisms with significant ecological and commercial significance.
- Enzymes: Biological molecules that catalyze biochemical reactions.
- Biosynthesis: Production of complex molecules within living organisms.
- Plant-mediated synthesis: Synthesizing nano-compounds using plant-based methods like plant juices and extracts.
- Microbial synthesis: Production of substances using microorganisms.

These methods are cost-effective, eco-friendly, and produce nanomaterials with unique properties. Additionally, using natural products in green synthesis reduces the risk of toxicity associated with conventional methods. Green synthesis is a promising approach to nanomaterial synthesis that is gaining momentum in the scientific community due to its numerous benefits [74]. In Fig 2.18, the synthesis of nanoparticles by plant extraction is shown.

Green synthesis methods are eco-friendly, cost-effective, and produce less waste than conventional chemicals. These methods utilize natural resources such as plant extracts, microorganisms, and biomolecules to produce nanomaterials. Green synthesis methods also produce biocompatible nanomaterials with unique properties, making them ideal for biomedical applications. Additionally, green synthesis methods have the potential to produce nanomaterials in large quantities, making them suitable for commercial applications. Green synthesis methods offer a sustainable and environmentally friendly approach to producing nanomaterials.

2.2.3.1 Plant Extract-Based Synthesis

Plant extracts have been increasingly used to synthesize nanomaterials due to their eco-friendliness and potential for large-scale production. The use of plant extracts as reducing and capping agent in synthesizing nanoparticles has been explored extensively, resulting in the development of various novel approaches [78]. One such approach is using plant extracts as a bio-template for synthesizing nanomaterials with unique morphologies. This method has shown promising results for producing

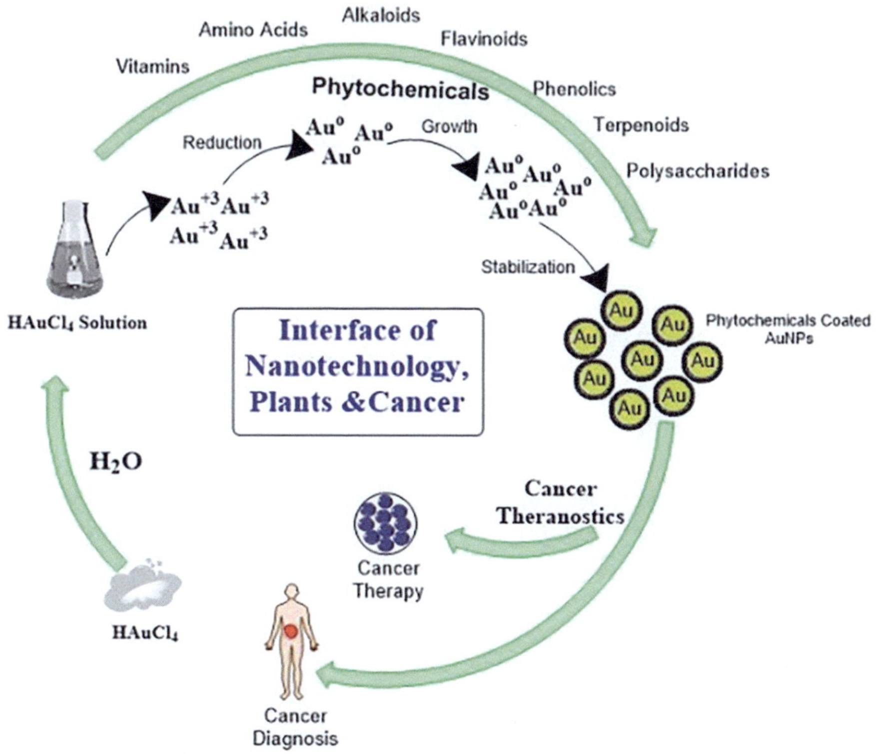

Fig. 2.18 Synthesis of nanoparticles by plant extract

nanomaterials with controllable sizes and shapes, which can be utilized in various applications such as drug delivery, catalysis, and biosensors.

2.2.3.2 Microbial Synthesis

Microbial nanomaterial synthesis involves using microbes such as bacteria, yeast and fungi, to produce nanomaterials through the reduction of metal ions present in a solution. Microbial synthesis is a cost-effective and eco-friendly approach to producing nanoparticles, making it an attractive alternative to traditional chemical methods. This method is also highly scalable, allowing for producing large quantities of nanoparticles. However, the size and shape of the nanoparticles produced through microbial synthesis can be difficult to control, making it challenging to produce nanoparticles with specific properties. Despite this limitation, microbial synthesis has shown much promise in nanotechnology [79].

2.2.3.3 Bacterial-Based Nanoparticle Synthesis

This method uses bacteria to synthesize nanoparticles with specific properties. It has gained attention recently due to its eco-friendliness and cost-effectiveness. Using bacteria makes it possible to produce high-quality nanoparticles in large quantities. Moreover, the nanoparticles produced through bacterial synthesis have improved biocompatibility and biodegradability properties, making them suitable for various biomedical applications [80].

2.2.3.4 Fungi-Based Nanoparticle Synthesis

Fungi-based nanoparticle synthesis is a promising research area that has gained much attention recently. These nanoparticles can be synthesized using various types of fungi, including yeasts and molds. One of the advantages of using fungi for nanoparticle synthesis is that they are readily available and can be grown in large quantities. Additionally, fungi can synthesize nanoparticles of different sizes and shapes, which makes them useful for various applications, such as drug delivery, imaging, and catalysis. Fungi-based nanoparticle synthesis is a fascinating research area that can potentially revolutionize many fields [81]. In Fig. 2.19, bacterial-assisted synthesis shell nanoparticles are given.

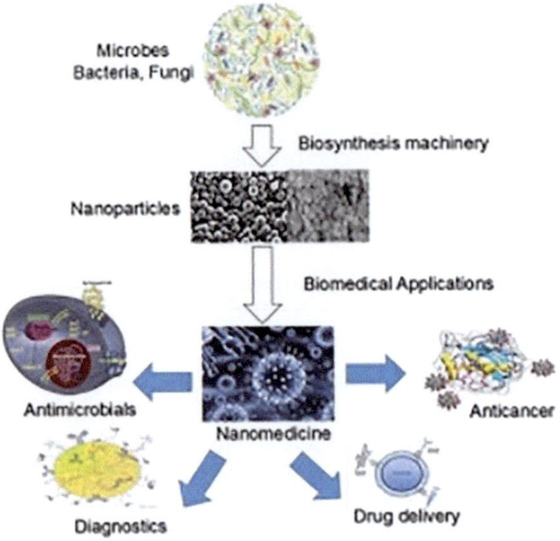

Fig. 2.19 Bacterial-assisted synthesis nanoparticles

2.2.3.5 Enzymes

Enzymes have unique properties that make them ideal for synthesizing nanomaterials. They can catalyze reactions with high specificity and efficiency, making them ideal candidates for controlling nanoparticle size, shape, and surface properties. Enzymes can also work under mild conditions, essential for preserving the integrity of the nanomaterials. With the help of enzymes, researchers can create nanomaterials with tailored properties that can be used in various applications, such as drug delivery, imaging, and sensing [82].

2.2.3.6 Biosynthesis

Biosynthesis of nanomaterials is an emerging field that uses biological entities such as bacteria, fungi, and plants to synthesize nanoparticles. This method is gaining popularity due to its eco-friendly and cost-effective nature. Biological entities allow to produce nanoparticles of different shapes and sizes, which can be used in various applications such as drug delivery, environmental remediation, and electronics. Some of the advantages of the biosynthesis of nanomaterials include scalability, reproducibility, and the ability to tune the properties of nanoparticles [83]. This method has benefits such as reduced environmental impact, lower energy consumption, renewable resources, and higher purity and quality. However, its disadvantages are limited availability, high cost of some renewable starting materials, and slower process requiring more steps.

2.2.4 Hybrid Method of Synthesis

Hybrid methods of preparing nanomaterials refer to using multiple techniques to synthesize nanoparticles. These methods usually involve combining two or more physical and chemical processes to obtain nanoparticles with specific properties or characteristics. The hybrid approach can improve nanomaterial synthesis's efficiency, yield, and reproducibility and provide more particle size, shape, and composition control. Examples of hybrid methods include sol–gel synthesis combined with hydrothermal treatment, microwave-assisted synthesis followed by calcination, and electrospinning combined with plasma treatment [84].

There are various types of hybrid synthesis methods, some of which are listed below

1. Sol–gel and hydrothermal synthesis
2. Chemical vapor deposition and electrospinning
3. Electrochemical and solvothermal synthesis
4. Microwave- and ultrasound-assisted synthesis
5. Coprecipitation and sono chemical synthesis
6. Solvothermal and sol–gel synthesis
7. Hydrothermal and sono chemical synthesis
8. Electrochemical and hydrothermal Synthesis
9. Sol–gel and electrospinning
10. Microwave and sol–gel synthesis.

Table 2.1 gives the classification of several synthesis techniques, the associated limitations, and the process's strength.

2.3 Conclusion

In conclusion, the synthesis of nanomaterials can be effectively achieved through various methods, each tailored to meet specific material requirements and desired applications. Physical processes, such as ball milling, PVD, plasma technology, and laser ablation, provide high control over particle size and morphology with applications where mechanical or thermal stability is crucial. Chemical methods, including sol–gel, hydrothermal, and microwave-assisted synthesis, fine-tune chemical composition and crystallinity, making them suitable for nanomaterials with specialized electronic or magnetic properties. Green synthesis methods present eco-friendly alternatives that use biological agents, reducing environmental impact while enabling the formation of biocompatible nanomaterials. Hybrid synthesis combines aspects of physical, chemical, and biological approaches to further refine nanomaterial characteristics for enhanced performance in specific applications. Each synthesis approach has its unique advantages, offering a comprehensive toolkit for developing nanomaterials with tailored properties and advancing the field of EM absorbers and other functional materials.

Table 2.1 Methods of nanoparticle synthesis [85]

S. No.	Technique	Type	Limitations	Strengths
1	Solution combustion	Chemical method, bottom-up	Agglomeration, evolving of toxic gases	It's a quick process, cheap, and best. Many nanoparticles can be created; they are pure, versatile, and easy to process
2	Sol–gel	Chemical method, bottom-up	Filtration of the product will take a long time	Cheap, various morphology of the product
3	Co-ppt	Chemical method, bottom-up	Filtration of the product will take a long time	Easy, cheap, various morphology of various nanomaterials can be prepared
4	Reduction	Chemical/biological method, bottom-up	Filtration of the product will take a long time	Suitable for metal nanoparticle preparation, quick, pure
5	Chemical vapor deposition (CVD) method	Chemical approach, bottom-up approach	Necessary high process temperatures, as well as complex, hazardous, and gas-producing processes, costly method	It is expensive, coats both organic and inorganic surfaces, and deposits compounds that are difficult to evaporate; it is ideal for producing CNTs, Buckyball, graphene, etc.
6	Mechano-chemical	Top-down Hybrid method	Grinding time	Simple, no sophisticated instrument required, and can be done with pestle and mortar
7	The inert gas condensation method	Physical method, bottom-up approach	Agglomeration, costly setup	Ultrafine-grained materials and excellent resistance to grain growth
8	Plasma arc discharge method	Physical method bottom-up approach	Required the right amount of space and temperature. Numerous factors, such as current, buffer, and gas, "also influence" the technique's results	Cost-effective

(continued)

Table 2.1 (continued)

S. No.	Technique	Type	Limitations	Strengths
9	Ion sputtering technique	Physical method top-down approach	The temperature distribution of a molten metal is frequently irregular. As a result, precise replication is challenging	They are extensively employed in thin-film technologies
10	Laser pyrolysis method	Physical method, bottom-up approach	The vapor phase needs to contain the necessary components	Affordable, quick reaction times, easy-to-use equipment, and high-purity goods
11	Hydrothermal synthesis method	Chemical, top-down approach	Among other factors, pH, temperature, and the concentration of the reactants influence the crystallized particles' structure and dimensions	Sensible control over crystallinity, size distribution, and form
12	Solvothermal synthesis method	Chemical, top-down approach	High temperature is necessary	Reasonable control over size, shape, and crystallinity compared to the hydrothermal method
13	Aerosol-based process method	Chemical methods. Bottom-up approach	Complex process	Provides reasonable control of particle size
14	Cryo chemical synthesis method	Chemical method	The shape of the nanocrystals, which guarantees their stability at room temperature and lack of changes upon heating, is another factor influencing the effect of temperature on particle size	Cost-effective, safe, and highly efficient

(continued)

Table 2.1 (continued)

S. No.	Technique	Type	Limitations	Strengths
15	Thermal plasma jets technique	Bottom-up approach	The problem of low power efficiency is severe	It is often employed in the synthesis of NPs. This technique's current application is "known and recognized" in microbiology, technology, waste treatment, and medicine
16	Laser ablation	Physical method top-down approach	Cost	Simple, safe, low porosity, minimal reliance on a chemical solution, "the capacity to produce materials with complex stoichiometry, narrower particle size distribution, and high purity"

References

1. Ahmadi, Y., Iqbal, S., Ahmad, S.: Chapter 4—Significance of naturally derived materials for potential EMI shields. In: Joseph, K., Wilson, R., George, G. (eds.) Materials for Potential EMI Shielding Applications, pp. 63–72. Elsevier (2020)
2. Green, M., Chen, X.: Recent progress of nanomaterials for microwave absorption. J. Materiomics **5**, 503–541 (2019)
3. Wang, T., Wang, B., Li, L., Liu, H., Zhang, K., Wu, X., Yu, K.: Coordinated multi-band stealth based on phase change materials. Int. J. Therm. Sci. **202**, 109101 (2024)
4. Li, Z., Yu, C.: Chapter 5—Synthesis methods and paradigms of nanomaterials. In: Li, Z., Yu, C. (eds.) Nanostructured Materials, pp. 145–183. Elsevier (2024)
5. Wei, J., Zhang, Y., Li, X., Zhang, H., Guo, Y., Wang, T., Qiao, X., Lei, W.: Recent progress in synthesis, growth mechanisms, and electromagnetic wave absorption properties of silicon carbide nanowires. Ceram. Int. **48**, 35966–35985 (2022)
6. Kulkarni, S.K.: Synthesis of nanomaterials—II (chemical methods). In: Nanotechnology: principles and practices, pp. 77–109. Springer International Publishing, Cham (2015)
7. Chandrakala, V., Aruna, V., Angajala, G.: Review on metal nanoparticles as nanocarriers: current challenges and perspectives in drug delivery systems. Emergent Materials **5**, 1593–1615 (2022)
8. Capek, I.: 1—Nanotechnology and nanomaterials. In: Capek, I. (ed.) Nanocomposite Structures and Dispersions (Second Edition), pp. 1–93. Elsevier, Amsterdam (2019)
9. Paras, Yadav, K., Kumar, P., Teja, D.R., Chakraborty, S., Chakraborty, M., Mohapatra, S.S., Sahoo, A., Chou, M.M.C., Liang, C.-T., Hang, D.-R.: A review on low-dimensional nanomaterials: nanofabrication, characterization and applications. **13**, 160 (2023)
10. Revathy, R., Sajini, T., Augustine, C., Joseph, N.: Iron-based magnetic nanomaterials: sustainable approaches of synthesis and applications. Results Eng. **18**, 101114 (2023)
11. Jie, L., Gao, X., Cao, X., Wu, S., Long, X., Ma, Q., Su, J.: A review of CdS photocatalytic nanomaterials: morphology, synthesis methods, and applications. Mater. Sci. Semicond. Process. **176**, 108288 (2024)
12. Kulkarni, G.K., Jadhav, S.A., Patil, K.T., Patil, P.S., Puri, V.R.: α-MnO2 nanorods-polyaniline (PANI) nanocomposites synthesized by polymer coating and grafting approaches for screening EMI pollution. Ceram. Int. **47**, 15044–15051 (2021)
13. Sahoo, R., Sundara, R., Venkatachalam, S.: Morphology dependent EMI shielding performance of Ag-Ni core-shell nanowires. J. Alloy. Compd. **981**, 173693 (2024)
14. Basa Reddy, S., Praveen, M., Basavarajappa, T., Bapu Tandale, V., Pachipulusu, R., Hegde, S.: EMI shielding performance of electroless coated iron nanoparticles on graphite in low-density polyethylene composite for X-band applications. J. Indian Chem. Soc. **100**, 100962 (2023)
15. Huang, W., Tong, Z., Li, W., Ma, M., Bi, Y., Liao, Z., Wang, R., Ma, Y.: Fabrication of flower-like $ZnFe_2O_4$@SiO_2@C@NiO nanochains/reduced graphene oxides as a high-performance microwave absorber. J. Alloy. Compd. **849**, 156658 (2020)
16. Dj, B.G., Wessley, G.J.J.: A study on EMI shielding in aircraft: introduction, methods and significance of using electrospun nanocomposites. J. Space Saf. Eng. **11**, 150–160 (2024)
17. Rani, P., Basha, S.K.J., Deshmukh, K., Pasha, S.K.K.: Chapter 29—Functionalized nanofibers for EMI shielding and microwave absorption applications. In: Deshmukh, K., Pasha, S.K.K., Barhoum, A., Mustansar Hussain, C. (eds.) Functionalized Nanofibers, pp. 831–866. Elsevier (2023)
18. Radoń, A., Ciuraszkiewicz, A., Hawełek, Ł, Kubacki, J., Kądziołka-Gaweł, M., Łukowiec, D., Gębara, P., Włodarczyk, P., Kolano-Burian, A.: $NiZnFe_2O_4$ core-shell nanoflowers as effective EMI shielding material dedicated to the absorption of electromagnetic waves in the UHF band and wide temperature range. Scripta Mater. **244**, 116020 (2024)
19. Yang, L., Peng, X., Zou, J., Deng, W., Liu, S., Li, C., Jiang, L., Zhong, C., Wang, Y.: Nanophotonic broadband infrared antireflection coatings based on dielectric Si_3N_4 nano-pillar arrays. Optics Commun. **552**, 130094 (2024)

20. Zhang, Z.-Q., Zhong, J., Ye, H.-F., Cheng, G.-G., Ding, J.-N.: Structural optimization and shear performances of the nanopins based on Y-junction carbon nanotubes. Physica B **504**, 13–22 (2017)
21. Praveen, M., Karthikeya, G.S., Krishna, R.H., Mamatha, G.M., Manjunatha, C., Khosla, A., Nagabhushana, B.M.: The role of magnetic nano $CoFe_2O_4$ and conductive MWCNT/graphene in LDPE-based composites for electromagnetic interference shielding in X-band. Diam. Relat. Mater. **130**, 109501 (2022)
22. Pontes, J.M., Pinto, A.K.M., Gomes, D.S., Dantas, M.A.L., Guerra, T., Azevedo, S.: Characterization of the structural, electronic, and magnetic properties of graphene and boron nitride nanoribbons with hexagonal doping. Physica E **161**, 115953 (2024)
23. Feng, C., Zhong, Z., Lin, S., Liu, Y., Hu, J.: Synthesis of silver nanorings via a facile one-pot method. Inorg. Chem. Commun. **130**, 108703 (2021)
24. Gordel-Wójcik, M., Pietrzak, M., Kołkowski, R., Zych, E.: Silica-coated gold nanoshells: surface chemistry, optical properties and stability. J. Lumin. **270**, 120565 (2024)
25. Wang, X., Shi, S., Zhang, F., Li, S., Tan, J., Su, B., Cheng, Q., Gou, Y., Zhang, Y.: Application of a nanotip array-based electrochemical sensing platform for detection of indole derivatives as key indicators of gut microbiota health. Alex. Eng. J. **85**, 294–299 (2023)
26. Althobaiti, M.G., Alosaimi, M.A., Alharthi, S.S., Alotaibi, A.A., Badawi, A.: Tailoring the optical performance of sprayed NiO nanostructured films through cobalt doping for optoelectronic device applications. Opt. Mater. **151**, 115341 (2024)
27. Yu, W., Chamkouri, H., Chen, L.: Recent Advancement on quantum dot-coupled heterojunction structures in catalysis: a review. Chemosphere, 141944 (2024)
28. Ucar, E., Teksoz, S., Ichedef, C., Kilcar, A.Y., Medine, E.I., Ari, K., Parlak, Y., Sayit Bilgin, B.E., Unak, P.: Synthesis, characterization and radiolabeling of folic acid modified nanostructured lipid carriers as a contrast agent and drug delivery system. Appl. Rad. Isotopes **119**, 72–79 (2017)
29. Ji, Z., Wang, H., She, X.: A novel CdS quantum dots decorated 3D $Bi_2O_2CO_3$ hierarchical nanoflower with enhanced photocatalytic performance. **10**, 1046 (2020)
30. Noor, U., Farid, M.F., Sharif, A., Saleem, A., Nabi, Z., Mughal, M.F., Abbas, K., Ahmed, T.: Functionalized carbon 1D/2D nanomaterials for effective water desalination: synthesis, applications and cost issues. An overview. Desalination **571**, 117086 (2024)
31. Lin, S.-Q., Tan, H., Fu, P.-H., Liu, J.-F.: Interaction-driven Chern insulating phases in the $\alpha-T3$ lattice with Rashba spin-orbit coupling. iScience. **26**, 107546 (2023)
32. Kulkarni, S.K.: Introduction to Quantum Mechanics, Nanotechnology: Principles and Practices, pp. 1–29. Springer International Publishing, Cham (2015)
33. Lee, K.W., Lee, Y.J., Lee, C.E.: Electron spin resonance study of mixed Bi_2Se_3 topological insulator–magnetic insulator nanoparticles. J. Phys. Chem. Solids **182**, 111554 (2023)
34. Takehana, K., Takamasu, T., Kido, G., Henini, M., Eaves, L.: Anomalous quantum Hall effect induced by nearby quantum dots. Physica E **34**, 148–151 (2006)
35. Saleh, T.A.: Trends in nanomaterial types, synthesis methods, properties and uses: toxicity, environmental concerns and economic viability. Nano-Struct. Nano-Objects **37**, 101109 (2024)
36. Qiu, L.-P., Zhang, Y.-R., Gao, S.-L., Zheng, Q.-H., Zhang, T.-T., Cheng, G.-T., Cao, S.-Z., Hang, W.-P., Ramakrishna, S., Long, Y.-Z.: Fabrication and magnetic properties of SiO_2 nanoparticles-doped BSCCO superconducting nanofibers by solution blow spinning. Physica C: Supercond. Appl. **608**, 1354251 (2023)
37. Chen, X., Sun, D., He, Z., Kang, S., Miao, Y., Li, Y.: Ferrite bismuth-based nanomaterials: from ferroelectric and piezoelectric properties to nanomedicine applications. Colloids Surf. B **233**, 113642 (2024)
38. Chkhartishvili, L., Rukhadze, L., Margiev, B., Tsagareishvili, O., Darchiashvili, M.: 6—Carbon and isostructural boron nitride nanomaterials doped with ferromagnetic clusters. In: Hussain, C.M., Patankar, K.K. (eds.) Fundamentals and Industrial Applications of Magnetic Nanoparticles, pp. 165–233. Woodhead Publishing (2022)
39. Carbonaro, C.M., Engelbrecht, L., Olla, C., Cappai, A., Casula, M.F., Melis, C., Stagi, L., Laaksonen, A., Mocci, F.: 13—Graphene quantum dots and carbon nanodots: modeling of

zero-dimensional carbon nanomaterials. In: Joseph, K., Wilson, R., George, G., Appukuttan, S. (eds.) Zero-Dimensional Carbon Nanomaterials, pp. 411–482. Woodhead Publishing (2024)
40. Mukherjee, N., Maity, P., Ranu, B.C.: Chapter 6—Use of ball milling for the synthesis of biologically active heterocycles. In: Brahmachari, G. (ed.) Green Synthetic Approaches for Biologically Relevant Heterocycles (Second Edition), pp. 167–187. Elsevier (2021).
41. Marks, M.R., Cheong, K.Y., Hassan, Z.: A review of laser ablation and dicing of Si wafers. Precis. Eng. **73**, 377–408 (2022)
42. Modi, P.D., Mehta, V.N., Prajapati, V.S., Patel, S., Rohit, J.V.: Chapter 2—Bottom-up approaches for the preparation of carbon dots. In: Kailasa, S.K., Hussain, C.M. (eds.) Carbon Dots in Analytical Chemistry, pp. 15–29. Elsevier (2023)
43. Pottathara, Y.B., Grohens, Y., Kokol, V., Kalarikkal, N., Thomas, S.: Chapter 1—Synthesis and processing of emerging two-dimensional nanomaterials. In: Beeran Pottathara, Y., Thomas, S., Kalarikkal, N., Grohens, Y., Kokol, V. (eds.) Nanomaterials Synthesis, pp. 1–25. Elsevier (2019)
44. Haque, F., Rahman, M.Z.: Carbon-based nanomaterials and nanocomposites synthesis, characterization, properties and applications: a review. In: Reference Module in Materials Science and Materials Engineering. Elsevier (2024)
45. Saleh, T.A., Gupta, V.K.: Chapter 6—Synthesis of nanomaterial–incorporated membranes by physical methods. In: Saleh, T.A., Gupta, V.K. (eds.) Nanomaterial and Polymer Membranes, pp. 161–186. Elsevier (2016)
46. Lee, J.-S., Park, K., Kang, M.-I.L., Park, I.L.W., Kim, S.-W., Cho, W.K., Han, H.S., Kim, S.: ZnO nanomaterials synthesized from thermal evaporation of ball-milled ZnO powders. J. Cryst. Growth **254**, 423–431 (2003)
47. Verma, A., Kumar, U., Chaudhary, P., Yadav, B.C.: Investigation on structural and optical properties of porous SnO_2 nanomaterial fabricated by direct liquid injection chemical vapour deposition technique. Solid State Commun. **348–349**, 114723 (2022)
48. Xie, Z., Zhang, D., Yang, B., Qu, T., Liang, F.: Regulation of high value-added carbon nanomaterials by DC arc plasma using graphite anodes from spent lithium-ion batteries. Waste Manage. **174**, 88–95 (2024)
49. Davis, E.M., Ribeiro, E.L., Kamali, S., Wen, J., Johnson, C., Johnson, J., Khomami, B., Mukherjee, D.: Fe-carbide/Fe-oxide-based nanocomposites synthesized as magnetic nanomaterials via laser ablation synthesis in solution (LASiS). Appl. Surf. Sci. **658**, 159682 (2024)
50. Ngo, T.T., Viaña, J.M., Romero, M., Calvo, M.E., Lozano, G., Míguez, H.: Enhancement of upconversion photoluminescence in phosphor nanoparticle thin films using metallic nanoantennas fabricated by colloidal lithography††. Electronic supplementary information (ESI) available. Mater. Adv. **4**, 6381–6388 (2023). https://doi.org/10.1039/d3ma00775h
51. Mattox, D.M.: Handbook of physical vapor deposition (PVD) processing. William Andrew (2010)
52. Arora, N., Sharma, N.N.: Arc discharge synthesis of carbon nanotubes: comprehensive review. Diam. Relat. Mater. **50**, 135–150 (2014)
53. Borane, N., Aralekallu, S., Boddula, R., Singh, J., Kurkuri, M.D.: Chapter 4—Recent trends in the "bottom-up" and "top down" techniques in the synthesis and fabrication of myriad carbonaceous nanomaterials. In: Biswas, K., Mohanta, Y.K., Mohanta, T.K., Saravanan, M. (eds.) Carbon-Based Nanomaterials in Biosystems, pp. 91–120. Academic Press (2024)
54. Zhou, G., Wu, H., Deng, Y., Miao, R., Lai, D., Deng, J., Zhang, J., Chen, Q., Shao, Q., Shao, C.: Synthesis of high-quality multi-walled carbon nanotubes by arc discharge in nitrogen atmosphere. Vacuum **225**, 113198 (2024)
55. Russo, R.E., Mao, X., Liu, H., Gonzalez, J., Mao, S.S.: Laser ablation in analytical chemistry—a review. Talanta **57**, 425–451 (2002)
56. Van Minh, N., Van Hieu, D., Van Tuan, P., Dung, N.D., Manh Hoang, C.: Characteristics of silicon nano-pillars fabricated by nano-sphere lithography and metal assisted chemical etching. Mater. Sci. Semicond. Process. **142**, 106483 (2022)
57. Sharma, A., Jaiswal, A., Jaiswal, R.P.: Fabrication of a tailorable polystyrene nanoscale mesh of honeycomb morphology using nano sphere lithography. Thin Solid Films **763**, 139567 (2022)

58. Ishikawa, A., Tanaka, T.: Metamaterial absorbers for infrared detection of molecular self-assembled monolayers. Sci. Rep. **5**, 12570 (2015)
59. Abid, N., Khan, A.M., Shujait, S., Chaudhary, K., Ikram, M., Imran, M., Haider, J., Khan, M., Khan, Q., Maqbool, M.: Synthesis of nanomaterials using various top-down and bottom-up approaches, influencing factors, advantages, and disadvantages: a review. Adv. Coll. Interface. Sci. **300**, 102597 (2022)
60. Armarego, W.L.F.: Chapter 7—Nanomaterials and nanotechnology. In: Armarego, W.L.F. (ed.) Purification of Laboratory Chemicals (Eighth Edition), pp. 1065–1106. Butterworth-Heinemann (2017)
61. Kim, S., Kang, S., Baek, S., Song, J., Mun, N.-E., Kwon, H., Kwon, H.-G., Pu, Y.-J., Lee, T.-W., Yoo, S., Oh, J.-M., Park, J., Kim, S.-W.: Highly thin film with aerosol-deposited perovskite quantum dot/metal oxide composite for perfect color conversion and luminance enhancement. Chem. Eng. J. **441**, 135991 (2022)
62. Mekuye, B., Abera, B.: Nanomaterials: an overview of synthesis, classification, characterization, and applications. **4**, 486–501 (2023)
63. Harish, V., Ansari, M.M., Tewari, D., Gaur, M., Yadav, A.B., García-Betancourt, M.-L., Abdel-Haleem, F.M., Bechelany, M., Barhoum, A.: Nanoparticle and nanostructure synthesis and controlled growth. Methods **12**, 3226 (2022)
64. Mimani, T., Patil, K.: Solution combustion synthesis of nanoscale oxides and their composites. Mater. Phys. Mech. (Russia) **4**, 134–137 (2001)
65. Parauha, Y.R., Sahu, V., Dhoble, S.J.: Prospective of combustion method for preparation of nanomaterials: a challenge. Mater. Sci. Eng. B **267**, 115054 (2021)
66. Gore, S.K., Jadhav, S.S., Tumberphale, U.B., Raut, S.D.: Chapter 2—Sol-gel technology for the synthesis of metal oxide nanostructures. In: Mane, R., Jadhav, V., Al-Enizi, A. (eds.) Solution Methods for Metal Oxide Nanostructures, pp. 19–38. Elsevier (2023)
67. Dobkin, D., Zuraw, M.K.: Principles of chemical vapor deposition. Springer Science & Business Media (2003)
68. Nikam, A.V., Prasad, B.L.V., Kulkarni, A.A.: Wet chemical synthesis of metal oxide nanoparticles: a review. CrystEngComm **20**, 5091–5107 (2018)
69. Rodriguez-Sanchez, L., Blanco, M.C., López-Quintela, M.A.: Electrochemical synthesis of silver nanoparticles. J. Phys. Chem. B **104**, 9683–9688 (2000)
70. Li, G.-R., Xu, H., Lu, X.-F., Feng, J.-X., Tong, Y.-X., Su, C.-Y.: Electrochemical synthesis of nanostructured materials for electrochemical energy conversion and storage. Nanoscale **5**, 4056–4069 (2013)
71. Liu, H., Wang, X., Lan, Z.: Cryogenic ball milling synthesis of Ag_3PO_4/h-BN nanoparticles with increased performance for photocatalytic oxygen evolution reaction. Ceram. Int. **45**, 16682–16687 (2019)
72. Modan, E.M., Plăiașu, A.G.: Advantages and disadvantages of chemical methods in the elaboration of nanomaterials. In: The Annals of "Dunarea de Jos" University of Galati. Fascicle IX, Metallurgy and Materials Science, 43, pp. 53–60 (2020)
73. Tulinski, M., Jurczyk, M.: Nanomaterials synthesis methods. In: Metrology and Standardization of Nanotechnology: Protocols and Industrial Innovations, pp. 75–98 (2017)
74. Kulkarni, S.K.: Synthesis of Nanomaterials—III (Biological Methods), Nanotechnology: Principles and Practices, pp. 111–123. Springer International Publishing, Cham (2015)
75. Shobha, N., Nanda, N., Giresha, A.S., Manjappa, P., Dharmappa, K.K., Nagabhushana, B.M.: Synthesis and characterization of Zinc oxide nanoparticles utilizing seed source of Ricinus communis and study of its antioxidant, antifungal and anticancer activity. Mater. Sci. Eng. C **97**, 842–850 (2019)
76. Enamala, M.K., Chavali, M., Pamanji, S.R., Tangellapally, A., Dixit, R., Singh, M., Kuppam, C.: Chapter 5—Greener synthesis of enzymes from marine microbes using nanomaterials. In: Kharisov, B., Kharissova, O. (eds.) Handbook of Greener Synthesis of Nanomaterials and Compounds, pp. 109–136. Elsevier (2021)
77. Saleh, H.M., Hassan, A.I.: Chapter 21—Nanomaterials fabrication: bacteria-assisted synthesis. In: Kharisov, B., Kharissova, O. (eds.) Handbook of Greener Synthesis of Nanomaterials and Compounds, pp. 683–700. Elsevier (2021)

78. Jagessar, R.: Plant extracts based nanoparticles, a good perspective in the development of drugs in nanomedicine. Mod. Approaches Drug Des. **3** (2020)
79. Narayanan, K.B., Sakthivel, N.: Biological synthesis of metal nanoparticles by microbes. Adv. Coll. Interface. Sci. **156**, 1–13 (2010)
80. Priyadarshini, S., Gopinath, V., Priyadharsshini, N.M., MubarakAli, D., Velusamy, P.: Synthesis of anisotropic silver nanoparticles using novel strain, Bacillus flexus and its biomedical application. Colloids Surf. B **102**, 232–237 (2013)
81. Guilger-Casagrande, M., Lima, R.: Synthesis of silver nanoparticles mediated by fungi: a review. Fronti. Bioeng. Biotechnol. **7** (2019)
82. Sanket, S., Das, S.K.: Role of enzymes in synthesis of nanoparticles. In: Thatoi, H., Mohapatra, S., Das, S.K. (eds.) Bioprospecting of Enzymes in Industry, Healthcare and Sustainable Environment, pp. 139–153. Springer Singapore, Singapore (2021)
83. Karunakaran, G., Sudha, K.G., Ali, S., Cho, E.-B.: Biosynthesis of nanoparticles from various biological sources and its biomedical applications. Molecules **28**, 4527 (2023)
84. Nguyen Tri, P., Ouellet-Plamondon, C., Rtimi, S., Assadi, A.A., Nguyen, T.A.: Chapter 3—Methods for synthesis of hybrid nanoparticles. In: Mohapatra, S., Nguyen, T.A., Nguyen-Tri, P. (eds.) Noble Metal-Metal Oxide Hybrid Nanoparticles, pp. 51–63. Woodhead Publishing (2019)
85. Imran, H.J., Hubeatir, K.A., Aadim, K.A., Abd, D.S.: Preparation methods and classification study of nanomaterial: a review. J. Phys. Conf. Ser. **1818**, 012127 (2021)

Chapter 3
Microwave Characterization

Abstract Naturally occurring materials do not necessarily offer electromagnetic shielding. Hence, unique materials with various compositions must be synthesized to achieve shielding effectiveness. Once these materials are designed, they must be characterized in the frequency range of interest. This chapter delves into the various methods of microwave characterization. The chapter also explains multiple industry-standard characterization methods with real-world examples. A detailed analysis of material properties using the *S*-parameter system is established. These concepts would give insight to the material designer in designing materials with high absorption. The ideas of Shielding Effectiveness (SE) are also explored in this chapter, and numerous use cases and real-world examples are provided.

3.1 Introduction

The materials designed must be characterized for their respective electromagnetic performance [1]. Synthesis alone would not qualify materials for their respective EM performance [2]. Shielding Effectiveness (SE) is one of the critical performance metrics to qualify materials for EM absorption. Shielding effectiveness could be measured using several techniques [3] discussed in this chapter. The type and design of the material play an essential role in the choice of the measurement method.

Even though the electromagnetic spectrum runs from ELF (shallow frequency) 1 to 30 Hz to Gamma radiations, with frequencies in the range of 10^{12} Hz [4], not all radiations are fundamental in EM absorption. The reason is that EMI is encountered mainly through devices and systems in the consumer market [5]. Hardware designed in the ELF band would require insanely huge antennas for transmission; the data rate would be pathetically low in addition to this challenge [6]. Furthermore, ELF bands are reserved for niche communications of low-data rate applications, such as the transmission of control signals. The question of shielding would not arise.

On the other hand, Gamma radiations would be radiated during nuclear reactions. These applications also seem to be niche and restricted to limited geographies. In other words, the commercial adoption of Gamma rays to design wireless systems

S. K. Koul et al., *Microwave Absorbers*, Engineering Materials,
https://doi.org/10.1007/978-981-96-9596-6_3

is poor. Even though adopting Gamma rays would deliver high data rates, their harmful ionizing nature prevents their use in wireless data transfer. It is well-known that lead would act as an absorber for Gamma rays [6]. It must also be noted that the researchers' interest in developing shielding materials for extremely high-frequency radiation is relatively low.

With the onset of 5G communication systems, millimetre waves are helpful for high data throughput systems. Millimetre waves typically are in the frequency range of 30 to 300 GHz; these carrier waves offer high data rates due to the relatively large available bandwidth compared to their previous wireless generations [7]. In other words, as companies design wireless transceivers centered around the 30 GHz bands, the EMI in these bands will also increase. To mitigate the EMI arising from the millimetre wave 5G devices, it is essential to designing specific materials that deliver high shielding effectiveness in the 5G bands and beyond [8].

On the other end of the commercially viable spectrum is the terahertz bands. Several hardware architectures involving THz have been deployed in a variety of applications. THz radiation would have a very low wavelength in the 30 μm to 3 mm range. This feature inhibits its propagation. Nevertheless, researchers have designed and characterized laboratory prototypes for THz shielding [9].

Most commercial wireless systems are designed for the microwave region. For instance, the ubiquitous Wi-Fi is in the 2.4 and 5.2 GHz bands. On the other hand, 4G long-term evolution (LTE) transceivers are built in the 0.7 to 3 GHz bands. Most commercially operating satellite bands are in the X-band, 8 to 12 GHz. These points illustrate that the potential EMI sources could be spread across the spectra above, especially in the 1 to 20 GHz range. Hence, it would be logical to design and characterize materials in the microwave bands.

Typically, absorbers could be designed to cover the entire microwave spectrum. These are normally the carbon granules impregnated pyramidal absorbers. These absorbers can achieve broad impedance matching and deliver superior shielding in the 60 to 100 dB range. The fundamental problem with these absorbers is their physical size, which inhibits their applications in consumer devices. It must also be noted that these absorbers are effective in one direction, inhibiting their application in bi-directional applications. Typically, these absorbers are commercially used to design RF-shielded anechoic chambers and EMI/Electromagnetic Compatibility (EMC) chambers [10].

Hence, material design for microwave absorption is an essential field of research, especially for commercial devices. Numerous methods characterize the microwave performance of materials. The choice of characterization method depends on the operation frequency and the application at hand. This chapter presents a glimpse of significant characterization methods and their technical nuances.

3.2 Measurement Techniques

There are several techniques to characterize the Shielding Effectiveness (SE) of a given material. Each of these techniques will be discussed in this section.

3.2.1 Waveguide Method

In the waveguide method of microwave characterization, the sample is placed within the sample holder of an empty or air-filled waveguide; a representative schematic of the method is sketched in Fig. 3.1. The primary measurement instrument is a Vector Network Analyzer (VNA). This instrument gives the *S*-parameters of the device under test (DUT). It must be noted that for material characterization, a two-port VNA is essential [11]. The ports of the VNA are typically connected to standard cables, which in turn would be connected to the coax-to-waveguide adapter; this adapter converts the coaxial mode to the dominant rectangular waveguide mode. Commercially available adapters are typically frequency-dependent. For instance, an S-band waveguide adapter, WR340, operates in the 2.2 to 3.3 GHz band. On the other hand, the X-band waveguide adapter, WR 900, operates in the 8.2 to 12.4 GHz band. The dimensions of the waveguide aperture typically decide the operating frequency. Most waveguide apertures are designed with the broad wall to be half wavelength in free space, wherein the wavelength is computed at the lowest operating frequency. The narrow wall of the aperture would be half the value of the broad wall. The sample must be placed between two identical waveguides as illustrated in the schematics, Fig. 3.1.

The operating spectrum decides the geometry of the sample. The material engineer must anticipate the operating frequency range before synthesizing the material. The material under test must be mechanically stable, as this sample would be placed within the waveguide adapters. Table 3.1 presents the typical dimensions required for microwave characterization for popular bands; the thickness of the sample must be at least $\lambda/4$, wherein λ is calculated at the center frequency.

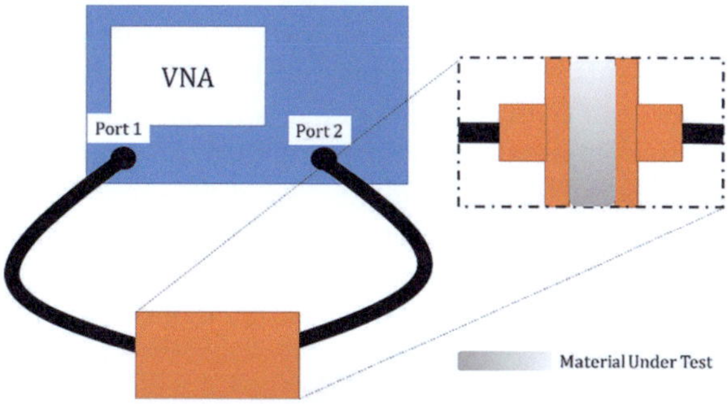

Fig. 3.1 Schematic for waveguide method

Table 3.1 Sample dimensions for various frequency bands with adapter nomenclature

Waveguide adapter	Frequency range (in GHz)	Sample dimensions (in mm)
WR1800	0.45–0.63	457.2 × 228.6 × 150
WR340	2.2–3.3	86.36 × 43.18 × 27.7
WR90	8.2–12.4	22.86 × 10.16 × 7.5
WR28	26.5–40	7.11 × 3.55 × 2.67
WR12	60–90	3.098 × 1.54 × 1.25

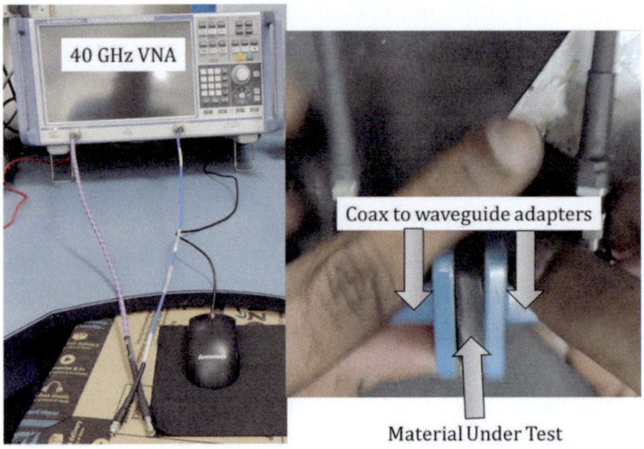

Fig. 3.2 Photograph of a 40 GHz VNA and X-band waveguide sample holder

If the sample is electrically thick, the ensemble of the waveguide adapters with the sample will radiate laterally. On the other hand, if the sample is electrically too thin, the absorption phenomenon would not be observed in the VNA. Figure 3.2 shows a photograph of a 40 GHz VNA and X-band Waveguide Sample holder.

The cables must be calibrated to the reference plane of the waveguide adapter. The sample photographed in Fig. 3.2 is prepared by compression molding. The sample contains different nanofillers, which deliver high shielding effectiveness through absorption. If the calibration seems improper, the material engineer can initially measure the S-parameters of the waveguides without any sample. Once the air-filled waveguide adapters are connected to the VNA, the VNA must show a flat line of 0 dB for $|S_{21}|$ throughout the expected operating frequency range. The reflection coefficients in both ports must be less than − 60 dB across the band; if the $|S_{21}|$ is lower than 0 dB, the calibration is faulty, or the waveguide sample is radiative. If the transmission loss is non-zero, then the values of $|S_{21}|$ in the empty waveguides could be subtracted from the $|S_{21}|$ of the sample to eliminate the error. Waveguide measurements are popular for narrowband frequency spectra. If a waveguide response is required, the engineer must recast the sample per the dimensional requirements of the respective waveguides. This method of characterization involves a lot of material in the lower frequencies.

3.2.2 *Free Space Method*

The free space method could characterize materials over multi-octave bandwidths, such as 1 to 40 GHz or 10 to 400 GHz. Figure 3.3 illustrates a generalized measurement topology of the free space method. A properly calibrated VNA is connected to two wideband planar antennas [12]. The material under test is placed in the middle to characterize it. It must be placed with an air gap along the axis of the radiating antennas.

It may be noted that the material must be electrically large to capture the absorption characteristics. Typically, the material under test must be at least $10\lambda \times 10\lambda$, wherein λ is calculated at the lowest frequency of operation. Table 3.2 gives the typical sizes of the material under test required for characterization at different frequency bands [13].

The antenna used for measurements must operate in the intended frequency range; typically, wideband planar Vivaldi-type antennas would suffice for this application [12]. Care must be taken to ensure the antenna's gain is high with unidirectional radiation patterns in the operating spectrum. The VNA must be connected to the antennas without the material under test; the transmission loss for this case must be

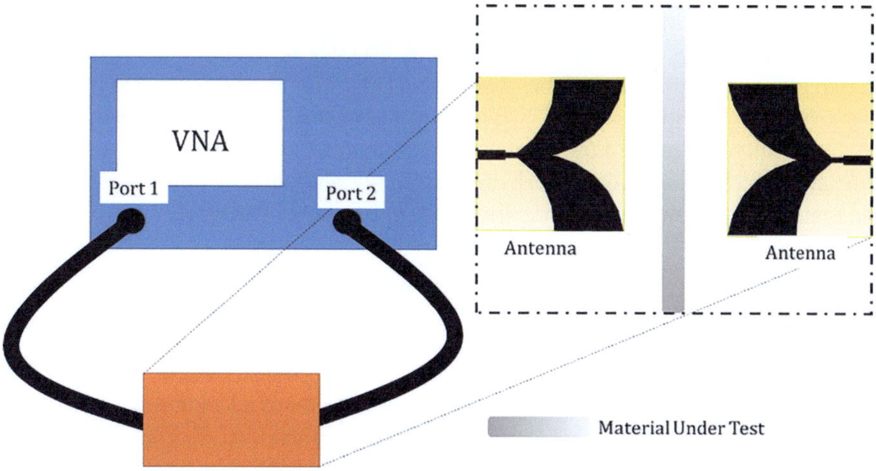

Fig. 3.3 Schematics for free space method

Table 3.2 Sample dimensions for various frequency bands

Band	Frequency range (in GHz)	Sample dimensions (in mm)
S-band	2–4	1500 × 1500
X-band	8–12	375 × 375
Ku-band	12–18	250 × 250
Ka-band	26.5–40	113 × 113

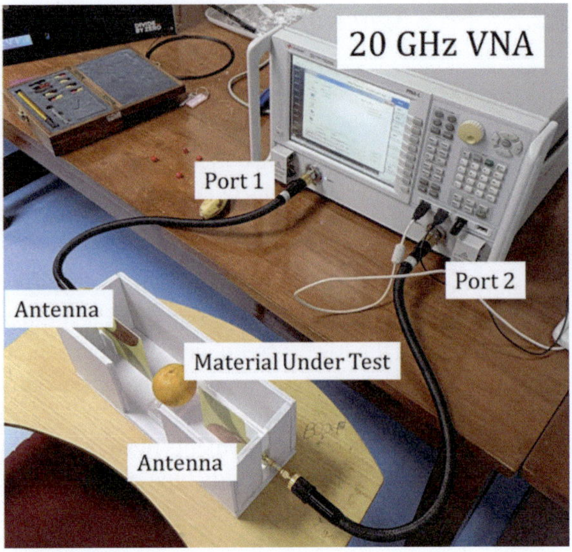

Fig. 3.4 Photograph of the setup for free space method

noted as $|S_{21}|$ free space. This value must be subtracted from the transmission loss measured with the material under test, $|S_{21}|$ material.

Figure 3.4 shows the setup photograph. A 20 GHz benchtop VNA is connected to two identical antipodal wideband antennas [14]. The antennas operate in the 2 to 10 GHz band. Sample material is placed in the center, and a 3D-printed case has been specially designed to mount the antennas and the material under investigation. This characterization method works for extremely wide bandwidths [15]. However, the material size requirements are also high, as shown in Table 3.2.

3.2.3 Coaxial Airline Method

A coaxial airline method could be used to solve the material requirement problem of the free space method. A general schematic is illustrated in Fig. 3.5. Here, a calibrated VNA is connected to a coaxial airline operating in a specific band [16]. As the propagation is Transverse Electromagnetic (TEM), the bandwidth of this characterization method is very high [14]. The coaxial line does not have a cut-off frequency; hence, the size of the coaxial setup is very compact, which starkly contrasts with the waveguide method. An impedance-matched air-filled coaxial line is mounted with the material under test, as shown in the inset of Fig. 3.5. Due to the material characteristics, the transmission and reflection coefficients are altered. The material under test must be molded into a cylinder that matches the inner dimensions of the coaxial airline. Air gaps between the material under test and the coaxial line must be minimal to mitigate measurement errors. The geometry of the material is a function of the test jig used for measurements and is independent of the frequency

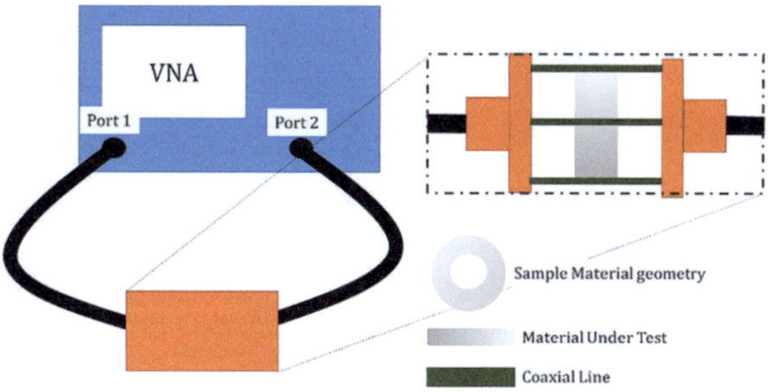

Fig. 3.5 Schematics for coaxial airline method

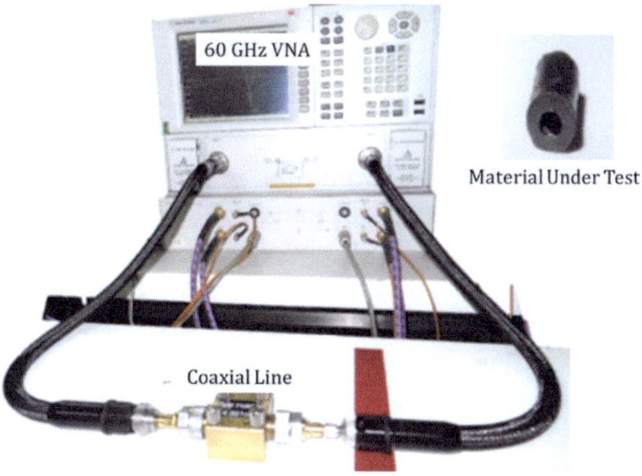

Fig. 3.6 Photograph for coaxial airline method

of operation. The MUT's thickness could be 10 mm for a 1 to 20 GHz frequency sweep.

Figure 3.6 shows a photograph of the coaxial airline measurement. A 60 GHz calibrated VNA is connected to the coaxial airline setup. The compression molded material sample, designed with nanocomposite fillers, is shown in the inset of the photograph. The coaxial airline method can work with extremely wide bandwidths [17]. The only catch is that the material sample's design must match the coaxial airline's dimensions.

3.2.4 Dielectric Measurements

The microwave characterization methods mentioned in Sects. 3.2.1 to 3.2.3 give the reflection and transmission coefficients. These values must be processed to extract the material's shielding effectiveness and dielectric constant under test. To avoid this additional step, an open-ended probe method could be used to estimate the dielectric properties of the material under test. A generalized schematic is illustrated in Fig. 3.7a. Here, a calibrated VNA is connected to a solid coaxial probe of known

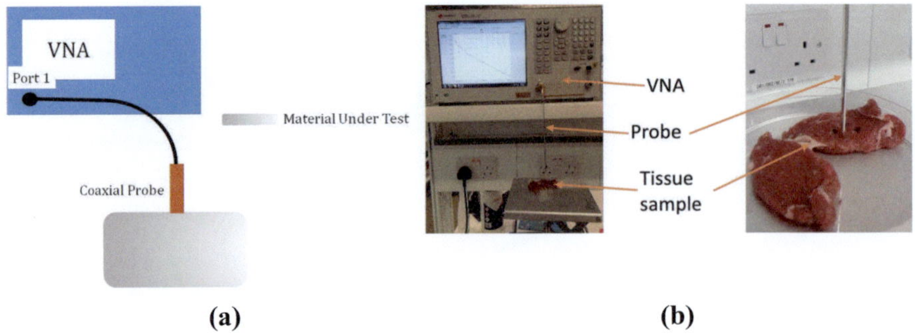

Fig. 3.7 Open-ended coaxial probe method: **a** schematics; **b** photograph of a measurement setup [21]

length. This probe, in turn, is mounted with the material under test. The calculator within the VNA would plot the real and imaginary parts of the material's dielectric constant under test in the specified frequency range [18]. The variation of the dielectric constant across the frequency range would give an estimate to the material designer to decide on the composition and nature of the absorber.

Absorbers typically have a high dielectric constant and a high dielectric loss tangent, aiding EM energy absorption. The open-ended coaxial method is suited for choosing the material or nanocomposite before the absorber design [19]. It must be noted that the coaxial probe method does not deliver the material's magnetic properties under test.

The absorptive and reflective characteristics of the MUT must be inferred through calculations using the experimental values of the permittivity. Often, this shielding effectiveness (SE) estimation method might be partial, as this method completely misses out on the losses arising from the magnetic materials. Certain classes of absorbers discussed in this book have both electric and magnetic fillers, which contribute to absorption. This method is quite popular for investigating the dielectric properties of biological tissues; one such example is illustrated in Fig. 3.7b [20].

3.3 Relationship Between *S*-Parameters and Material Characteristics

S-parameters are one of the common performance metrics for characterizing Radio Frequency (RF) and microwave devices in the context of absorber design. *S*-parameters deliver information about the attenuation and reflection happening exclusively due to the material [22]. A simulation model establishes a relationship between the fundamental material characteristics and *S*-parameters. The topology of the simulation model is illustrated in Fig. 3.8. Here, ports are used, and the material under test is inserted between the ports.

Fig. 3.8 Generalized setup for material characterization using waveguide method

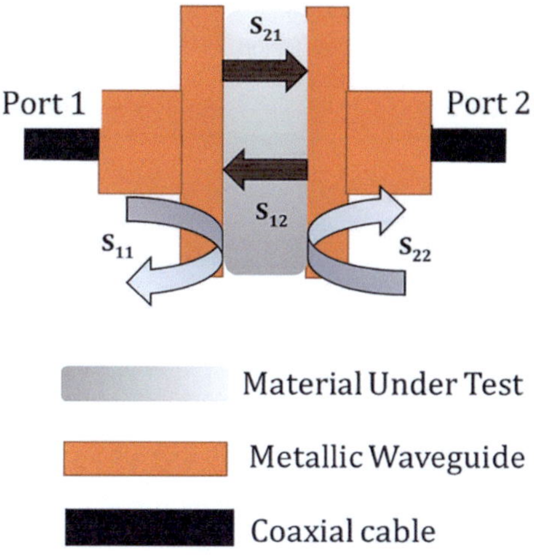

The schematic indicates the various S-parameters in the context of material absorber design. They are as follows: $|S_{21}|$ is forward transmission of the material under test in dB. $|S_{12}|$ is reverse transmission of the material under test in dB. $|S_{11}|$ is reflection coefficient at the interface of the material under test at Port 1 in dB. $|S_{22}|$ is reflection coefficient at the interface of the material under test at Port 2 in dB. It must be noted that all the S-parameters are in the logarithmic scale, dB, as it is the industry standard. The relationship between linear and dB scale is presented in (3.1) and (3.2) [23]:

$$|S_{21}| \text{ (in dB)} = 20\log_{10}|S_{21}|\text{linear} \tag{3.1}$$

$$|S_{11}| \text{ (in dB)} = 20\log_{10}|S_{11}|\text{linear} \tag{3.2}$$

As a rule of thumb, the $|S_{21}|$ forward transmission must be less than -20 dB. In addition to this criterion, the reflection coefficient, $|S_{11}|$, must be less than -20 dB [21]. Unless otherwise specified, these values are expected to be true for the reverse transmission loss, $|S_{12}|$, and the reverse reflection coefficient, $|S_{22}|$. Most synthesized materials typically have reciprocal characteristics, i.e., $|S_{12}| = |S_{21}|$ and $|S_{11}| = |S_{22}|$. If the material is nonreciprocal under certain circumstances, then the absorption phenomenon might be restricted to one of the directions of power flow. It must be noted that the transmission loss alone will not suffice to demonstrate absorption. For instance, a reflective type of material also exhibits very low $|S_{21}|$ close to $-\infty$ dB, but the reflection coefficient would be high, close to 0 dB, demonstrating reflection. The conditions above would be applicable irrespective of the microwave measurement method, as discussed in Sect. 3.2.

The industry-standard software Ansys HFSS® (High-Frequency Structure Simulator) builds and designs a simulation model, illustrated in Fig. 3.9. Here, an X-band

Fig. 3.9 Simulation setup for material characterization using waveguide method

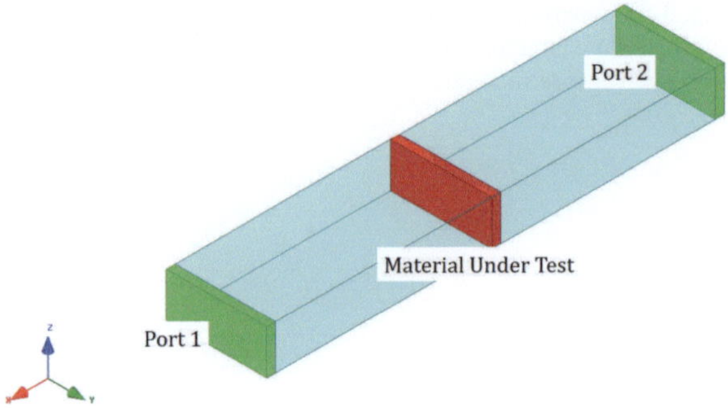

waveguide operating in the 8 to 12 GHz band is assumed. The waveguide's aperture matches the dimensions mentioned in Table 3.1.

The waveguide is assumed to be air-filled and 60 mm long. The length of the waveguide ensures only the dominant mode of propagation between the two ports. The software assumes a semi-infinite waveguide impinging linearly polarized EM waves to the waveguide. The polarization of the incoming electromagnetic wave is along the z-axis, concerning Fig. 3.9. The material under test is placed at the center of this waveguide. Various properties of the materials, such as permittivity (ε), dielectric loss tangent ($\tan\delta_e$), permeability (μ), and magnetic loss tangent ($\tan\delta_m$), are varied to illustrate the context of the ideal set of parameters for absorption.

The first simulation is when ε_r is varied with $\mu_r = 1$, $\tan\delta_e = \tan\delta_m = 0$. In this case, relative permittivity characterizes material behavior exclusively. The reflection coefficient for this case is illustrated in Fig. 3.10.

Here $\varepsilon_r = 1$, a non-existent material, the waveguide is empty, and electromagnetic power flows from Port 1 to Port 2. The transmission loss for this case is shown in Fig. 3.11. When $\varepsilon_r = 1$, there is no loss within the system as relative permittivity

Fig. 3.10 Input reflection coefficient for case 1

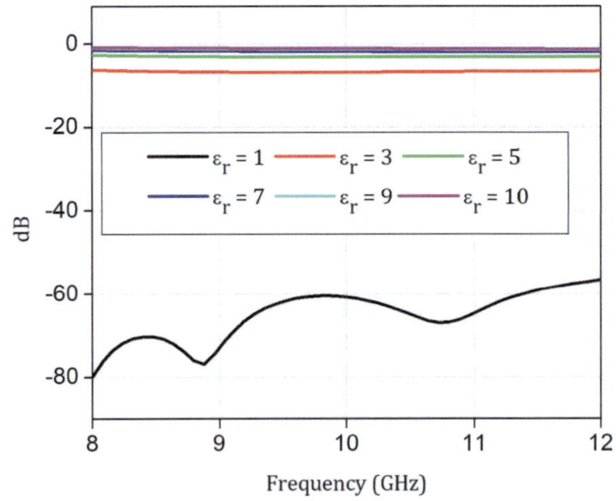

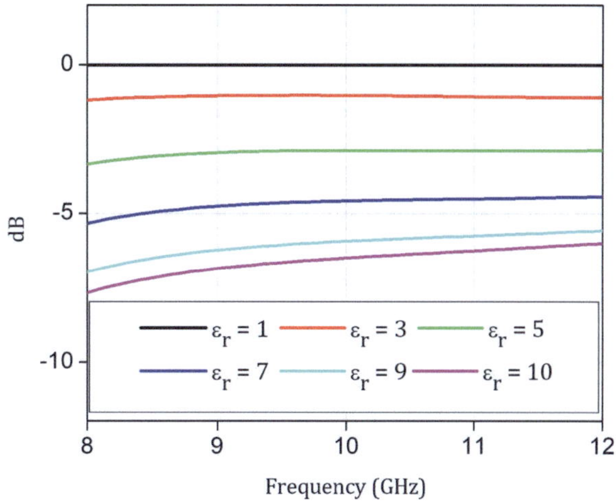

Fig. 3.11 Transmission loss for case 1

is increased. The reflective nature of the material is evident when $\varepsilon_r = 3$, when the material is moderately lossy.

The second simulation is to characterize the dielectric loss tangent, tan δ_e exclusively; the other parameters are fixed as follows: $\mu_r = 1$ (non-magnetic) and tan $\delta_m = 0$ (no magnetic loss). The input reflection coefficient and transmission loss are illustrated in Figs. 3.12 and 3.13, respectively. When the dielectric loss tangent is zero, the power flow is continuous through the material. Typically, polymer metrics exhibit low dielectric loss tangents and moderate relative permittivity. As the dielectric loss tangent increases, the material exhibits reduced transmission. The fundamental definition of loss tangent is tan $\delta_e = \sigma/\omega\varepsilon$, where σ is the conductivity of the material under test, ω is the angular frequency, and ε is the permittivity of the material. According to this expression, the conductivity increases inherently, exhibiting a reflective nature as the loss tangent is enhanced within the material.

The third simulation is to study the effects of ε_r and tan δ_e, with fixed permeability and magnetic loss tangent ($\mu_r = 1$ and tan $\delta_m = 0$). The input reflection coefficient for this case is shown in Fig. 3.14. It is evident that when the $\varepsilon_r = 1$ and tan $\delta_e = 0$, the propagation is highest.

Hence, $|S_{11}|$ is below $-$ 60 dB, with high transmission. The insertion loss for the condition above is illustrated in Fig. 3.15. Here, the transmission decreases with an increase in the electric loss tangent, but this cannot be attributed to the absorption phenomenon. The high reflection coefficient is due to the enhanced intrinsic conductivity of the material under test. These two figures show that the permittivity and electric loss tangents alone would not be sufficient to create an absorber.

In the fourth simulation studies, the effects of the material under the test's permeability on the transmission and reflection characteristics are studied. In this simulation, the permittivity, ε_r, is fixed at 1, tan $\delta_e = 10$, and tan $\delta_m = 0$. The reflection coefficient is illustrated in Fig. 3.16.

Fig. 3.12 Input reflection coefficient for case 2

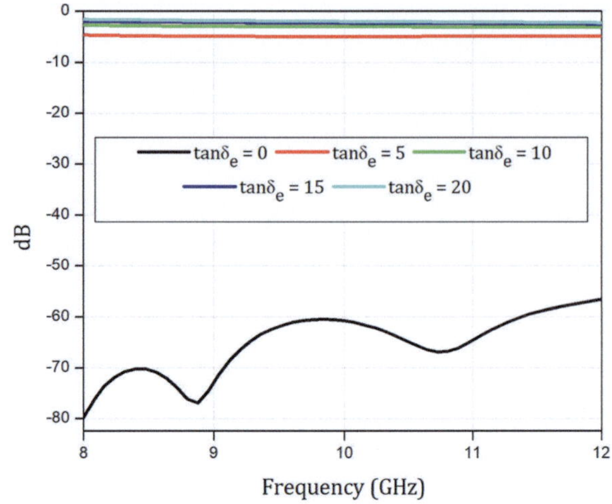

Fig. 3.13 Transmission loss for case 2

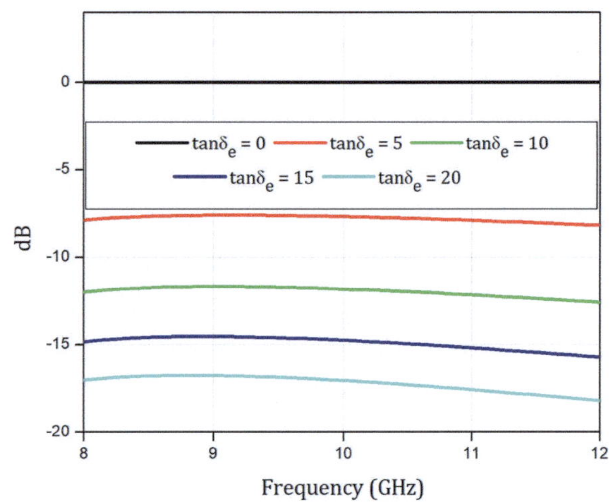

Figure 3.17 illustrates the transmission loss for this case. These graphs clearly show that permeability alone will not be able to deliver absorption. As μ_r increases, a slight increase in the reduction of the transmission coefficient is observed. On the other hand, the reflective nature of the material also increases, prohibiting absorption. The curves are similar to their electric counterparts depicted in Figs. 3.10 and 3.11.

The fifth simulation exclusively investigates the effects of magnetic loss. The other parameters are fixed: $\mu_r = 1$, $\varepsilon_r = 1$, and tan $\delta_e = 0$. The input reflection coefficient and the insertion loss curves for various cases are displayed in Figs. 3.18 and 3.19, respectively.

Like the dielectric loss tangent effects, the transmission coefficient goes below − 10 dB when the magnetic loss tangent coefficient increases to 15. On the other

Fig. 3.14 Input reflection coefficient for case 3

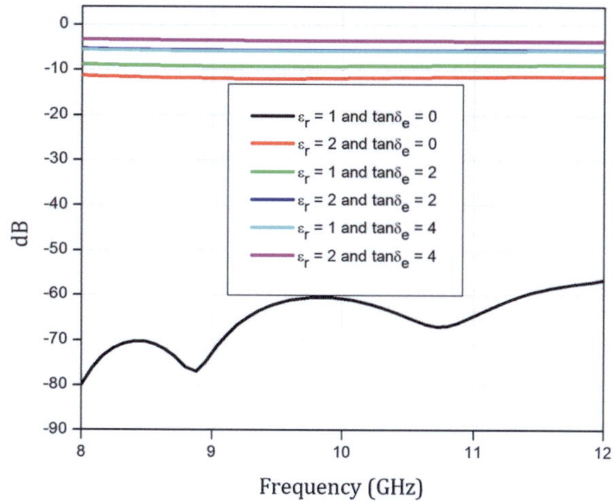

Fig. 3.15 Transmission loss for case 3

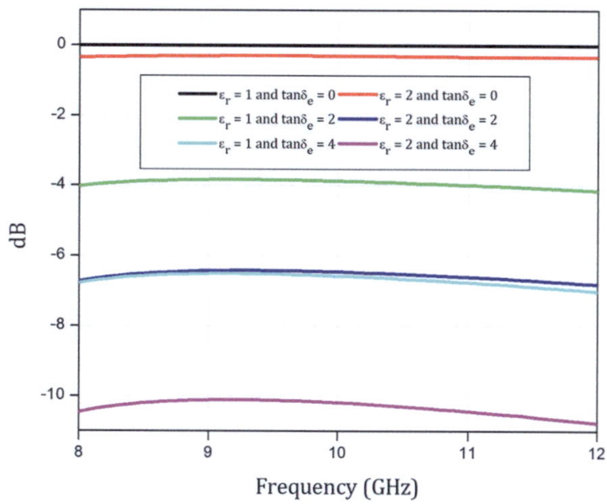

hand, the material becomes reflective and delivers a reflection coefficient close to 0 dB. The fourth and fifth simulations indicate that permeability and magnetic loss tangents alone would not be sufficient to create absorption within the material.

The last simulation to characterize the material under test is to investigate the effects of ε_r, μ_r, tan δ_e, and tan δ_m in the same material. Figures 3.20 and 3.21 illustrate the input reflection coefficient and insertion losses for various cases.

The true absorptive phenomenon is observed when $\varepsilon_r = \mu_r = 2$ and tan $\delta_e =$ tan $\delta_m = 5$. Here, the transmission is below − 30 dB throughout the operating spectrum. In addition to this feature, the reflection coefficient is below − 10 dB throughout the operating spectrum. All six simulations indicate that altering ε_r and μ_r exclusively would not produce an absorption phenomenon within the material. However, the right

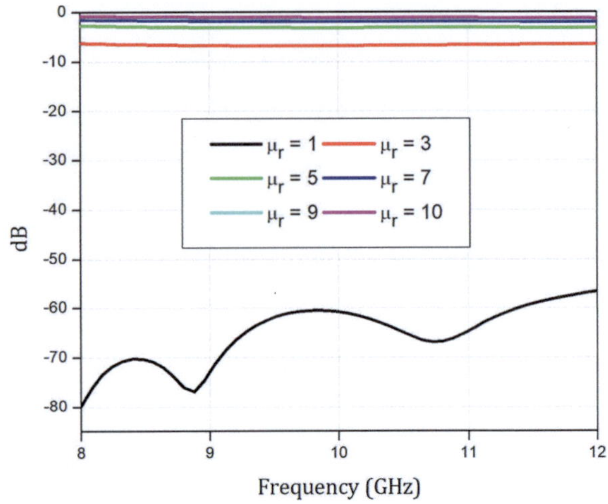

Fig. 3.16 Input reflection coefficient for case 4

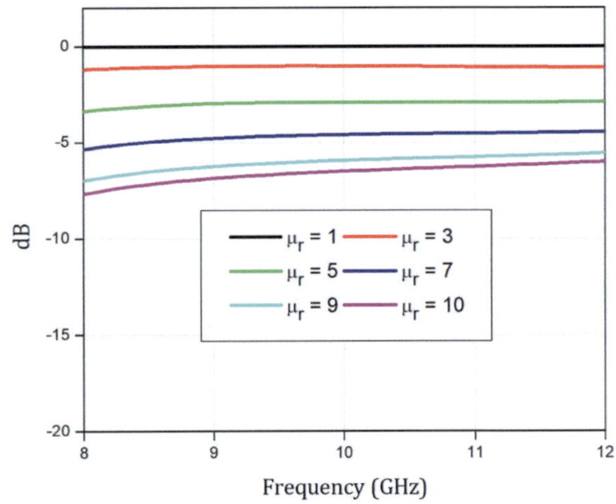

Fig. 3.17 Transmission loss for case 4

combination of ε_r, μ_r, tan δ_e, and tan δ_m would create absorption in the context of material design. The nanocomposites within the primary polymer matrix must adhere to the above conditions to achieve the absorption phenomenon. To understand the reflection and absorption phenomenon, the E-field plots at 10. GHz are presented.

When the material under test is a perfect absorber, $|S_{21}|$ is less than $-$ 30 dB, and $|S_{11}|$ is less than $-$ 20 dB, traveling waves are absorbed between Port 1 and the absorber. This phenomenon is captured in Fig. 3.22. The EM energy would be one-directional. These kinds of materials offer superior absorption. If this absorber is applied to various consumer devices, the energy penetration would be minimal.

Figure 3.23 illustrates the E-field propagation of a perfect reflector placed within an X-band waveguide. Here, the ideal electric conductor (PEC) would reflect the

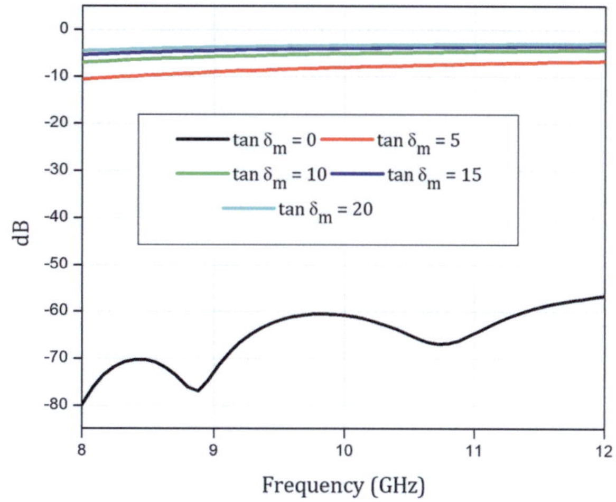

Fig. 3.18 Input reflection coefficient for case 5

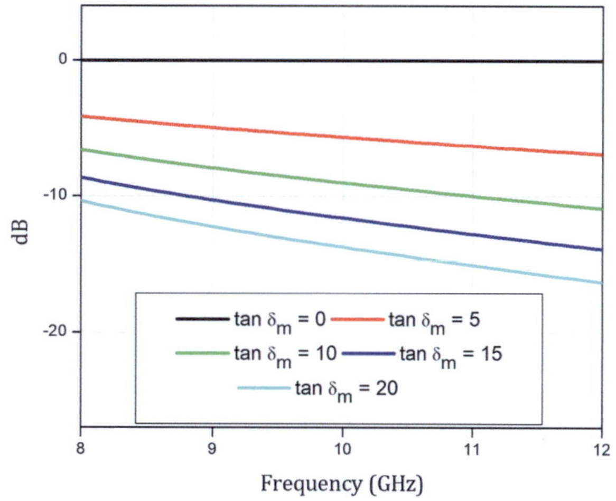

Fig. 3.19 Transmission loss for case 5

incoming waves to Port 1, thus creating standing waves in between. This type of material also offers shielding, but in a reflective way.

Most of the materials designed are partial absorbers, whose E-field propagation is illustrated in Fig. 3.24. Hybrid waves are formed between Port 1 and the material, meaning the material under test is partially reflective and absorptive.

Fig. 3.20 Input reflection coefficient for case 6

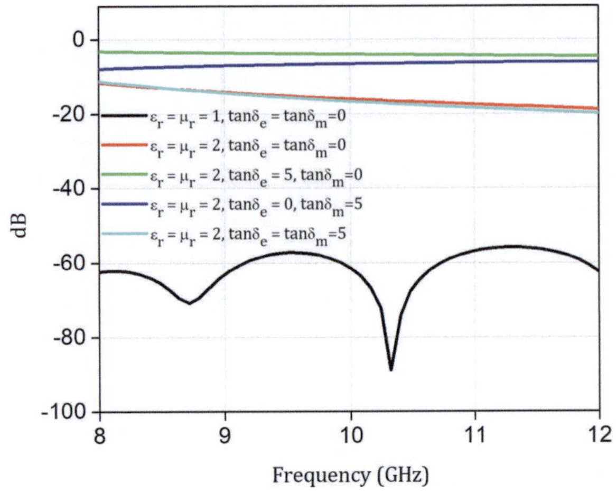

Fig. 3.21 Transmission loss for case 6

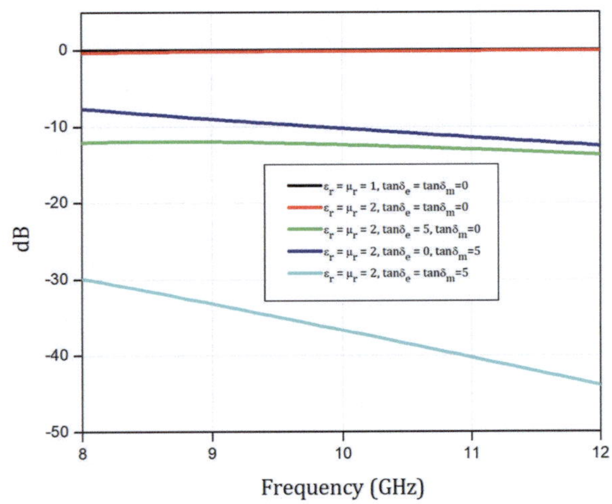

Fig. 3.22 E-field propagation of a perfect absorber at 10 GHz

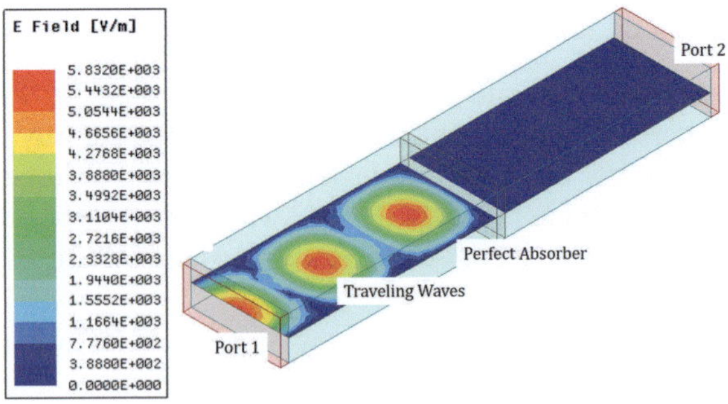

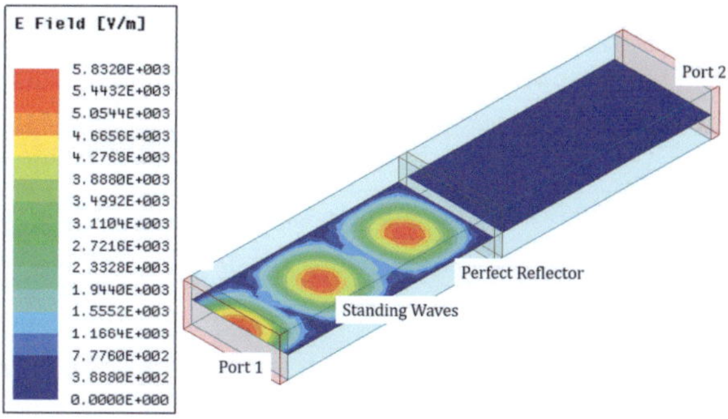

Fig. 3.23 E-field propagation of a perfect reflector at 10 GHz

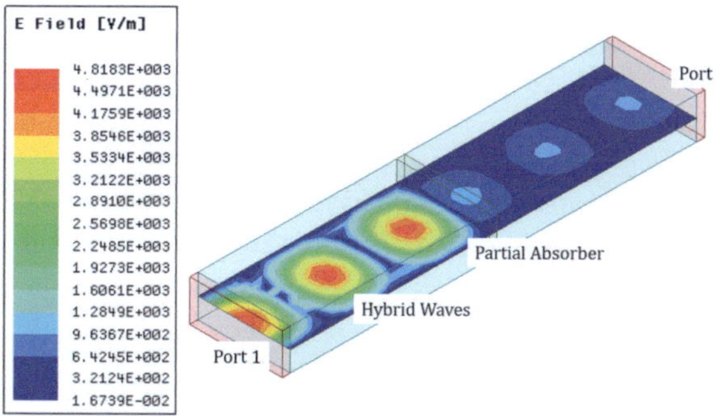

Fig. 3.24 E-field propagation of a partial absorber at 10 GHz

3.4 Shielding Effectiveness and S-Parameters

The measured S-parameters alone might not adequately characterize the material in the operating frequency comprehensively. Hence, the term shielding effectiveness is introduced. The first formula is the shielding effectiveness due to reflection SE_R. The relationship is presented in (3.3) as

$$SE_R = -10\log_{10}\left(1 - |S_{11}|^2\right) \text{ in dB} \tag{3.3}$$

Here, readers must note that the values for $|S_{11}|$ must be on a linear scale, not a logarithmic scale. Typically, instruments such as VNA export $|S_{11}|$ on a dB scale. These values must be converted to a linear scale using (3.2). The values for $|S_{11}|$ could range from 0 to 1. In other words, SE_R could range from 0 to $+\infty$ dB. When the material is purely reflective, $|S_{11}|$ would be close to 1 or 0 dB, which, in turn, would lead to SE_R being plus $+\infty$ dB. Similarly, shielding effectiveness due to absorption SE_A could be defined as (3.4):

$$SE_A = 10\log_{10}\left(1 - |S_{11}|^2/|S_{21}|^2\right) \text{ in dB} \tag{3.4}$$

Table 3.3 Shielding effectiveness of a generic class of absorbers

Type of material	$\lvert S_{11}\rvert$ in dB	$\lvert S_{21}\rvert$ in dB	SE_R in dB	SE_A in dB	SE_T in dB
Ideal absorber	$-\infty$	$-\infty$	0	$+\infty$	$+\infty$
Good absorber	< -20	< -20	< 0.045	> 19.95	> 20
Moderate absorber	< -10	< -10	< 0.46	> 9.54	> 10

The numerator indicates the actual transmitted power into the material. Hence, $\lvert S_{11}\rvert^2$ is subtracted from unity. In other words, the reflected component is eliminated to compute SE_A, and the denominator indicates the power flow from Ports 1 to 2. For a good absorber, $\lvert S_{21}\rvert$ should be close to zero or $-\infty$ dB, and $\lvert S_{11}\rvert$ also should be close to zero or $-\infty$ dB. This would lead to SE_A being $+\infty$ dB. The total shielding effectiveness SE_T is presented by (3.5) as follows:

$$SE_T = SE_A + SE_R \text{ in dB} \tag{3.5}$$

A good absorber would have a high SE_A and low SE_R. Table 3.3 gives an overview of typical, expected values from different types of absorbers.

3.5 Conclusion

Microwave characterization is essential for materials with shielding applications. Even though electromagnetic radiation has an extremely wide range of frequencies, only a certain portion of the spectrum is used to design and deploy for most commercial wireless systems. Hence, microwave characterization methods are essential to quantizing the performance metrics of absorber materials. This chapter explored numerous characterization methods and discussed shielding effectiveness regarding *S*-parameters.

References

1. Geetha, S., Satheesh Kumar, K.K., Rao, C.R.K., Vijayan, M., Trivedi, D.C.: EMI shielding: methods and materials—a review. J. Appl. Polym. Sci. **112**, 2073–2086 (2009)
2. Wang, X.-Y., Liao, S.-Y., Wan, Y.-J., Zhu, P.-L., Hu, Y.-G., Zhao, T., Sun, R., Wong, C.-P.: Electromagnetic interference shielding materials: recent progress, structure design, and future perspective. J. Mater. Chem. C **10**, 44–72 (2022)
3. Micheli, D., Pastore, R., Vricella, A., Delfini, A., Marchetti, M., Santoni, F.: Chapter 9—Electromagnetic characterization of materials by vector network analyzer experimental setup. In: Thomas, S., Thomas, R., Zachariah, A.K., Mishra, R.K. (eds.) Spectroscopic Methods for Nanomaterials Characterization, , pp. 195–236. Elsevier (2017)
4. J. Norgard, G.L.: Best, The electromagnetic spectrum. In: National Association of Broadcasters Engineering Handbook, pp. 3–10. Routledge (2017)

5. Bangerter, B., Talwar, S., Arefi, R., Stewart, K.: Networks and devices for the 5G era. IEEE Commun. Mag. **52**, 90–96 (2014)
6. Oglat, A.A.: Studying the radiation absorption and scattering of gamma rays by using different absorbers. Radiat. Phys. Chem. **176**, 109072 (2020)
7. Rappaport, T.S., Sun, S., Mayzus, R., Zhao, H., Azar, Y., Wang, K., Wong, G.N., Schulz, J.K., Samimi, M., Gutierrez, F.: Millimeter wave mobile communications for 5G cellular: it will work! IEEE Access **1**, 335–349 (2013)
8. Choudhary, A., Pal, S., Sarkhel, G.: Broadband millimeter-wave absorbers: a review. **15**, 347–363 (2023)
9. Cheng, X., Huang, R., Xu, J., Xu, X.: Broadband terahertz near-perfect absorbers. ACS Appl. Mater. Interfaces **12**, 33352–33360 (2020)
10. Nornikman, H., Soh, P.J., Azremi, A.A.H., Anuar, M.S.: Performance simulation of pyramidal and wedge microwave absorbers. In: 2009 Third Asia International Conference on Modelling & Simulation, pp. 649–654 (2009)
11. Manjappa, P., Rajan, H.K., Mahesh, M.G., Sadananda, K.G., Channegowda, M., Shivashankar, G.K., Mutt, N.B.: Effective attenuation of electromagnetic waves by synergetic effect of α-Fe_2O_3 and MWCNT/graphene in LDPE-based composites for EMI applications. Materials (2022)
12. Gazit, E.: Improved design of the Vivaldi antenna. IEE Proc. H (Microw. Ant. Propag.), 89–92 (1988)
13. Ozturk, T., Morikawa, O., Ünal, İ, Uluer, İ: Comparison of free space measurement using a vector network analyzer and low-cost-type THz-TDS measurement methods between 75 and 325 GHz. J. Infrared Millimeter Terahertz Waves **38**, 1241–1251 (2017)
14. Keam, R.B., Williamson, A.G.: Broadband design of coaxial line/rectangular waveguide probe transition. IEE Proc. Microw. Ant. Propag. **141**, 53–58 (1994)
15. Gorai, A., Karmakar, A., Pal, M., Ghatak, R.: A super wideband Chebyshev tapered antipodal Vivaldi antenna. AEU-Int. J. Electron. C. **69**, 1328–1333 (2015)
16. Venkatarayalu, N.V., Yuan, C.J.: Eliminating errors due to position uncertainty in coaxial airline based measurement of material parameters, pp. 464–467. IEEE
17. Scott, M.M., Morris, A.P., Reid, D.R., Bean, J.A.: Broadband isotropic and anisotropic permittivity determination using partially filled coaxial airlines. Microw. Opt. Technol. Lett. **57**, 1864–1868 (2015)
18. Marsland, T.P., Evans, S.: Dielectric measurements with an open-ended coaxial probe. IEE Proc. H (Microw. Ant. Propag.), 341–349 (1987)
19. Dilman, I., Akinci, M.N., Yilmaz, T., Çayören, M., Akduman, I.: A method to measure complex dielectric permittivity with open-ended coaxial probes. IEEE Trans. Instrum. Meas. **71**, 1–7 (2022)
20. La Gioia, A., O'Halloran, M., Porter, E.: Challenges of post-measurement histology for the dielectric characterisation of heterogeneous biological tissues. Sensors **20**, 3290 (2020)
21. Micheli, D., Pastore, R., Apollo, C., Marchetti, M., Gradoni, G., Primiani, V.M., Moglie, F.: Broadband electromagnetic absorbers using carbon nanostructure-based composites. IEEE Trans. Microw. Theory Tech. **59**, 2633–2646 (2011)
22. Abbas, S.M., Chandra, M., Verma, A., Chatterjee, R., Goel, T.C.: Complex permittivity and microwave absorption properties of a composite dielectric absorber. Compos. A Appl. Sci. Manuf. **37**, 2148–2154 (2006)
23. Yusoff, A.N., Abdullah, M.H., Ahmad, S.H., Jusoh, S.F., Mansor, A.A., Hamid, S.A.A.: Electromagnetic and absorption properties of some microwave absorbers. J. Appl. Phys. **92**, 876–882 (2002)

Chapter 4
Material Characterization

Abstract Material characterization is a vital aspect of materials science and engineering, as it involves understanding the properties and structures of materials at a microscopic level. The importance of material characterization, various techniques used for characterization, and the significance of accurate material characterization in research and development are discussed here. This chapter comprehensively covers the characterization techniques used for materials in EMI shielding applications. Most research scientists worldwide use these modern, standardized, advanced techniques. Material characterization is crucial in engineering as it allows researchers to determine the behavior and performance of materials for different applications. Understanding the properties of materials at a microscopic level helps develop new materials with improved properties and performance metrics. The techniques used for material characterization in EMI shielding applications are constantly evolving to keep up with technological advancements. Accurate material characterization is essential in research and development, as it ensures the reliability and effectiveness of the materials used in various engineering applications.

4.1 Introduction

Microscopy and spectroscopy characterize materials, analyze microstructural features using optical, SEM, and TEM techniques, and analyze chemical composition using NMR, Raman, FTIR, XRD, XRF, and SEM (EDS). Efficient and robust characterization techniques are crucial for future research and development, as they help confirm practical work on materials. These techniques are essential for understanding bulk properties and their microscopic structure. They can be used to characterize various materials for different applications, and their outputs are studied using various detectors to comment on the material structure. Two standard techniques for material characterization study are classification and interaction of EM waves with the material. Two standard techniques used for material characterization, namely classification of material characterization and interaction of EM wave with material are shown in Fig. 4.1.

S. K. Koul et al., *Microwave Absorbers*, Engineering Materials,
https://doi.org/10.1007/978-981-96-9596-6_4

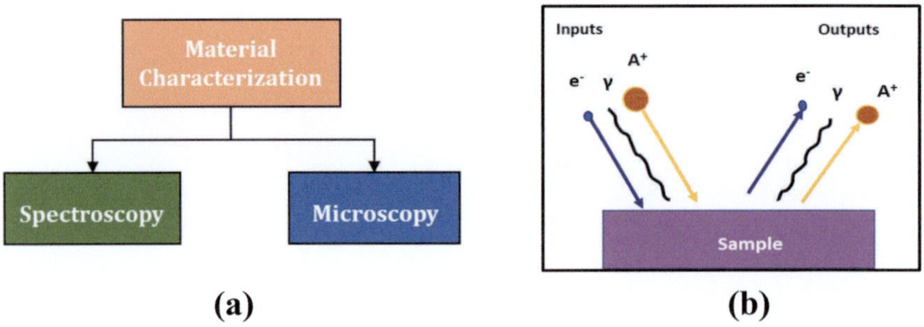

Fig. 4.1 **a** Classification of material characterization and **b** interaction of EM waves with material

Material characterization methods are essential for understanding various materials' properties and behavior. These methods involve analyzing materials' structure, composition, and properties at different scales, from the atomic to the macroscopic levels [1, 2]. By combining techniques such as microscopy, spectroscopy, and mechanical testing, researchers can gain valuable insights into the performance and durability of materials in different applications [3].

The data obtained from these characterization methods can help design optimization, quality control, and failure analysis of materials. Material characterization methods are crucial in advancing materials science and engineering by providing insightful information for developing new materials with improved properties and performance metrics [4]. Examination and comprehension of materials' physical, electrical, and chemical properties are characterizations [5]. It is a crucial component of materials engineering study that aids in determining whether materials are appropriate for a particular application [6].

This chapter includes a variety of topics for performing material characterization [7]. By applying these methods, engineers might improve their understanding of the material properties and optimize their designs for increased performance and reliability. This, in turn, aids in their decision-making regarding the application of customized materials and in developing safe and effective goods with certain behaviors [7, 8].

Materials science and engineering involve characterization and analyzing materials' visible and invisible properties to understand their behavior. This process collects information about chemical, physical, and mechanical properties, aiding in the design of composite materials with high shielding effectiveness [2].

This section will discuss various methods of characterizing materials. Furthermore, advanced imaging techniques, including transmission electron microscopy (TEM), scanning electron microscopy (SEM), and X-ray diffraction, will be presented, along with mechanical tests of materials like tensile and hardness tests [6].

4.2 Techniques for Material Characterization

Several techniques are used to characterize materials in various aspects so that the materials exhibit desired physical and chemical properties [9]. Spectroscopic techniques are popular in the industry to extract useful engineering information about the material. These techniques comprise extracting material data by letting the material interact with the EM waves and, following the contact, collecting the reflected, refracted, or transmitted data using relevant technology and software. The raw data can be depicted in graphical form that professionals at various levels can comprehensively analyze [4].

Spectroscopy is the dispersion of light into its component colors; it studies the interaction of these EM waves with matter. It is a technique to determine the intensity of light that passes through a test material [10, 11]. The light illuminated onto the test material would be composed of various frequencies. In other words, spectroscopy delivers information about the frequency response of the test material.

According to analytical science, every element or compound has its characteristic spectrum. Light is absorbed and dispersed by each chemical within a specific wavelength range. While spectroscopy is functional in Chemistry for studying or analyzing different chemical compounds or elements, it is also used in Physics to ascertain the composition of planets' atmospheres [4].

Spectroscopic techniques can be classified based on different criteria, such as the type of radiation used, the nature of the interaction of incident light with the test material, and the information obtained [10]. Here is a classification based on these criteria.

Electromagnetic Spectrum-Based Classification:

- **Optical Spectroscopy**: Involves using visible, ultraviolet, and infrared light.
- **Infrared Spectroscopy**: Focuses on the absorption, emission, or reflection of infrared radiation, which is beneficial for studying molecular vibrations.
- **UV–Visible Spectroscopy**: Deals with the absorption, emission, or scattering of ultraviolet and visible light.
- **X-Ray Spectroscopy**: Determines the electronic structure of atoms and molecules.
- **Raman Spectroscopy**: Investigates molecules' vibrational and rotational modes through the inelastic scattering of monochromatic light, typically laser light. The electromagnetic spectrum is shown in Fig. 4.2.

4.2.1 Fourier Transform Infrared Spectroscopy (FTIR)

Fourier Transform Infrared Spectroscopy (FTIR) analyzes materials in the infrared part of the electromagnetic spectrum. Infrared radiation has a longer wavelength and lower frequency than visible light. Absorption spectroscopy is fundamental in

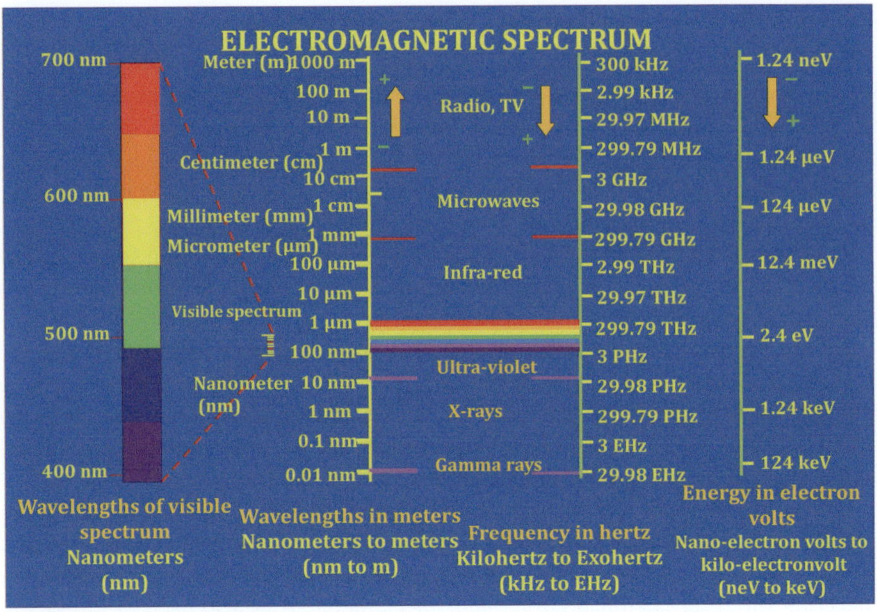

Fig. 4.2 Electromagnetic (EM) spectrum

infrared spectroscopy, with the infrared region ranging from 700 nm to 1 mm or 10^{12}–10^{14} Hz. The infrared spectrum includes near-infrared (NIR), mid-infrared (mid-IR or MIR), and far-infrared (far-IR or FIR) regions. NIR spans 12,500–4000 cm^{-1} (800–2500 nm) and features overtone and combination bands, lacking fundamental bands. NIR bands come from vibrational transitions with lower probability, resulting in weaker absorption coefficients (ε), about 100 times smaller than basic bands. The split into sub-regions is based on physical and practical considerations [11].

The exact molecular vibration pertains to atom movement independent of the center (center) of mass or rotation. These vibrations, known as "normal modes," are specific to molecules. Normal modes result from stretching and bending molecular bonds in nonlinear molecules, sometimes with asymmetrical or scaly properties. Not all normal mode vibrations of a molecule will be active in response to infrared light, as the molecule's symmetry affects the location and quantity of infrared-active peaks [12, 13].

FTIR spectroscopy can quantify vibrations in key environmental substances like organic acids and mineral phases in the mid-IR region. It is used to study bonding mechanisms in solids and surfaces [14]. Molecular vibrations can be tied directly to molecules' symmetry. Hence, the FITR spectrum precisely gives information about the molecule's behavior on surfaces [15]. The image of FTIR spectroscopy instrumentation is illustrated in Fig. 4.3.

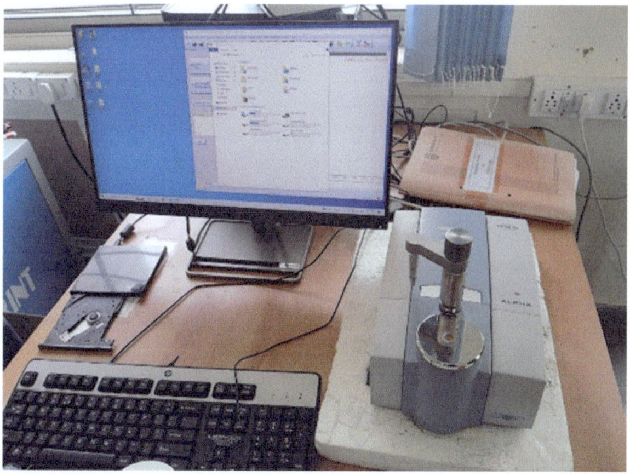

Fig. 4.3 Instrumentation of FTIR spectroscopy

4.2.2 UV–Visible (UV–Vis)

UV–Vis's spectroscopy is a straightforward, adaptable, inexpensive, and safe analytical method for detecting light absorption or transmission in various fields such as agriculture, food, pharmaceuticals, and the environment. It is adaptable, inexpensive, and safe for multiple organic and inorganic substances [16]. This methodology has undoubtedly resulted in significant discoveries across various fields. UV–Vis measures sample properties in the UV–Vis light range (190–900/1100 nm) without sample destruction [17]. The UV–visible spectrometer will measure absorbance, reflectance transmission, and energy/bandgap associated with the substance over a UV–vis frequency range [18, 19]. The photograph of a UV–visible spectrometer is shown in Fig. 4.4.

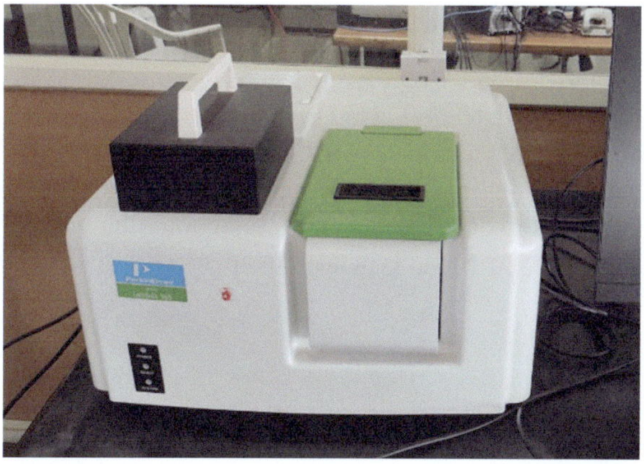

Fig. 4.4 UV–visible spectrometer

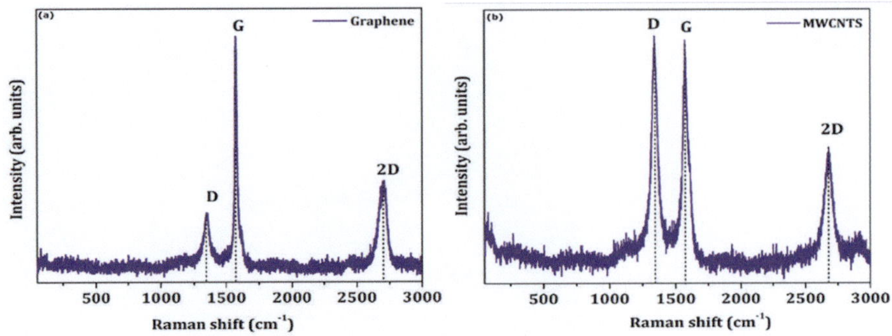

Fig. 4.5 Raman spectra of **a** GNP and **b** MWCNT [24]

4.2.3 Raman Spectroscopy

Raman spectroscopy is an essential spectroscopic method named after its inventor, Prof. C. V. Raman, an Indian physicist. This technique analyzes a system with low-frequency modes of rotation, vibration, and other phenomena. Raman spectroscopy is commonly used in chemistry to identify molecules by producing a unique fingerprint. It is based on Raman scattering, which involves inelastic scattering of monochromatic light [20].

Interaction of light with molecules or phonons causes laser photon energy to oscillate, forming vibrational modes in objects. Various advanced Raman spectroscopy techniques include surface-enhanced Raman, resonance Raman, tip-enhanced Raman, polarized Raman, stimulated Raman, transmission Raman, spatially offset Raman, and hyper Raman [21]. Raman spectroscopy sends monochromatic radiation through a sample, causing it to reflect, absorb, or scatter the photons. Due to vibration and rotation changes in the material, the scattered photons have a different frequency than the incident photons, leading to a wavelength shift analyzed in IR spectra [22].

The laser source bandwidth influences Raman spectrum resolution. Raman-scattered light is analyzed using spectrographs or interferometers for Fourier Transform (FT) detection. Modern Raman spectrometers utilize array detectors such as charged-coupled devices (CCDs) tailored for specific wavelength ranges [23].

FT-Raman is frequently paired with NIR lasers and requires detectors such as Germanium or InGaAs. A laser rejection filter with notch or long-pass filters is used for quality spectra. Typical Raman spectra of GNP and MWCNT are displayed in Fig. 4.5.

4.2.4 X-Ray Diffraction (XRD)

X-ray diffraction uses X-rays to analyze materials' crystalline structure, helping evaluate composition and characteristics. It's non-destructive and provides detailed

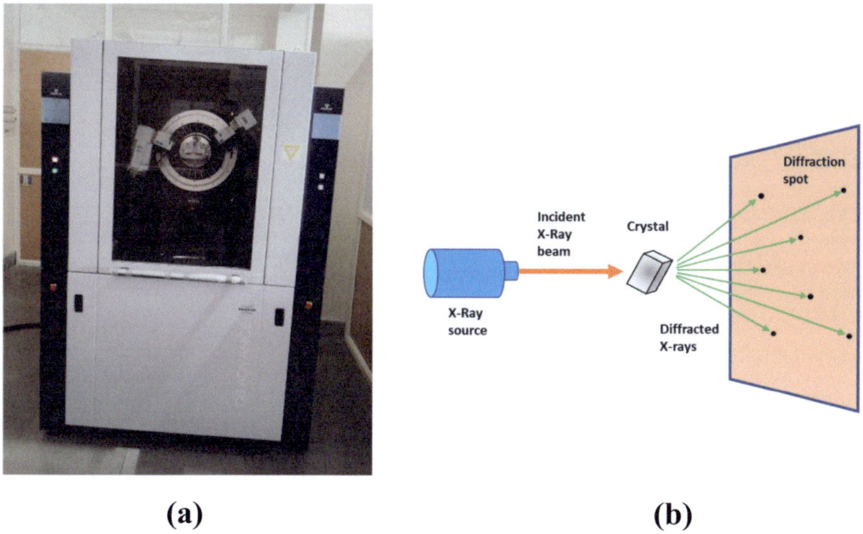

Fig. 4.6 **a** XRD spectrometer instrument and **b** XRD mechanism

information on crystallographic structure, chemical composition, and physical properties using constructive interference of monochromatic X-rays and a crystalline sample [25].

X-ray diffraction (XRD) is utilized to analyze nanomaterials' crystalline structures, measuring properties such as stress, particle size, orientation, and defects. The equipment includes a source, sample components, and optics, with cathode rays directed toward a detector for X-ray spectra processing. XRD is a safe and noninvasive technique that produces diffractograms displaying antiphase boundaries, faults, and crystallite sizes, and it is widely used in various material studies. Figure 4.6 shows XRD spectrometer instrument and its mechanism.

In materials research, scientists have several analytical issues linked to materials' chemical composition and crystalline constitution. X-ray diffraction is the sole non-destructive laboratory technique for obtaining crucial information on crystal structure, orientation, composition, and more. Therefore, researchers employ XRD to examine various materials, from powders to solids, thin films, and nanomaterials. Typical XRD spectra of (a). $SrFe_{12}O_{19}$, (b). GNP, and (c). MWCNT are shown in Fig. 4.7.

4.2.5 Scanning Electron Microscopy (SEM)

This method uses an electron microscope to examine the structure and composition of materials at a tiny scale. It assesses a material's homogeneity, lattice cleanliness, and structure. A scanning electron microscope (SEM) is a tool that can be used to see

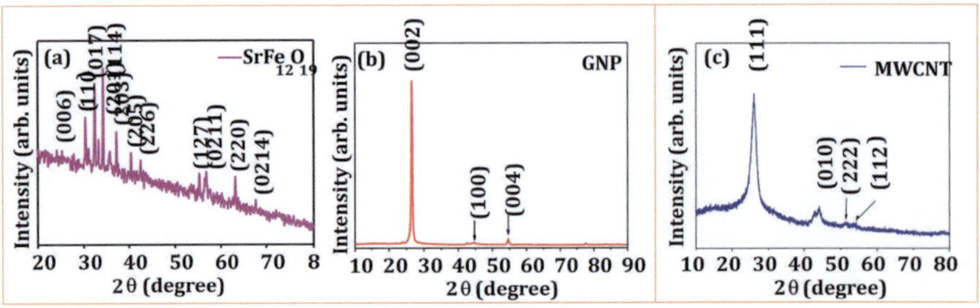

Fig. 4.7 XRD spectra of **a** $SrFe_{12}O_{19}$, **b** GNP, and **c** MWCNT [24]

nanomaterials prepared in the laboratory on a nanometer scale. This SEM is different from the conventional microscope [26].

In a traditional microscope, lenses are used with different reflective indices and focal lengths involved in the objective and eyepiece. This SEM has an electromagnetic coil instead of the lens. In the SEM, an electron beam is used instead of light as an illumination source. The microscope cannot observe material as small as a nanoscale since it has limitations with the wavelength of light, whereas, in the SEM, the electron ray is the source [27]. It has the advantage of resolution up to the Angstrom 10^{-6} (Å) level.

Nanomaterials and their structures are visible in the SEM, where a concentrated beam of energetic electrons is directed onto the specimen's surface. This generates signals to analyze the specimen's properties. In a scanning electron microscope, electrons focused on the specimen's surface have high kinetic energy [28].

Electrons have much shorter wavelengths than visible light, allowing electron microscopes to achieve much higher resolution than light microscopes. While the theoretical resolution of an electron microscope can approach fractions of an angstrom due to these short wavelengths, practical limitations—such as lens aberrations and electron scattering—restrict actual resolution to about 0.1 nm (1 Å). The resolution of any microscope is fundamentally limited by the wavelength of the imaging medium and is defined as the minimum distance at which two objects can be distinguished as separate.

Theoretically, this means it is possible to image species as small as 0.3 Å using this setup. Most atoms have a size ranging from 2 to 3 angstroms (Å). If the sample doesn't carry electricity, electrons will build up on it, called a charging or capacitance effect. Because of this, the picture will get too bright [29]. Therefore, gold sputtering will transform insulative samples into conductive ones. Gold will be deposited on the surface of the insulative sample, thereby generating conduction on the surface. Subsequently, the complementary electrical charges will be exhausted. Gold will be sputtered to a thickness of a few nanometers, resulting in the deposition of 2–3 atomic layers of Gold. Therefore, sputtering is unnecessary when the sample is conductive [30]. The scanning electron microscopy instrument is shown in Fig. 4.8.

While focusing on the sample, the electromagnetic lens will try to diverge/converge the electron rays, which will be utilized to magnify the sample. The change

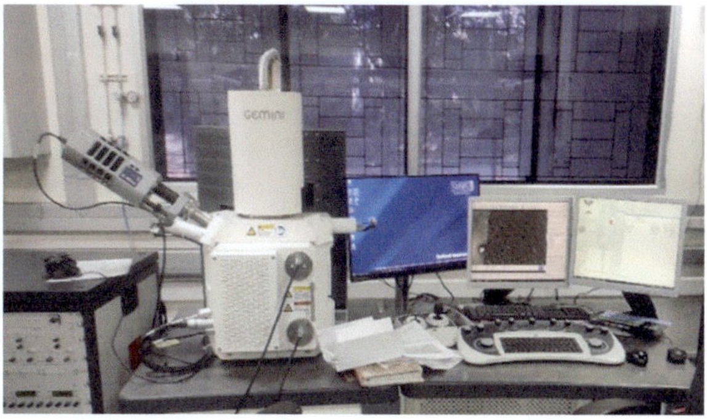

Fig. 4.8 Scanning electron microscopy instrument

in the current in the electromagnetic coil leads to a change in a magnetic field, which is essential for magnification and focusing on the sample's surface. Several current and voltage combinations will work for proper sample magnification so that the user can get a clear, well-focused image of high quality. A good view of the sample's surface morphology, structure, shape, and size could be realized by SEM, in principle [31].

A modern SEM instrument has energy-dispersive X-ray spectroscopy (EDX) built into it. EDX is another powerful tool that arbitrarily determines material composition and quantity. Energy-dispersive X-ray spectroscopy (EDS, EDX, or XEDS) is a method of analysis employed to determine a material's elemental composition or chemical properties. The process depends on examining the interaction between an X-ray excitation source and a sample. It can be utilized for qualitative and quantitative analysis, allowing users to determine the presence of elements and the precise proportion of each element's concentration in the sample [32].

EDX analysis is a noninvasive method that uses electron beams to analyze atoms, ejecting electrons from the inner shell and creating a positively charged electron hole. This process is different from conventional SEM as it requires minimal sample preparation and does not require any invasive techniques [33].

Electron transitions within atoms emit X-rays when moving from higher to lower energy shells, with unique emission patterns for each element. Elemental mapping along with the EDX is possible, giving the color to each element type in the sample. This is very helpful in studying the elemental distribution in the sample in the region of view [33, 34]. Figure 4.9 shows sample images taken by SEM spectroscopy.

4.2.6 Thermogravimetric Analysis (TGA)

Thermogravimetric analysis is a technique used to measure the change in a material's weight when subjected to a change in temperature [35]. This method evaluates the thermal resistance and stability of a material. Thermogravimetric analysis is used to ascertain the thermal profile and stability of the sample [36]. Different phases of the material can be understood with the help of TGA. As an illustration, Polyethene

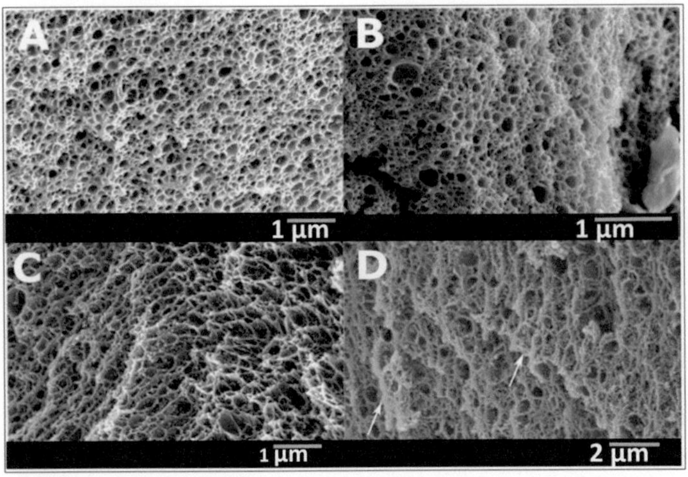

Fig. 4.9 Sample images taken by SEM spectroscopy [34]

exists in a solid state at room temperature. If the temperature climbs to around 80 °C, it will become smooth and soft [37].

Once the temperature reaches between 120 and 130 °C, it will undergo a phase change and become a liquid. In this process, the solid substance will transform into a fluid state. The substance evaporates at 160 °C, producing polyethylene fumes [38]. Once the temperature surpasses this threshold, the object will ignite if exposed to high temperatures. After 300 to 400 °C, it will instantaneously burn out, and polyethylene is a hydrocarbon substance taken from a petroleum source and decompressed to CO_2 [39].

Thermal gravimetric analysis (TGA) will analyze all temperature versus weight factors. By employing the differential scanning calorimetry (DSC) approach, we may ascertain if the combustion reaction is characterized by absorption or release of heat. This procedure will create two small crucibles of Al_2O_3, Silica, Platinum, and other materials. These crucibles will be carefully measured, and the sample will be placed in one of them. The chamber will be cleansed with ultra-high-purity nitrogen, with a concentration of 99.997% [35]. Figure 4.10 shows TGA of pure LDPE and 50:5:40:5 composite.

4.2.7 Electrical Characterization

Examining the dielectric properties is essential for describing any material-related electrical behavior. Samples may be available in various types, such as liquids, powders, polymers, gels, crystals, and minerals. The instrument is furnished with a heating furnace, which may deliver any amount of heat to the sample, and the

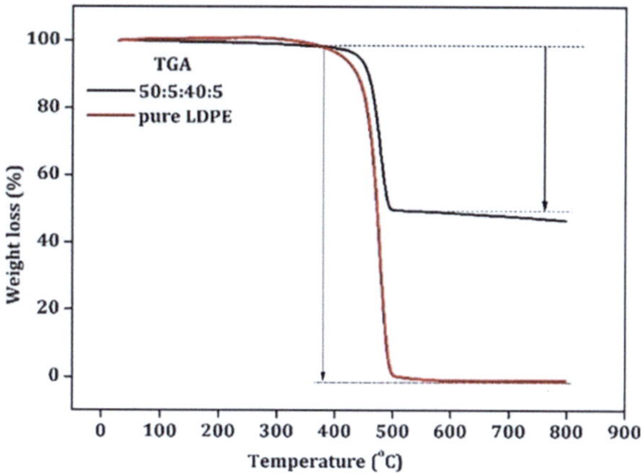

Fig. 4.10 TGA of pure LDPE and 50:5:40:5 composite [24]

performance of the sample at that temperature will be measured over the frequency range.

Silver is the most electrically conductive element, followed by copper and gold; while carbon, in the form of graphite,also conducts electricity effectively. Due to affordability and corrosion resistance, copper and gold are more commonly used in electrical applications. Factors like cross-sectional area, conductor length, and temperature affect conductivity and resistivity. Table 4.1 displays the electrical resistivity and conductivity of copper, gold, platinum, and glass materials. Electrical resistivity, shown by ρ (rho), indicates a material's resistance to electric current flow. Lower resistivity allows greater charge flow, while electrical conductivity, denoted by σ (sigma), κ (kappa), or γ (gamma), measures a material's ability to conduct an electric current.

Researchers have devised methods to produce conducting polymer composites and hybrids with enhanced properties, including CP nanocomposites with carbon nanospecies. Improvements involve increased conductivity through enhanced structural order of CP chains and promoting charge carrier delocalization. These composites offer a broad range of conductivities, from insulating to metallic, with high mechanical stability, flexibility, and conductivity, making them valuable for diverse technology applications.

Conductive polymers (CPs) and electrically conductive polymer composites are used due to their unique properties in applications such as electrochromic devices, batteries, LEDs, transistors, solar cells, and sensors. By adding ions and conductive nanofillers, the properties of CPs can be altered to emit different light colors. CP LEDs have advantages like easy patterning, low DC voltage operation, and uniform light distribution, making them suitable for backlighting and display applications despite not replacing fluorescent bulbs in terms of efficiency.

In non-conductive polymer composites, conductive polymers facilitate the flow of electricity due to their π-bond system's conjugation. This conjugation leads to electron delocalization, where π-bonding electrons are spread across multiple atoms

Table 4.1 Resistivity and conductivity of different materials at 20 °C

Material	ρ (Ω·m) at 20 °C resistivity	σ (S/m) at 20 °C conductivity
Silver	1.59×10^{-8}	6.30×10^{7}
Copper	1.68×10^{-8}	5.96×10^{7}
Annealed copper	1.72×10^{-8}	5.80×10^{7}
Gold	2.44×10^{-8}	4.10×10^{7}
Aluminum	2.82×10^{-8}	3.5×10^{7}
Calcium	3.36×10^{-8}	2.98×10^{7}
Tungsten	5.60×10^{-8}	1.79×10^{7}
Zinc	5.90×10^{-8}	1.69×10^{7}
Nickel	6.99×10^{-8}	1.43×10^{7}
Lithium	9.28×10^{-8}	1.08×10^{7}
Iron	1.0×10^{-7}	1.00×10^{7}
Platinum	1.06×10^{-7}	9.43×10^{6}
Tin	1.09×10^{-7}	9.17×10^{6}
Carbon steel	(10^{10})	1.43×10^{-7}
Lead	2.2×10^{-7}	4.55×10^{6}
Titanium	4.20×10^{-7}	2.38×10^{6}
Grain-oriented electrical steel	4.60×10^{-7}	2.17×10^{6}
Manganin	4.82×10^{-7}	2.07×10^{6}
Constantan	4.9×10^{-7}	2.04×10^{6}
Stainless steel	6.9×10^{-7}	1.45×10^{6}
Mercury	9.8×10^{-7}	1.02×10^{6}
Nichrome	1.10×10^{-6}	9.09×10^{5}
GaAs	5×10^{-7} to 10×10^{-3}	5×10^{-8} to 10^{3}
Carbon (amorphous)	5×10^{-4} to 8×10^{-4}	1.25 to 2×10^{3}
Carbon (graphite)	2.5×10^{-6} to 5.0×10^{-6} // basal plane 3.0×10^{-3} ⊥basal plane	2 to 3×10^{5} //basal plane 3.3 $\times 10^{2}$ ⊥basal plane
Carbon (diamond)	1×10^{12}	~ 10^{-13}
Germanium	4.6×10^{-1}	2.17
Sea water	2×10^{-1}	4.8
Drinking water	2×10^{1} to 2×10^{3}	5×10^{-4} to 5×10^{-2}
Silicon	6.40×10^{2}	1.56×10^{-3}
Wood (damp)	1×10^{3} to 4	10^{-4} to 10^{-3}
Deionized water	1.8×10^{5}	5.5×10^{-6}
Glass	10×10^{10} to 10×10^{14}	10^{-11} to 10^{-15}
Hard rubber	1×10^{13}	10^{-14}
Wood (oven-dry)	1×10^{14} to 16	10^{-16} to 10^{-14}

(continued)

Table 4.1 (continued)

Material	ρ (Ω·m) at 20 °C resistivity	σ (S/m) at 20 °C conductivity
Sulfur	1×10^{15}	10^{-16}
Air	1.3×10^{16} to 3.3×10^{16}	3×10^{-15} to 8×10^{-15}
Paraffin wax	1×10^{17}	10^{-18}
Fused quartz	7.5×10^{17}	1.3×10^{-18}
PET	10×10^{20}	10^{-21}
Teflon	10×10^{22} to 10×10^{24}	10^{-25} to 10^{-23}

rather than being confined between two atoms. This delocalization enables faster electron flow, contributing to electrical conductivity. When introducing oxidative dopants, they remove an electron from the π backbone of the polyacetylene chain, generating a positive charge or "hole" on a carbon atom, forming a positively charged radical cation known as a polaron. This configuration allows the remaining lone electron in the π-bond to move freely, causing the double bond to shift along the molecule. With further oxidation, a bipolar configuration can form. If polyacetylene undergoes heavy oxidation, polarons pair to create mobile charged solitons, which provide unique "spinless" conductivity specific to polyacetylene. Polyacetylene structure with conjugated double bonds is shown in Fig. 4.11.

The π-bond shifts along the carbon chain, allowing electrons to move, enabling conductivity. Heeger et al. [40] combined chemistry and physics to enhance polymer conductivity. They used a halogen dopant to create electron holes, allowing charge flow. Various theories explain how conducting polymer conductivity changes, involving solitons, polarons, and bi-polarons, as charges move through the material.

An element's electrical conductivity is mainly influenced by its electronic structure, and energy band theory distinguishes conductors, insulators, and semiconductors based on the bandgap. In conductors, valence and conduction bands overlap,

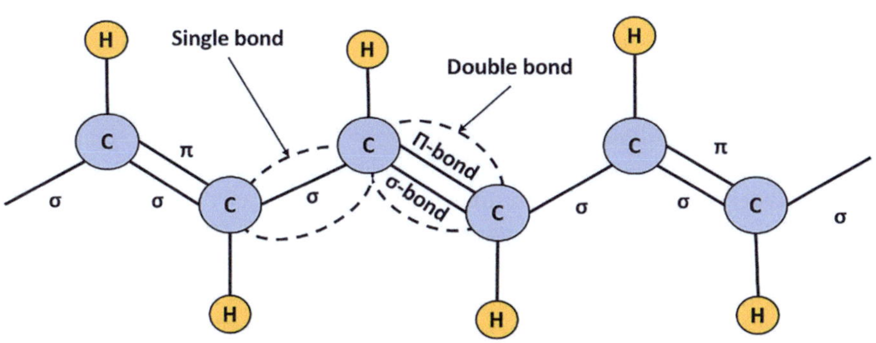

Fig. 4.11 The structure of polyacetylene

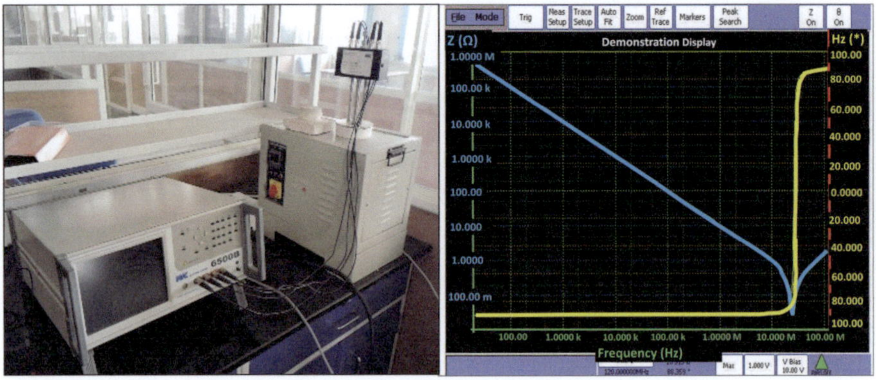

Fig. 4.12 Computer-controlled impedance analyzer operates in direct *I–V* mode with an attached test fixture

allowing electron movement. Semiconductors have a small bandgap enabling electron and hole currents, while insulators have a wide bandgap preventing electrical conduction.

Dielectric consistency is crucial in EMC testing to assess materials' effectiveness in managing electromagnetic interference. The dielectric constant, known as relative permittivity, characterizes how a polymer composite responds to an electric field. Understanding the dielectric constant is essential for mitigating EMI in EMC testing [36]. A computer-controlled impedance analyzer in direct *I–V* mode uses a test fixture shown in Fig. 4.12.

4.2.8 Mechanical Characterization

Mechanical characterization involves testing the physical properties of materials used in motor vehicles to ensure safety and regulatory compliance. This testing covers a range of materials like plastics, metals, rubbers, paints, and coatings, evaluating static and dynamic properties from − 40 to 400 °C. Tests include tensile, compression, bending, torsion, tear strength, shearing, peeling, creep, fatigue, endurance, pendulum shock, hardness, micro and ball indentation, and pressure testing supported by specialized equipment and expertise.

Evaluation of mechanical and physical properties is vital for determining the strength and rigidity of polymer composites, following set standards. Tests like tensile, flexural, and impact tests gauge a material's ability to withstand stress. Fiber-reinforced polymer composites' mechanical properties are notably impacted by fiber type, polymer type, and fiber/matrix bonding. Physical properties such as water absorption, hardness, shear, void content, and density also control the suitability of composites for different applications.

Mechanical and physical testing of composites is crucial for various engineering fields such as construction, furniture, automotive, and electrical devices. These tests are essential for collecting valuable data on polymers and composites, especially when integrating materials like steel, cement, and concrete with polymer composites in industries like buildings, railway interiors, and storage solutions.

Practical applications necessitate mechanical stability and durability, including tolerance for vibrations, wear, flexibility, rigidity, adhesion, and load-bearing capacity. The strength of materials calculates stresses and strains in structures' beams, columns, and shafts.

Predicting a structure's response to loading and vulnerability to failure involves material properties such as yield strength, ultimate strength, Young's modulus, and Poisson's ratio. Geometric features like dimensions, boundary constraints, and shape changes also impact behavior and stability under load [41].

Nanomaterials and nanocomposites require nanoindentation and micro-stress analysis. The mechanical strength of polymer, fibers, laminates, and metal composites can be tested using a tensiometer and universal testing machine (UTM) [42]. During product design research and development, this device evaluates materials' tensile and compressive strength. It can determine a material's loads, strengths, and tensile strength while universally assessing standard properties. Parameters provided include stress, strain, modulus, strength, vibration resistance, abrasion resistance, compression strength, flexural strength, elongation at breaks, plasticity, and engineering strength [43]. The sample will adhere to ASTM standards, including flat dog bone, double shapes, and flat samples of uniform thickness for mechanical strength testing [44]. The sample will be attached to the holding fixture to prevent slippage during evaluation with the tensiometer instrument, ensuring a secure and tight fit [45].

A tensiometer comprises two grips holding sample material for testing. It applies a pulling force using a screw to measure the material's strength, making it portable and versatile. The device can determine Young's modulus, the initial slope of the material's stress–strain curve. This helps to identify the material's elastic limit, beyond which irreversible damage can occur. Evaluating these properties is crucial for ensuring the material is suitable for its intended purpose. Figure 4.13 depicts the tensiometer instrument during mechanical sample testing.

This aids the material scientists in organizing the material for appropriate technical use. What is the material's capacity in any device, lifespan, and the load it can sustain in that stress state? The nanomaterials under examination have the potential to be utilized in various electronic applications, such as constructing components and producing EMI and EMC-compliant sheets and coatings [46, 47].

4.2.9 Atomic Force Microscope (AFM)

AFM methods for studying topographical, electrical, and mechanical properties can utilize various Raman spectrometer laser sources or external lighting such as a solar

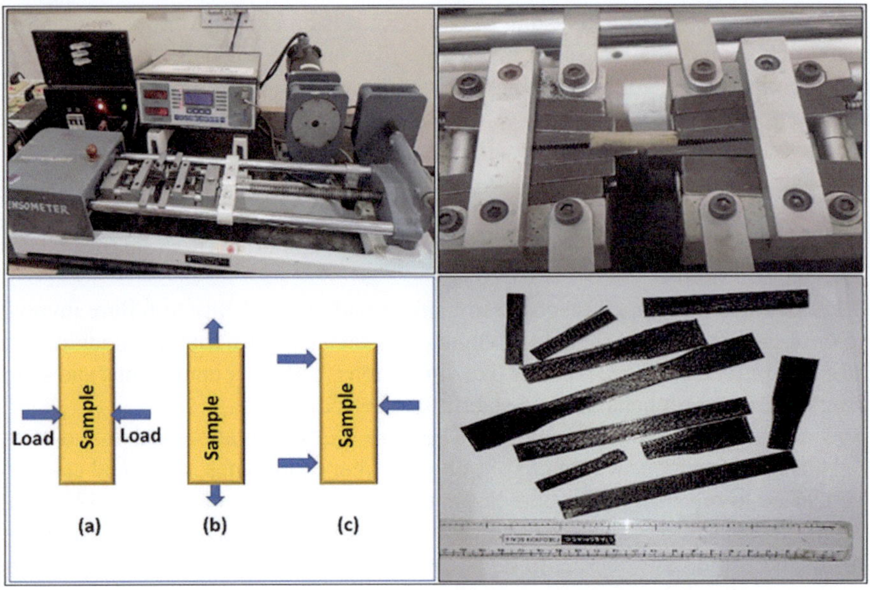

Fig. 4.13 Material being loaded in **a** compression, **b** tension, and **c** shear

simulator. The AFM was invented by IBM scientists in 1985, while the STM was created by Gerd Binnig and Heinrich Rohrer at IBM Research—Zurich in the early 1980s, resulting in a Nobel Prize in Physics in 1986 [48].

That year, Gerd Binning, Calvin Quate, and Christoph Gerber invented the atomic force microscope (AFM), which transformed nanoscale characterization and measurement. AFM is a form of scanning probe microscopy (SPM) using a cantilever probe with an unconstrained tip [49].

AFM analyzes nanocoating, film, and other material surface morphology. Nanoscience research uses dynamic, robust methods. AFM was made with fewer tools. A cantilever, a thin and pointed tool, is used to investigate the specimen‘s surface. A laser beam measures cantilever deviation, and a photodetector converts it to electricity. During scanning, the cantilever moves grid-like across the sample surface. Laser light bounces off the cantilever and hits the photodetector. The photodetector records the laser location as the cantilever moves. AFM operates in different modes. There are three modes of operation for an Atomic Force Microscope (AFM): Contact Mode, Tapping Mode, and Non-contact Mode. AFM is used for surface roughness investigation, nanoscale biological sample imaging, and material mechanical property characterization. Contact Mode involves constant tip-sample contact, Tapping Mode oscillates the tip, and Non-contact Mode scans without touching the sample [50].

Atomic Force Microscopy (AFM) is a versatile technique that can characterize various properties such as mechanical, electrical, magnetic, and optical spectroscopic properties. In addition to imaging, AFM can manipulate surfaces, write patterns, and

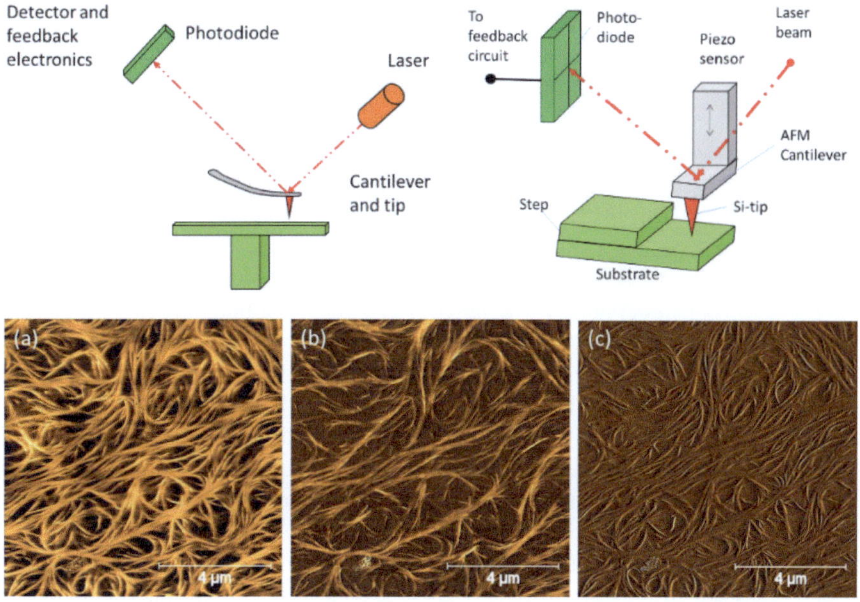

Fig. 4.14 Working principle of AFM and some sample images of nanostructures taken with AFM [52]

exert forces in experiments like lithography and molecular pulling. AFM offers high resolutions down to the nanoscale and is effective in diverse environments, from ultra-high vacuum to liquids. It has become a widely used technique in material characterization, complementing optical and electron microscopy in physics, chemistry, biology, and materials science [51]. The working principle of AFM and some sample images of nanostructures taken from AFM are shown in Fig. 4.14 [52].

The resolution achieved by AFM, which is in the order of fractions of a nanometer, exceeds the optical diffraction limit by more than 1000 times. The surface structure of the nanocoating will be recorded by the interaction of a mechanical microscale probe, which will function via the piezoelectric effect. The cantilever operation of the probe will work in two modes: continuous and tapping [53]. The constant mode operates with the probe and substrate's continuous scanning contact. This procedure would be preferred for the homogeneous and smooth surface. At the same time, the tapping mode is applied to the sample with a non-uniform surface since the probe is likely to get damaged in continuous mode [54, 55].

4.2.10 Vibrating-Sample Magnetometer (VSM)

Modern society relies heavily on magnetic materials like computer disks, credit cards, and speakers. Lodestone, mainly made of magnetite, was the first permanent magnet in ancient Greece. Many magnetic materials have been developed since then, leading to the growth of industries around them [56]. A vibrating-sample magnetometer (VSM), developed by Simon Foner at MIT Lincoln Laboratory in 1955, studies magnetic properties using Faraday's Law of Induction, referenced in 1956 by G. W. Van Oosterhout and P. J. Flanders [57, 58]. A sample is placed in a magnetic field which aligns its magnetization if magnetic. The sample's magnetic dipole moment changes a field as it moves up and down [59].

Piezoelectric material generates an electric field in VSM pickup coils via an alternating magnetic field. Higher current signifies greater sample magnetization proportionally. Consequently, a hysteresis curve is recorded and analyzed [60, 61]. Understanding the magnetic properties of solids is crucial, as it provides insights into the fundamental structure of various metallic and non-metallic materials. A vibrating-sample magnetometer (VSM) measures the magnetic moment of solid samples, an essential factor in magnetism. A dipole moment is created when a sample is placed in a uniform magnetic field related to its susceptibility and field strength. This generates an electrical signal in pickup coils if the sample vibrates, with the signal being proportional to the magnetic moment, amplitude, and vibration frequency. The device reports the magnetic moment in electromagnetic units (e.m.u.) [62]. VSM instruments and working principle are depicted in Fig. 4.15.

The working principle of a vibrating-sample magnetometer involves suspending a sample in a magnetic field and subjecting it to small amplitude vibrations. This allows the measurement of the sample's magnetic properties by detecting the changes in the induced voltage as the sample vibrates in the magnetic field.

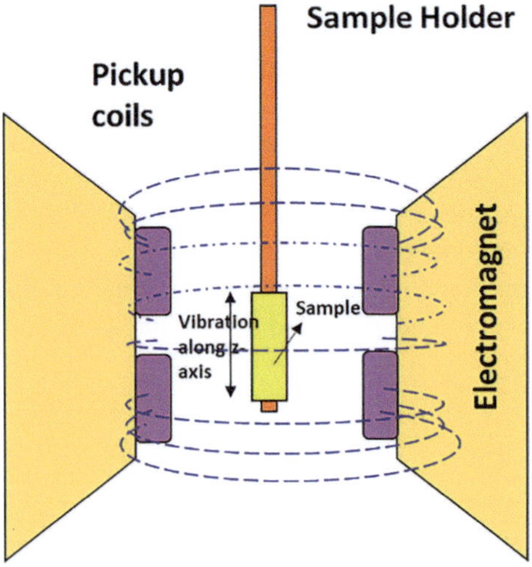

Fig. 4.15 VSM instrument and working principle

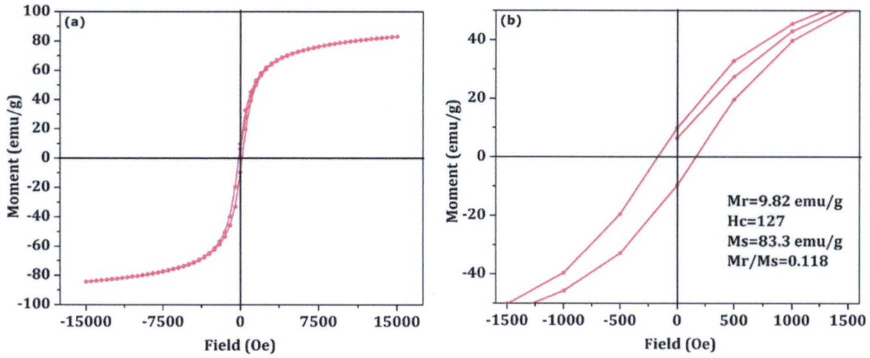

Fig. 4.16 VSM plots of $SrFe_{12}O_{19}$ nanoparticles at room temperature

The VSM comprises an amplifier, control chassis, meter, computer interface, sensor coils, vibration exciter, sample holder, and Hall probe to optimize signal induction. Coil setup may vary based on the studied material [57]. VSM plots of $SrFe_{12}O_{19}$ nanoparticles at room temperature are shown in Fig. 4.16.

Advanced method uses SQUID-VSM to analyze magnetic properties in materials. Unlike traditional VSMs, it uses a highly sensitive Superconducting Quantum Interference Device to detect magnetic flux. It can identify minute amounts of quantum flux smaller than a billionth of Earth's magnetic field [59].

4.2.11 Transmission Electron Microscopy (TEM)

TEM uses an electron beam to examine internal material structure, like a microscope. Interpretation of TEM images can be more complex than Scanning Electron Microscopy (SEM) images, which focus on external solid morphology. Though basic TEM imaging principles stay consistent, modern TEM microscopes have added various analytical features [63, 64]. In addition to essential imaging, modern TEMs include X-ray and electron spectroscopy, electron diffraction, and other features. The electron beam in SEM is much longer and thinner than the light beam in optical microscopy, limiting dispersion even at a distance from the source. This focused beam allows SEM to maintain focus on objects with significant topographical relief, providing a depth of field 10 to 100 times larger than that of an optical microscope [65–67]. TEM image of Mixed-Phase TiO_2–ZrO_2 nanoparticles [68] is shown in Fig. 4.17.

Transmission Electron Microscopy (TEM) is a highly effective method employed to visualize the interior composition of materials at the nanoscale. Transmission electron microscopy (TEM) passes a highly narrow beam of electrons through a specimen, which then interacts with the object to generate a picture. Under the far shorter wavelength of electrons, this technique enables the observation of minute entities that

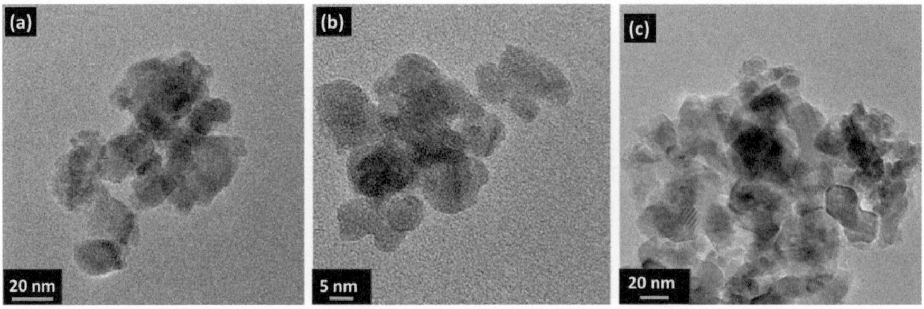

Fig. 4.17 TEM image of mixed-phase TiO_2–ZrO_2 nanoparticles [68]

are not achievable with light microscopy. Transmission electron microscopy (TEM) finds extensive utility in diverse domains, including materials science, biology, and nanotechnology, providing comprehensive data on materials' crystalline structure, flaws, and morphology. It is an essential instrument for researchers and scientists. To summarize, Transmission Electron Microscopy (TEM) is an invaluable instrument for investigating nanoscale structures and has been crucial in enhancing our comprehension of the molecular realm [64].

The operational mechanism of a transmission electron microscope (TEM) entails using an electron beam to visualize the specimen. As the electrons traverse the material, their interaction generates a high-resolution picture, enabling meticulous analysis of the specimen's interior structure. The intricate arrangement of electromagnetic lenses and detectors in the TEM enables the successful capture and magnification of the picture, therefore offering significant insights into the morphology and composition of the material [69].

The fundamental components of a transmission electron microscope (TEM) include a high-voltage electron gun for generating an electron beam and electromagnets for focusing and directing the electron beam. A sequence of electromagnetic lenses is used to focus the electron beam further, a specimen stage is used to clamp the sample, and a fluorescent screen or photographic plate is used to capture the magnified shadow image of the specimen. A vacuum system facilitates the unhindered movement of electrons through the column and their interaction with the specimen. The implementation of this approach also serves to mitigate contamination and safeguard both the equipment and the specimen.

Selective Area Electron Diffraction (SAED) pattern is a crucial method employed in transmission electron microscopy (TEM) to investigate material crystal structure. It is acquired by choosing a limited sample region for examination using the electron beam. The diffraction pattern obtained from such analysis offers insights into the crystallographic planes and orientations within the sample, therefore enabling the identification of its crystal structure. The transmission electron microscope (TEM) is an indispensable instrument for examining the interior composition of objects under extremely high levels of magnification [63].

This technique employs an electron beam to represent the specimen visually. This methodology is especially advantageous for analyzing the microstructure of materials and biological samples at a highly minute level. The selected area electron diffraction (SAED) technique is frequently employed with transmission electron microscopy (TEM). Electron microscopy directly applies an electron beam to a designated specimen region and then analyzes the resulting diffraction pattern. The technique mentioned above yields significant insights into the crystal structure and orientation of the specimen. Transmission electron microscopy (TEM) and scanning electron microscopy (SAED) are essential instruments for researchers and scientists in many disciplines like materials science, nanotechnology, biology, etc. [70]. Table 4.2 gives the characteristics and properties of the instruments used in materials characterization.

Table 4.2 Characteristic properties of the instruments used in materials characterization

S. No.	Characteristics (properties)	Instrument used
1	Crystal dimension and structural behavior of nanomaterials	XRD, TEM, electron diffraction
2	Shape and morphology	SEM, TEM, AFM,
3	Agglomeration state	DCS, UV–Vis, SEM, TEM
4	Elemental-chemical composition	SEM with EDX, XRD,
5	Chemical structure and chemical bonds	FTIR, UV, Raman, XPS
6	Single-particle properties	HRTEM
7	Size distribution	DCS
8	Structural defects	HRTEM
9	Density	DCS
10	3D visualization	AFM, SEM
11	Magnetic properties	Vibrating-sample magnetometers and superconductor quantum interference devices (SQUIDs), vibration sample magnetometry (VSM), magnetic susceptibility
12	Size	SEM, TEM, XRD, AFM
13	Concentration of nanoparticles in a media	UV–Vis, DCS
14	Nano-size analysis	TEM, SEM, magnetic susceptibility
15	Chemical and oxidation state	X-ray absorption spectroscopy (XAS), thermogravimetric analysis
16	Optical characterization	UV–Vis-NIR, STEM
17	Nanoparticles dispersion	TEM, SEM, AFM

4.3 Conclusion

Material characterization methods are crucial in understanding materials' properties and their behavior. X-ray diffraction, scanning electron microscopy, and thermal analysis provide information on materials' structure, composition, and performance. By combining these methods, researchers can understand material properties comprehensively, enabling them to make informed decisions in material selection and design. Material characterization methods are essential for advancing research and development in various industries, including aerospace, automotive, and electronics.

References

1. Nalluri, A., Dogiparthi, L.K., Chakravorty, A., Rather, G.A., Gangadhar, L., Sana, S.S.: Chapter 14—Physicochemical characterization of nanomaterials for production of biofuel and bioenergy. In: Kumar, R.P., Bharathiraja, B. (eds.) Nanomaterials, pp. 309–329. Academic Press (2021)
2. Holbrook, R.D., Galyean, A.A., Gorham, J.M., Herzing, A., Pettibone, J.: Chapter 2—Overview of nanomaterial characterization and metrology. In: Baalousha, M., Lead, J.R. (eds.) Frontiers of Nanoscience, pp. 47–87. Elsevier (2015)
3. Solangi, N.H., Karri, R.R., Mubarak, N.M., Mazari, S.A.: Mechanism of polymer composite-based nanomaterial for biomedical applications. Adv. Industr. Eng. Polym. Res. **7**, 1–19 (2024)
4. Mourdikoudis, S., Pallares, R.M., Thanh, N.T.K.: Characterization techniques for nanoparticles: comparison and complementarity upon studying nanoparticle properties. Nanoscale **10**, 12871–12934 (2018)
5. Joudeh, N., Linke, D.: Nanoparticle classification, physicochemical properties, characterization, and applications: a comprehensive review for biologists. J. Nanobiotechnol. **20**, 262 (2022)
6. Mekuye, B., Abera, B.: Nanomaterials: an overview of synthesis, classification, characterization, and applications. **4**, 486–501 (2023)
7. T.A. Saleh, Chapter 7 - Characterization and description of adsorbents and nanomaterials, in: T.A. Saleh (Ed.) Interface Science and Technology, Elsevier2022, pp. 199–232.
8. Catita, J.A.M.: Challenges in nanomaterial characterization—from definition to analysis. In: Louro, H., Silva, M.J. (eds.) Nanotoxicology in Safety Assessment of Nanomaterials, pp. 3–17. Springer International Publishing, Cham (2022)
9. Canevari, T.C.: Chapter 5—Fabrication routes for metallic nanostructured electrochemical biosensors. In: Manjunatha, J.G. (ed.) Novel Nanostructured Materials for Electrochemical Bio-Sensing Applications, pp. 79–96. Elsevier (2024)
10. Mahmoudi, M.: The need for robust characterization of nanomaterials for nanomedicine applications. Nat. Commun. **12**, 5246 (2021)
11. Isah, U.A., Rashid, M.I., Lee, S., Kiman, S., Iyodo, H.M.: Correlations of coal rank with the derived Fourier Transform Infra-Red (FTIR) spectroscopy structural parameters: a review. Infrared Phys. Technol., 105456 (2024)
12. Kosel, J., Legan, L., Bračič, M., Ropret, P.: Cationic silicone based paper impregnation agent for the protection of historical paper against fungal growth. Detection and quantification of fungi using micro-invasive fluorescence microscopy and non-invasive FTIR spectroscopy. J. Cult. Heritage **68**, 195–204 (2024)
13. Gong, Y., Chen, X., Wu, W.: Application of Fourier transform infrared (FTIR) spectroscopy in sample preparation: material characterization and mechanism investigation. Adv. Sample Prep. **11**, 100122 (2024)

14. Rosalina, R., Kamwilaisak, K., Sutthanut, K., Srisongkram, T., Weerapreeyakul, N.: Probing the stability and quality of the cellulose-based Pickering emulsion containing sesamolin-enriched sesame oil by chemometrics-assisted ATR-FTIR spectroscopy. Food Chem. **452**, 139555 (2024)
15. Chu, C., Wang, H., Luo, X., Fan, Y., Nan, L., Du, C., Gao, D., Wen, P., Wang, D., Yang, Z., Yang, G., Liu, L., Li, Y., Hu, B., Zunongjiang, A., Zhang, S.: Rapid detection and quantification of melamine, urea, sucrose, water, and milk powder adulteration in pasteurized milk using Fourier transform infrared (FTIR) spectroscopy coupled with modern statistical machine learning algorithms. Heliyon **10**, e32720 (2024)
16. Beć, K.B., Grabska, J., Huck, C.W.: Chapter 1—Physical principles of infrared spectroscopy. In: Cozzolino, D. (ed.) Comprehensive Analytical Chemistry, pp. 1–43. Elsevier (2022)
17. Gariglio, S., Malegori, C., Menżyk, A., Zadora, G., Vincenti, M., Casale, M., Oliveri, P.: Determination of time since deposition of bloodstains through NIR and UV–Vis spectroscopy—a critical comparison. Talanta **278**, 126444 (2024)
18. León, A.Y., Núñez-Méndez, K.S., Salas-Chia, L.M., Orozco-Agamez, J.C., Peña-Ballesteros, D.Y., Martínez-Vertel, J.J., León, P.A., Molina-Velasco, D.R.: Prediction of some physicochemical properties in Colombian crude oils upgraded by catalytic aquathermolysis using UV–VIS spectroscopy associated with chemometric methods. Spectrochim. Acta Part A Mol. Biomol. Spectrosc. **310**, 123965 (2024)
19. Khalid, K., Ishak, R., Chowdhury, Z.Z.: Chapter 15—UV–Vis spectroscopy in non-destructive testing. In: Otsuki, A., Jose, S., Mohan, M., Thomas, S. (eds.) Non-Destructive Material Characterization Methods, pp. 391–416. Elsevier (2024)
20. Xia, G., Pan, Q., Pan, Q., Li, Y., Yu, D., Zhu, J.: Dual-channel Raman spectrometer device based on multi-field-of-view spectrometer for on-line detection. Vib. Spectrosc. **134**, 103732 (2024)
21. Zheng, C., Zhang, L., Huang, H., Wang, X., Van Schepdael, A., Ye, J.: Raman spectroscopy: a promising analytical tool used in human reproductive medicine. J. Pharm. Biomed. Anal. **249**, 116366 (2024)
22. Kavitha, C.: Chapter 10—Graphene oxide in surface enhanced Raman spectroscopy for sensing and imaging applications. In: Kailasa, S.K. (ed.) Comprehensive Analytical Chemistry, pp. 267–301. Elsevier (2024)
23. Marchetti, M., Diliberto, C., Mechling, J.-M., Kang, J.W., Bourson, P., Offroy, M., Gouadec, G.: Chapter 14—Raman spectroscopy—part two. In: Otsuki, A., Jose, S., Mohan, M., Thomas, S. (eds.) Non-Destructive Material Characterization Methods, pp. 353–390. Elsevier (2024)
24. Praveen, M., Karthikeya, G.S., Krishna, R.H., Mamatha, G.M., Manjunatha, C., Khosla, A., Nagabhushana, B.M.: The role of magnetic nano $CoFe_2O_4$ and conductive MWCNT/graphene in LDPE-based composites for electromagnetic interference shielding in X-band. Diam. Relat. Mater. **130**, 109501 (2022)
25. Adam, M.R., Mokhter, M.A., Aziz, M.H.A., Kamaludin, R., Ahmad, N.A., Azmi, M.F.I.: 12—Chemical analysis of photocatalytic membrane (FTIR, XRD, UV-vis/optical, XPS, and zeta potential). In: Othman, M.H.D., Rahman, M.A., Matsuura, T., Adam, M.R., Mohd Makhtar, S.N.N. (eds.) Advanced Ceramics for Photocatalytic Membranes, pp. 295–315. Elsevier (2024)
26. de Assumpção Pereira-da-Silva, M., Ferri, F.A.: 1—Scanning electron microscopy. In: Da Róz, A.L., Ferreira, M., de Lima Leite, F., Oliveira, O.N. (eds.) Nanocharacterization Techniques, pp. 1–35. William Andrew Publishing (2017)
27. Abd Mutalib, M., Rahman, M.A., Othman, M.H.D., Ismail, A.F., Jaafar, J.: Chapter 9—Scanning electron microscopy (SEM) and energy-dispersive x-ray (EDX) spectroscopy. In: Hilal, N., Ismail, A.F., Matsuura, T., Oatley-Radcliffe, D. (eds.) Membrane Characterization, pp. 161–179. Elsevier (2017)
28. Lefkowitch, J.H.: Chapter 17—Electron microscopy and other techniques. In: Lefkowitch, J.H. (ed.) Scheuer's Liver Biopsy Interpretation (Tenth Edition), pp. 410–430. Elsevier, London (2021)
29. Makhtar, S.N.N.M., Hamed, N.K.A., Hamdan, M.A.H.b.: 10—Morphological analysis of photocatalytic membrane (SEM, FESEM, TEM). In: Othman, M.H.D., Rahman, M.A., Matsuura, T., Adam, M.R., Mohd Makhtar, S.N.N. (eds.) Advanced Ceramics for Photocatalytic Membranes, pp. 221–238. Elsevier (2024)

30. Huang, L., Gan, Y.: A review on SEM imaging of graphene layers. Micron **187**, 103716 (2024)
31. Ozcan, A., Meral, H.: Chapter 3—Introduction to fundamentals of nanoparticle synthesis, characterization, and properties. In: Pudake, R.N., Tripathi, R.M., Gill, S.S. (eds.) Nanotechnology for Abiotic Stress Tolerance and Management in Crop Plants, pp. 41–60. Academic Press (2024)
32. Vairaperumal, T., Selvaraj, D., Chakraborty, P.: Chapter 10—Microplastic in the environment: sources, workflow, identification techniques, and impacts on human health. In: Srivastav, A.L., Grewal, A.S., Markandeya, Pham, T.D. (eds.) Role of Green Chemistry in Ecosystem Restoration to Achieve Environmental Sustainability, pp. 91–103. Elsevier (2024)
33. Kievits, A.J., Duinkerken, B.H.P., Fermie, J., Lane, R., Giepmans, B.N.G., Hoogenboom, J.P.: Optical STEM detection for scanning electron microscopy. Ultramicroscopy **256**, 113877 (2024)
34. Muri, H.I., Hoang, L., Hjelme, D.R.: Mapping nanoparticles in hydrogels: a comparison of preparation methods for electron microscopy. Appl. Sci. **8**, 2446 (2018)
35. Aniza, R., Chen, W.-H., Kwon, E.E., Bach, Q.-V., Hoang, A.T.: Lignocellulosic biofuel properties and reactivity analyzed by thermogravimetric analysis (TGA) toward zero carbon scheme: a critical review. Energy Convers. Manage.: X **22**, 100538 (2024)
36. Zakiyan, S.E., Azizi, H., Ghasemi, I.: Effect of cell morphology on electrical properties and electromagnetic interference shielding of graphene-poly(methyl methacrylate) microcellular foams. Compos. Sci. Technol. **157**, 217–227 (2018)
37. Pirsaheb, M., Seifi, H., Gholami, T., Ganduh, S.H., Jasim, L.S., Mahdi, M.A., Salavati-Niasari, M.: Thermal analysis techniques for evaluating the thermal stability of battery materials: a comprehensive review. J. Anal. Appl. Pyrol. **174**, 106136 (2023)
38. Nath, B., Chen, G., Bowtell, L., Graham, E.: An investigation of thermal decomposition behavior and combustion parameter of pellets from wheat straw and additive blends by thermogravimetric analysis. Int. J. Thermofluids **22**, 100660 (2024)
39. Monteiro, S.N., Calado, V., Rodriguez, R.J.S., Margem, F.M.: Thermogravimetric stability of polymer composites reinforced with less common lignocellulosic fibers—an overview. J. Market. Res. **1**, 117–126 (2012)
40. Heeger, A.J., MacDiarmid, A.G., Shirakawa, H.: Macromolecules **35**(4), 1137–1139 (2002)
41. Sharma, B.B., Parashar, A.: Chapter 20—Effect of defects and functionalization on mechanical and fracture properties of two-dimensional nanomaterials. In: Thomas, S., Kalarikkal, N., Abraham, A.R. (eds.) Fundamentals and Properties of Multifunctional Nanomaterials, pp. 543–567. Elsevier (2021)
42. Fitch, E.C.: 9—Mechanical stability. In: Fitch, E.C. (ed.) Proactive Maintenance for Mechanical Systems, pp. 261–286. Elsevier, Oxford (1992)
43. Numata, K.: Chapter 6—Stability. In: Numata, K. (ed.) Biopolymer Science for Proteins and Peptides, pp. 143–178. Elsevier (2021)
44. Simitses, G.J., Hodges, D.H.: 2—Mechanical stability models. In: Simitses, G.J., Hodges, D.H. (eds.) Fundamentals of Structural Stability, pp. 19–46. Butterworth-Heinemann, Burlington (2006)
45. Guo, H., Wang, Y., Zhang, H., An, K.: Recent advances and strategies in mechanical stability of superhydrophobic surfaces. Prog. Org. Coat. **194**, 108595 (2024)
46. Vaidya, A., Pathak, K.: 17—Mechanical stability of dental materials. In: Asiri, A.M., Inamuddin, Mohammad, A. (eds.) Applications of Nanocomposite Materials in Dentistry, pp. 285–305. Woodhead Publishing (2019)
47. Pashley, D., Tay, F.: 5—Mechanical stability of resin–dentine bonds. In: Curtis, R.V., Watson, T.F. (eds.) Dental Biomaterials, pp. 112–161. Woodhead Publishing (2008)
48. Yu, T., Wang, J., Zhang, H., Sun, J.: Statistical analysis of aging micro-characteristics for asphalt binder AFM images based on neural network classification. Fuel **371**, 132080 (2024)
49. Sittl, S., Das, M., Helfricht, N., Petekidis, G., Papastavrou, G.: Direct force measurements between sub-micron rod-shaped colloids by AFM. Colloids Surf. A **697**, 134319 (2024)
50. Vargas, V., Castillo, J., Bouyssiere, B., Carrier, H., Radji, S.: Study of asphaltenes and subfraction nanoaggregates by AFM force spectroscopy. Coll. Surf. A: Physicochem. Eng. Aspects, 134638 (2024)

51. Nandee, R., Chowdhury, M.A., Hossain, N., Rana, M.M., Mobarak, M.H., Khandaker, M.R.: Surface topography and surface morphology of graphene nanocomposite by FESEM, EDX and AFM analysis. Nano-Struct. Nano-Objects **38**, 101170 (2024)
52. Matheson, A.B., Koutsos, V., Euston, S.R., Clegg, P.S.: Atomic force microscopy of phytosterol based edible oleogels. Gels **9**, 750 (2023)
53. Khan, M.K., Wang, Q.Y., Fitzpatrick, M.E.: 1—Atomic force microscopy (AFM) for materials characterization. In: Hübschen, G., Altpeter, I., Tschuncky, R., Herrmann, H.-G. (eds.) Materials Characterization Using Nondestructive Evaluation (NDE) Methods, pp. 1–16. Woodhead Publishing (2016)
54. Joshua, A.M., Cheng, G., Lau, E.V.: Soft matter analysis via atomic force microscopy (AFM): a review. Appl. Surf. Sci. Adv. **17**, 100448 (2023)
55. Emam-Djomeh, Z., Pure, A.E., Pure, M.E.: Chapter 5—Atomic force microscopy (AFM) of nanoencapsulated food ingredients. In: Jafari, S.M. (ed.) Characterization of Nanoencapsulated Food Ingredients, pp. 159–188. Academic Press (2020)
56. Characterization, M.: In: Wang, Z.L., Liu, Y., Zhang, Z. (eds.) Handbook of Nanophase and Nanostructured Materials, pp. 595–625. Springer, US, Boston, MA (2002)
57. Ghasemi, A.: Chapter 2—Ferrite characterization techniques. In: Ghasemi, A. (ed.) Magnetic Ferrites and Related Nanocomposites, , pp. 49–124. Elsevier (2022)
58. Rezaei, B., Mostufa, S., Wu, K.: Chapter 6—Characterization methods for magnetic nanoparticles. In: Wu, K., Wang, J.-P. (eds.) Magnetic Nanoparticles in Nanomedicine, pp. 155–177. Woodhead Publishing (2024)
59. Chapter 9—Magnetometer technology. In: You, Z. (ed.) Space Microsystems and Micro/nano Satellites, pp. 341–360. Butterworth-Heinemann (2018)
60. Agarwal, N., Nair, M.S., Mazumder, A., Poluri, K.M.: Chapter 3—Characterization of nanomaterials using nuclear magnetic resonance spectroscopy. In: Mohan Bhagyaraj, S., Oluwafemi, O.S., Kalarikkal, N., Thomas, S. (eds.) Characterization of Nanomaterials, pp. 61–102. Woodhead Publishing (2018)
61. Kobayashi, K., Fuller, M.D.: Vibration magnetometer. In: Collinson, D.W., Creer, K.M., Runcorn, S.K. (eds.) Developments in Solid Earth Geophysics, pp. 450–456. Elsevier (2013)
62. Helsley, C.E.: Design of spinner magnetometers 1 1Contr. No. 12, Geosciences Division, Graduate Research Center of the Southwest, Dallas, Texas (U.S.A.). In: Collinson, D.W., Creer, K.M., Runcorn, S.K. (eds.) Developments in Solid Earth Geophysics, , pp. 115–118. Elsevier (2013)
63. Kwiecińska, B., Pusz, S., Valentine, B.J.: Application of electron microscopy TEM and SEM for analysis of coals, organic-rich shales and carbonaceous matter. Int. J. Coal Geol. **211**, 103203 (2019)
64. Chen, L.J., Wu, W.W.: In situ TEM investigation of dynamical changes of nanostructures. Mater. Sci. Eng. R. Rep. **70**, 303–319 (2010)
65. De Yoreo, J.J.: In-situ liquid phase TEM observations of nucleation and growth processes. Prog. Cryst. Growth Charact. Mater. **62**, 69–88 (2016)
66. Wu, F., Yao, N.: Advances in windowed gas cells for in-situ TEM studies. Nano Energy **13**, 735–756 (2015)
67. Danino, D.: Cryo-TEM of soft molecular assemblies. Curr. Opin. Colloid Interface Sci. **17**, 316–329 (2012)
68. Kumari, P., Saha, R., Saikia, G., Bhujel, A., Choudhury, M.G., Jagdale, P., Paul, S.: Synthesis of mixed-phase TiO_2–ZrO_2 nanocomposite for photocatalytic wastewater treatment. Toxics **11**, 234 (2023)
69. Li, M.-H., Yang, Y.-Q., Huang, B., Luo, X., Zhang, W., Han, M., Ru, J.-G.: Development of advanced electron tomography in materials science based on TEM and STEM. Trans. Nonferr. Metals Soc. China **24**, 3031–3050 (2014)
70. Muhammad Sajeer, P., Simran, Nukala, P., Manoj, M.V.: TEM based applications in solid state nanopores: from fabrication to liquid in-situ bio-imaging, Micron **162**, 103347 (2022)

Chapter 5
Inorganic Materials for Electromagnetic Shielding

Abstract The use of inorganic materials for electromagnetic shielding purposes is discussed here. This includes examining the properties of inorganic materials that facilitate EM shielding in various bands. The chapter also touches upon recent advancements and research in the field of electromagnetic shielding using inorganic materials, highlighting key findings and potential future directions for this technology. A concise summary of the importance and relevance of inorganic materials in developing electromagnetic shielding solutions is also included in this chapter. As the demand for smaller, faster, and more efficient electronic devices continues to grow, the development of advanced electromagnetic shielding techniques using inorganic materials is crucial in meeting the needs of various industries.

5.1 Introduction

Electromagnetic shielding is essential in diverse industries to safeguard electronic devices from interference. Metals and metal oxides are frequently employed due to their advantageous characteristics [1]. Metal oxide nanomaterials have been found to have the advantages of providing a large surface area-to-volume ratio, good transport performance, and physical properties [2]. Current research endeavors to create innovative inorganic materials to improve electromagnetic shielding efficiency in contemporary applications [3].

This chapter explores electromagnetic shielding applications and emphasizes the significance of safeguarding electronic devices from interference. In this process, inorganic materials, specifically metals like copper and aluminum, and metal oxides, like ferrites, are crucial due to their conductivity and magnetic permeability qualities [4]. In exceptional cases, some researchers also used liquid metals [5]. These materials can be easily included in shielding structures using coatings, laminates, or composites. The current research in this area is centered on creating new inorganic materials to improve electromagnetic shielding effectiveness [6]. EMI shielding commonly employs inorganic materials with high magnetic permeability to efficiently divert and absorb electromagnetic radiation [7].

S. K. Koul et al., *Microwave Absorbers*, Engineering Materials,
https://doi.org/10.1007/978-981-96-9596-6_5

An approach to integrating inorganic elements into an electromagnetic shielding-compliant material involves the application of physical vapor deposition [8]. This process aids in forming a protective shield that redirects electromagnetic energy [9]. Nevertheless, a significant hindrance to the utilization of inorganic materials for shielding is their substantial weight and high cost, rendering them impractical for large-scale production [10–13]. However, current research is focused on creating lighter and more affordable options to enhance the accessibility of electromagnetic shielding for many purposes [14]. This chapter discusses the inorganic nanomaterials to improve their effectiveness in electromagnetic shielding applications [12].

5.2 Inorganic Ferrite Nanoparticle Composites

Ferrite beads are passive electronic components that suppress high-frequency noise in electronic circuits. They are made from ferrite, a ceramic material of iron oxide (Fe_2O_3) mixed with other metal oxides. The exact composition of ferrite beads can vary depending on the application and desired properties. Ferrite beads are composed of a nano-sized ferrimagnetic substance known as ferrite. This substance exhibits the characteristics of an inductor constructed from a spiral arrangement of conductive material. The appeal of this component lies in its compact size without compromising the high inductance value [15]. Ferrite is an insulating material that can be employed to diminish electromagnetic interference (EMI) in electronic systems using magnetic induction [16]. Ferrite components such as beads, cores, and core inductors have several applications in electromagnetic interference (EMI) filtering [17]. Ferrite beads can prevent high-frequency disturbances from entering a device. They can also function as an LC filter with a bypass capacitor and establish connections across distinct analog and digital ground/power planes. Ferrite cores can be utilized in cables to mitigate electromagnetic interference (EMI) by impeding low-frequency noise and minimizing high-frequency noise [16].

These devices are effective for keyboards, mice, and PC cables since they can lower standard mode current and radiation. Ferrite core inductors are utilized in power electronic systems to reduce the intensity of high-frequency harmonics. Ferrite cores mitigate noise interference in electrical devices [18]. Traditional chip ferrite beads have limited possibilities for both frequency range and attenuation. However, wire-wound ferrite nanoparticle beads offer a relatively higher magnitude of attenuation over a wide range [19].

Additionally, wire-wound ferrite beads have better current ratings and lower direct current resistance (DCR). Every ferrite bead has a maximum DC voltage at which it can no longer function without breaking down [20]. Heat significantly affects a ferrite bead; the rated current value drops with the temperature. Figure 5.1 shows a non-split non-crystalline and split-core sample (left) and a non-split ceramic and split-core sample (right) [21].

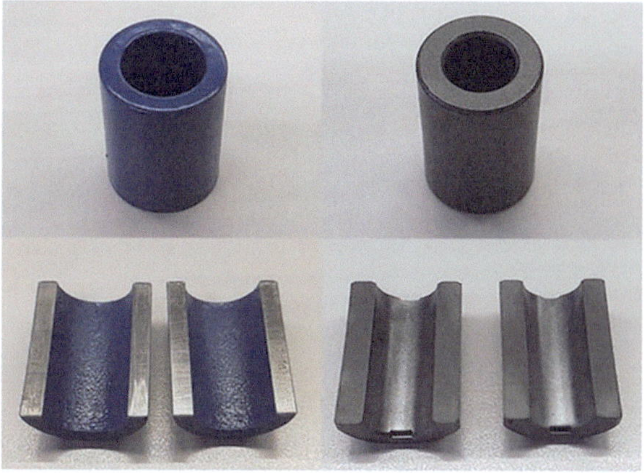

Fig. 5.1 Non-split non-crystalline and split-core sample (left) and non-split ceramic and split-core sample (right) [21]

5.3 Classification of Inorganic Nanomaterials Used for EMI/EMC Application

Inorganic nanomaterials, such as metallic nanoparticles, carbon-based nanomaterials, and ceramics, offer unique properties like high electrical conductivity, lightweight, and high surface area, which make them ideal for EMI shielding applications.

5.3.1 Metallic Nanomaterials

Metallic and conductive nanomaterials exhibit exceptional electrical conductivity and other unique properties due to their nanoscale dimensions. These materials are crucial in various applications, including electronics, energy storage, sensing, and EMI shielding. The classification of inorganic nanomaterials is shown in Fig. 5.2, which gives a detailed classification of inorganic nanomaterials under four sections: Metallic nanoparticles, metal oxide nanoparticles, non-metallic nanoparticles, and hybrid nanocomposites. Different nanoforms are classified, as shown in Fig. 5.3. Important metallic nanomaterials, including Ag, Cu, Pt, and Ni, are discussed in subsequent sections.

5.3.1.1 Silver-Based Nanomaterials

Silver is a commonly used inorganic nanomaterial in applications requiring high conductivity. Silver nanoparticles have unique plasmonic properties, which can be utilized in optical and electronic applications. Silver also has excellent thermal

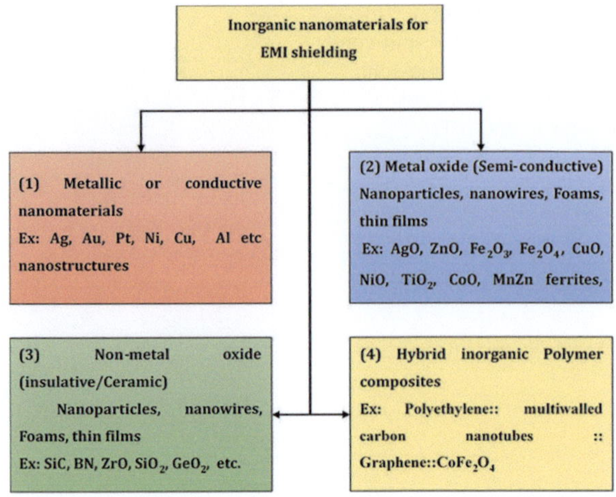

Fig. 5.2 Classification of inorganic nanomaterials for shielding applications

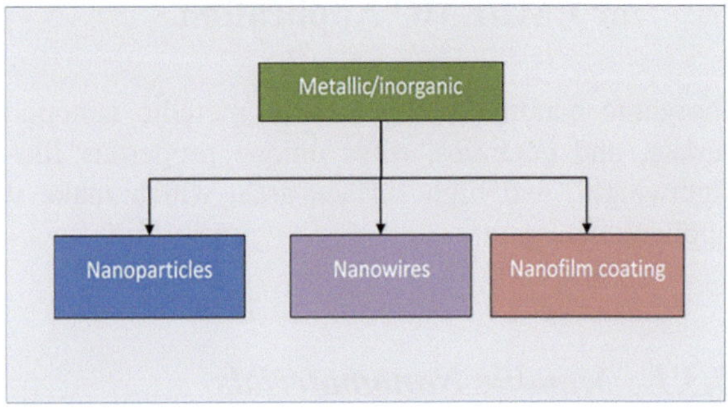

Fig. 5.3 Classification of metallic or inorganic nanostructure

conductivity, helping to dissipate heat generated during EMI absorption. These applications include EMI shielding. Silver is one of the best-suited inorganic nanomaterials, which can be used in nanoparticles of definite shape or irregular morphology, nanowires of silver metals, and thin-film coating of silver metal over any substrate [22], which will work for an EMI/EMC application. Silver nanowires' low sheet resistance and flexibility have made them highly sought-after material for use in flexible transparent conductive films (TCFs). Metal-based transparent conductive films (MTCFs) such as indium-tin-oxide (ITO) are essential components in optoelectronics and flexible electronics, combining transparency with electrical conductivity.These metal-based TCFs offer promising alternatives to ITO for next-generation flexible, stretchable, and wearable optoelectronic applications due to their mechanical robustness, tunable performance, and cost-effective manufacturing. These alternatives offer a combination of high conductivity and excellent sheet resistance, resulting in a high level of transmittance. These materials have high electrical conductivity and exceptional shielding efficiency [23].

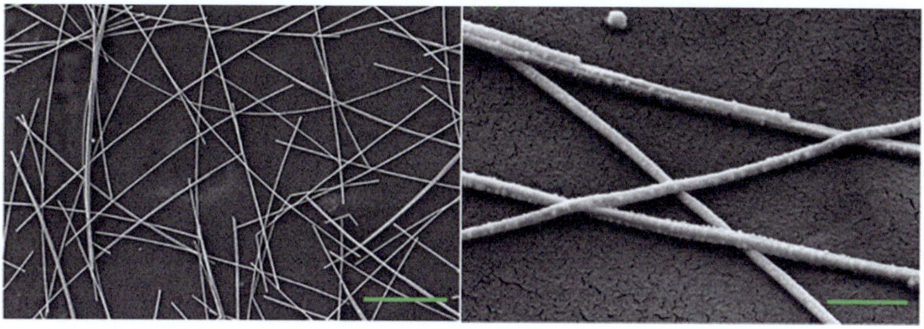

Fig. 5.4 Silver metallic nanowire under SEM [27]

Ag nanowires (NWs) with a 1D structure can attain percolation at low concentrations because of their high aspect ratio. Transparent conductive films (TCFs) are highly sought for their use in display devices, touch sensors, solar cells, organic photovoltaic devices, and light-emitting diodes. Indium-tin-oxide (ITO) thin films possess exceptional optoelectronic characteristics, but their fragility hinders their use in forthcoming flexible and stretchable electronics [24–26]. Figure 5.4 shows a silver nanowire under a scanning electron microscope (SEM) [27]. The silver nanoparticle finds its application in conductive inks, inkjet printers, screen-printing purposes, and conductive ink pens to build broken circuits and coat the inner portion of electronic device panels. The silver coating is also used in the plasma and physical vapor deposition (PVD) coating to deposit the thin conductive film on the substrate as shielding materials [28].

5.3.1.2 Copper Nanomaterials

Like Silver, Copper, Gold, and Platinum-derived nanomaterials also have high conductivity and are cost-effective compared to silver. They are used in ink and coatings for flexible and printable electronics. Copper Nanowires (CuNWs) have high electrical conductivity and are cost-effective compared to silver. They are used in transparent conductive films, flexible circuits, and EMI shielding coatings. Copper nanoparticles (CuNPs) have excellent electrical and thermal conductivity. They can be used in ink and coatings for printable electronics and EMI shielding [29, 30]. Figure 5.5 shows SEM images of copper nanowires with (a) dense nanowires, (b) nano-needles, and (c) nanoparticles [31]. In the SEM images of copper nanoparticles, we can observe different structures and morphology of the prepared copper nanoparticles.

Silver, Gold, Platinum, and Copper nanowires and nanoparticles are highly effective materials for electromagnetic interference (EMI) shielding due to their excellent electrical conductivity, high aspect ratio (for nanowires), and unique surface finishing properties. Copper is one of the best electrical conductors, and both nanowires and nanoparticles retain this property, making them highly effective for EMI shielding.

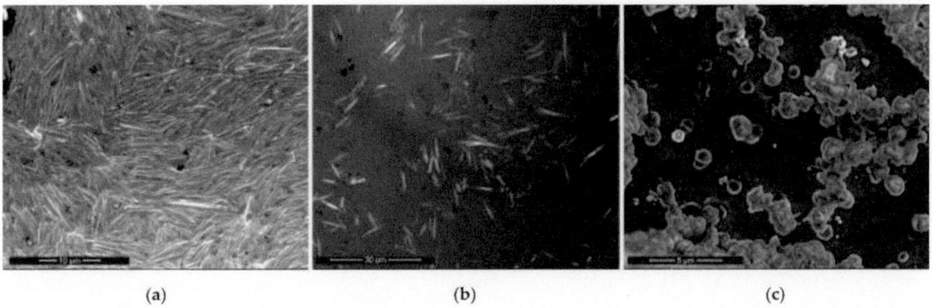

Fig. 5.5 SEM images of copper nanowires with **a** dense nanowires, **b** nano-needles, and **c** nanoparticles [31]

Silver nanowires have a high length-to-diameter ratio, which enhances their ability to form conductive networks [32].

Applications of Ag/Au/Pt/Cu Nanostructures

Metallic nanowires are employed in transparent conductive films in displays and touchscreens to shield against electromagnetic interference (EMI), enabling high conductivity and transparency. Ag/Au/Pt/Cu nanowires are used in intelligent windows and transparent panels to provide EMI shielding without obstructing visibility. In flexible electronics, Ag/Au/Pt/Cu nanoparticles and nanowires are incorporated into polymer composites to create flexible and lightweight EMI shielding materials for wearable electronics and portable devices [33]. Coatings and films are applied on surfaces to provide effective EMI shielding while maintaining flexibility. Metallic nanoparticles are used in conductive inks for printed electronics, enabling the creation of EMI shielding patterns and components through printing techniques. Figure 5.6 shows (a) inkjet printing digital display circuit and (b) digital display circuit with soldered components [34]. The image shows the conducting circuit successfully printed over a transparent, flexible plastic substrate. Many electronic components were soldered onto it.

Silver nanowire and nanoparticle coating protect sensitive components from EMI in electronic device enclosures and casings. They are also used in electronic enclosure panels for high-frequency applications to provide EMI shielding and enhance signal integrity. These materials have advantages compared to other nanomaterials, such as metallic nanostructures with high-shielding effectiveness. Excellent conductivity and a high aspect ratio (for nanowires) result in effective EMI shielding.

Transparency associated with silver nanowires can form transparent conductive films, making them suitable for optical applications. Flexibility is another aspect that can be incorporated into materials for wearable and portable electronics [35]. The metallic nanomaterials also have good thermal conductivity, which helps manage heat generated during EMI absorption. Silver is an expensive material, which can increase the cost of EMI shielding solutions. Silver nanoparticles agglomerate, making uniform dispersion in composites difficult. Silver can oxidize, affecting its conductivity and shielding effectiveness over time.

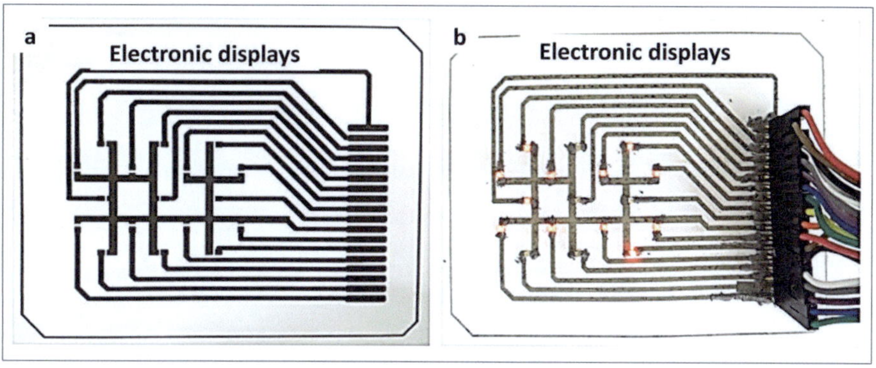

Fig. 5.6 **a** Inkjet printing digital display circuit and **b** digital display circuit with soldered components [34]

Similarly, copper nanoparticles possess high conductivity in nanostructure form, which is cost-effective compared to silver. They are used in ink and coatings for flexible and printable electronics. Furthermore, nickel nanoparticles have good magnetic properties and conductivity. They are often used in combination with polymers to enhance EMI shielding performance. Iron nanomaterials have high magnetic permeability and electrical conductivity. They enhance EMI shielding in composites.

Gold Nanoparticles (AuNPs) have exceptional conductivity and chemical stability despite their high cost. They are utilized extensively in specialized applications where cost is not a limitation. Aluminum nanoparticles are used for various purposes in the aerospace and automotive industries as a lightweight EMI shielding solution due to their excellent conductivity, corrosion resistance, and low weight. Aside from being lightweight, the nanomaterial made of aluminum also offers excellent conductivity and resistance to corrosion. Iron and steel's high permeability can benefit magnetic shielding [36]. Figure 5.7 shows the coating of conductive paint used for the EMI shielding effectiveness test [37, 38].

When combined with metal nanoparticles, nanostructures can also form metallic foams that contain nanocavities or nano-sized thick walls. These foams are light because relatively less material is used. Metal foams are also less expensive, more malleable, and have good conductivity.

5.3.1.3 Metal Foams

These are lightweight with good shielding properties. Due to their unique properties, metal foams are increasingly used in electromagnetic interference (EMI) shielding. Figure 5.8 shows SEM images of ultralight metal foams [38].

Metal foams are significantly lighter than solid metals, making them suitable for applications where weight is critical. The porous structure provides a large surface area, enhancing its interaction with electromagnetic waves. Despite their

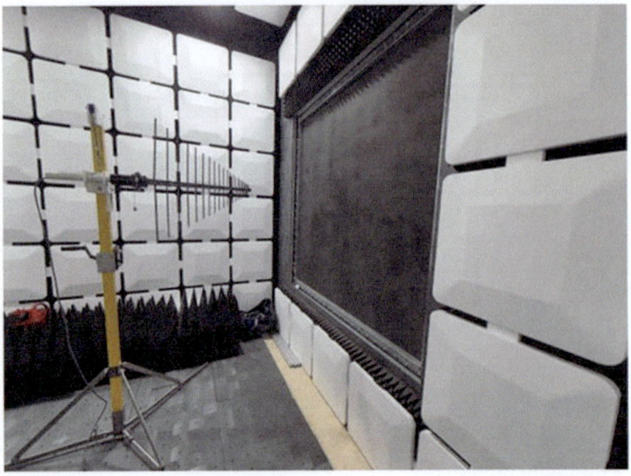

Fig. 5.7 Coating of conductive paint for EMI shielding effectiveness test [37]

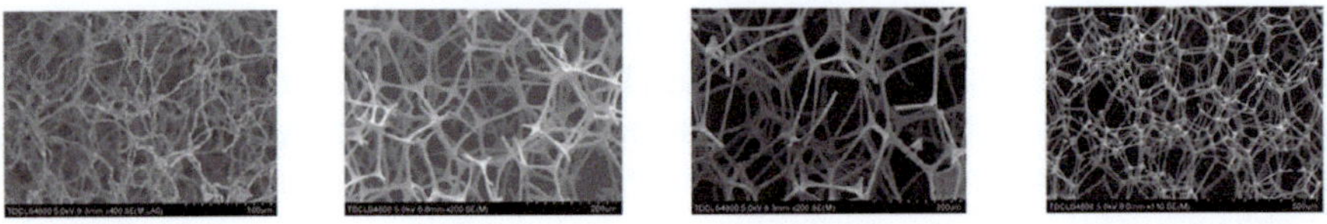

Fig. 5.8 SEM images of ultralight metal foams [38]

porous nature, metal foams maintain good electrical conductivity, which is crucial for effective EMI shielding. Metal foams retain good mechanical strength, providing structural support while offering shielding. They possess good thermal conductivity, helping them to dissipate heat generated during the absorption of electromagnetic waves.

Metal foams' porous structure increases the path length for electromagnetic waves, enhancing absorption and converting the energy into heat, leading to EM wave absorption. Metal foams' conductive nature causes them to reflect electromagnetic waves, preventing them from passing through the material and leading to reflection.

The internal porous structure leads to multiple reflections within the material, further attenuating the electromagnetic waves. Aluminum foam has unique properties, such as being lightweight, cost-effective, and suitable for electrical and thermal conductivity. Copper foam produces excellent electrical conductivity and is relatively expensive but effective in high-performance applications. It is used in electronics, telecommunications, and shielding enclosures.

Nickel foam materials have good electrical conductivity and corrosion resistance. They are suitable for harsh environments and are used in battery electrodes, fuel cells, and EMI shielding. Steel foam exhibits good mechanical properties and is

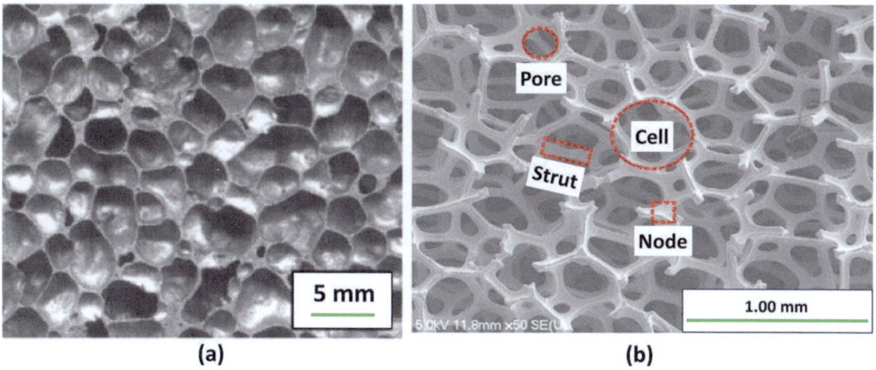

Fig. 5.9 The SEM images of closed-cell and open-cell metal foams: **a** closed-cell foam and **b** open-cell foam [39]

cost-effective. It is used in civil construction and automotive applications. Figure 5.9 shows the SEM images of closed-cell and open-cell metal foams [39].

Applications of Metal Foams Metal foams are used to construct enclosures and casings for electronic devices, providing lightweight yet effective EMI shielding solutions. Metal foam panels can be used in building walls and partitions to block electromagnetic interference in sensitive areas like data centers and laboratories. Metal foam gaskets and seals are employed in connectors and joints to ensure continuous EMI shielding. Metal foams can be used as EMI filters in various electronic and telecommunication devices. They are also used in vehicles and aircraft to shield electronic components from EMI while minimizing weight.

However, high-quality metal foam can be expensive and require additional design considerations. As manufacturing techniques improve and costs decrease, metal foams' use in EMI shielding is expected to expand.

Mu-metal is also gaining popularity among researchers who use metal-based shielding materials. It consists of a nickel–iron alloy with high magnetic permeability, which makes it effective for magnetic shielding.

Permalloy is another nickel–iron alloy that is effective for magnetic and electromagnetic shielding. Copper–nickel alloys have good corrosion resistance and are effective at various frequencies.

5.3.1.4 Metal-Coated Textile

Metal-coated fabrics are flexible and can be used in various applications. The metal coating provides electrical conductivity. Flexibility is retained from the fabric base. It is lightweight and lighter than most solid metal shields. Metal-coated textiles are resistant to wear and environmental conditions. The fabric absorbs electromagnetic energy, converting it to heat. Additional attenuation occurs due to the scattering of

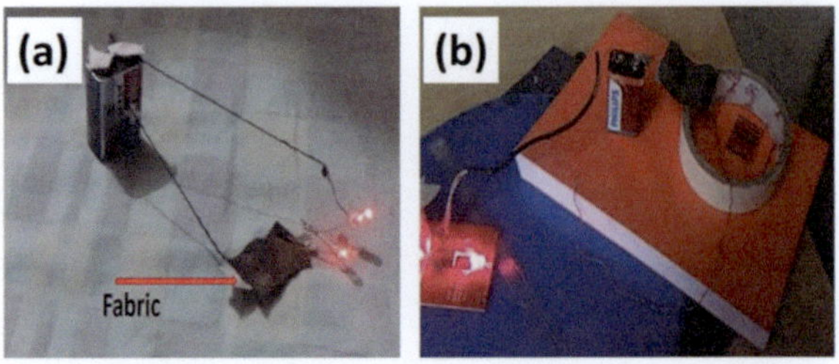

Fig. 5.10 Flow of electricity through **a** silver particles and **b** copper particle-coated fabrics [40]

electromagnetic waves by texture. The silver coating on fabrics is expensive, highly conductive, and antibacterial. Aside from being inexpensive, copper also offers good conductivity. Nickel coating on clothes is conductive and impervious to corrosion. Aluminum is lightweight and cost-effective.

Applications Conductive textiles find applications in wearable electronics, clothing, and accessories. In military and aerospace applications, lightweight shielding is necessary for equipment and personnel. In health care, shielding for medical devices and protective garments are used. In consumer electronics, these conductive textiles cover outer cases and linings for gadgets. In the automotive sector, shielding for electronic systems in vehicles is necessary to develop driverless and signal-free mobility.

Architecture zones (e.g., data centers) are applied to prevent the outside EM waves from interacting with the human beings inside the buildings. Metal-coated fabrics are suitable for conformable shielding, are easier to handle and integrate, have tailorable properties for specific needs, and resist various atmospheric conditions. These materials suffer from challenges that include cost, and high-quality coatings can be expensive. They ensure long-term adherence and effectiveness and require precise manufacturing control. Metal-coated fabrics offer versatile and effective EMI shielding, combining conductivity, flexibility, and lightweight properties for various applications. Advances in technology continue to improve their performance and broaden their use. Figure 5.10 illustrates the flow of electricity through (a) silver particles and (b) copper particle-coated fabrics [40].

5.3.1.5 Metal-Polymer Composites

These composites combine the properties of metals and polymers for versatile applications. They are materials composed of metal particles embedded in a polymer matrix. These composites are increasingly used for electromagnetic interference (EMI) shielding due to their unique properties, which provide effective shielding while maintaining flexibility and lightweight characteristics, as shown in Fig. 5.11. Fluorinated polyimide nano silver paste-coated material in a patterned manner is shown in this figure [41].

Fig. 5.11 Fluorinated polyimide nano silver paste-coated material in a patterned manner [41]

Properties of Metal-Polymer Composites The embedded metal particles provide high electrical conductivity, which might be crucial for reflecting and absorbing electromagnetic waves. The polymer matrix maintains the composite's flexibility, allowing it to be used in applications where rigid materials are unsuitable.

Metal-polymer composites are lighter than solid metals, which is advantageous for applications where weight is a concern. Depending on the choice of polymer, these composites can exhibit good thermal stability. They can be designed to resist wear, tear, and environmental conditions. The conductive metal particles reflect incoming electromagnetic waves, preventing them from passing through the composite. The composite absorbs electromagnetic waves and converts them into heat, reducing EMI [42].

Common Metals Used in Metal-Polymer Composites:

- **Silver**: Excellent conductivity. It is a high-cost material but very effective in high-performance applications.
- **Copper**: Good conductivity and cost-effectiveness. It is commonly used to balance performance and price.
- **Nickel**: Good conductivity and corrosion resistance. It is used in applications requiring durability.
- **Aluminum**: Lightweight and cost-effective. Suitable for applications where weight is critical.

Applications of Metal-Polymer Composites in EMI Shielding: They are used in electronic enclosures, creating lightweight, flexible, and effective EMI shields for electronic devices. Wearable electronics are incorporated into fabrics and flexible materials for wearable devices to protect against EMI. The automotive field shields electronic systems within vehicles to ensure reliable operation. Medical devices shield sensitive medical equipment to prevent interference from external electromagnetic sources.

In consumer electronics, such as covers and internal linings for smartphones, tablets, and laptops, such materials can be used. It is also suitable for applications requiring flexibility and movement. It is easier to handle and integrate into products than traditional metal shields. The properties of metal-polymer composites can be tailored by varying the type and amount of metal and polymer used. It can be designed to withstand various environmental conditions and mechanical stress. Uniform dispersion ensuring even distribution of metal particles within the polymer matrix can be challenging to achieve.

Manufacturing processes need to ensure consistent quality and performance. Metal-polymer composites are highly effective for EMI shielding, offering a blend of conductivity, flexibility, and lightweight properties. Their ability to reflect, absorb, and scatter electromagnetic waves makes them suitable for various applications, including electronics, automotive, medical devices, and wearable technology. As advancements in materials science and manufacturing techniques continue, the performance and application scope of metal-polymer composites in EMI shielding will likely expand.

5.3.2 Metal Oxides

Metal oxides have high dielectric loss tangents, which leads to high absorption characteristics. Ferrite nanoparticles can absorb electromagnetic waves, converting them into heat through magnetic losses (e.g., hysteresis, eddy currents, and domain wall resonance). Due to their high magnetic permeability, metal oxide nanoparticles can reflect and alter electromagnetic waves' direction and penetration power. As mentioned earlier, the nanoscale size and high surface area of ferrite nanoparticles enhance the material's absorption characteristics.

5.3.2.1 Magnetite Nanoparticles (Fe_3O_4)

They have superior magnetic properties. They are used in absorbent shielding materials and composite matrices. Hard ferrite nanoparticles are widely used in electromagnetic interference (EMI) shielding due to their excellent magnetic and dielectric properties. Here's an overview of the EMI shielding properties of ferrite nanoparticles:

Ferrites have high magnetic permeability, which allows them to absorb and dissipate electromagnetic waves effectively. They exhibit significant dielectric properties, contributing to their ability to attenuate EMI through polarization mechanisms. Ferrites generally have high electrical resistivity, which helps reduce eddy current losses and enhances their EMI shielding effectiveness.

5.3.2.2 Ferrites

Ferrite nanoparticles are stable at high temperatures, making them suitable for various environmental conditions. Common ferrite materials used in EMI shielding are manganese–zinc ferrite ($MnZnFe_2O_4$), which has high magnetic permeability and is effective at relatively lower frequencies (kHz to GHz range). It is commonly used in transformers, inductors, and EMI suppression cores.

Nickel–zinc ferrite ($NiZnFe_2O_4$) has a higher resistivity than MnZn ferrite and is effective at relatively higher frequencies (MHz to GHz range). It is used in high-frequency applications, including antennas and filters. Cobalt ferrite ($CoFe_2O_4$) has high coercivity and magnetic anisotropy. It is suitable for applications requiring strong magnetic fields, such as data storage and magnetic sensors.

Ferrite nanoparticles can be incorporated into paints and coatings to create surface EMI shielding layers. Polymer composites can be embedded in polymer matrices to produce lightweight, flexible EMI shielding materials. Ferrite nanoparticles are used to design inductors, transformers, and EMI suppression cores for electronic circuits. Enclosures and casings are used to construct enclosures for electronic devices to shield sensitive components from EMI. Ferrite nanoparticle-filled gaskets and seals provide EMI shielding in connectors and enclosures. The photograph of the ceramic nanoparticle paste is shown in Fig. 5.12.

These ceramic nanoparticle pastes and paints can be applied on various surfaces and easily cover the desired area. Additionally, it can protect the materials from corrosion and heat resistance.

Ferrite nanoparticles provide excellent attenuation of electromagnetic waves across a broad frequency range. They maintain their properties under high temperatures and harsh environmental conditions. It can be used in various forms and integrated into different materials and products. Ferrites are generally non-toxic and environmentally friendly.

High-quality ferrite materials are relatively expensive to produce. Composites and coatings containing ferrite nanoparticles might be complex to disperse evenly.

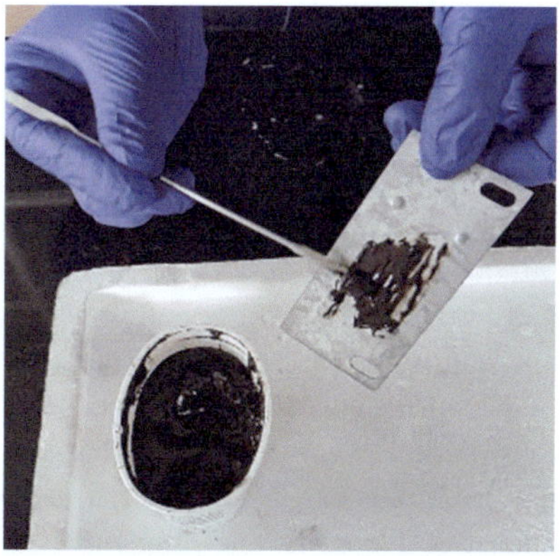

Fig. 5.12 Photograph of ceramic nanoparticle paste

Different ferrite materials are effective in other frequency ranges, requiring careful selection for specific applications. Ferrite nanoparticles offer robust solutions for EMI shielding, combining high magnetic permeability and dielectric properties with thermal stability and versatility [43].

Their ability to attenuate electromagnetic waves through absorption, reflection, and multiple reflections makes them valuable in protecting electronic devices and systems from electromagnetic interference. Like all the metal oxide nanomaterials, such as zinc oxide (ZnO), Silicon dioxide (SiO_2), and titanium dioxide (TiO_2), they have high electron mobility and good ultraviolet (UV) light shielding properties. They are used in combination with other materials to enhance EMI shielding. Due to their unique properties, zinc oxide (ZnO) and titanium oxide (TiO_2) nanoparticles have emerged as promising materials for electromagnetic interference (EMI) shielding [44].

They exhibit high magnetic permeability and demonstrate shielding effectiveness at high frequencies. They are used in coatings, composites, and porous materials to protect against electromagnetic interference (EMI). Even when applied to very thin material, M-type hexagonal ferrites demonstrate substantial reflection loss and improved absorption, making them a potential choice as the radar-absorber compound for electromagnetic interference (EMI) attenuators in weather monitoring, radar tracking, and air traffic management. Moreover, the synthesized magnetic Yttrium-doped Barium M-type hexagonal ferrites have potential commercial applications such as long-term magnetic data recording and information storage [3, 12, 45].

5.3.2.3 Zinc Oxide (ZnO)

ZnO nanoparticles exhibit semiconducting properties, which can be tailored for better conductivity by doping with other elements. ZnO particles' nanoscale size provides a large surface area, enhancing their interaction with electromagnetic waves. ZnO nanoparticles are stable at high temperatures, which benefits applications in various environments.

ZnO has excellent UV absorption and visible transparency, making it suitable for transparent shielding applications. These metal oxide nanoparticles may not directly contribute to the EMI shielding when designed alone but have efficient shielding performance when used along with metal structures. SiO_2, ZnO, and TiO_2 nanoparticles can scatter electromagnetic waves with multiple reflections, converting them into heat and reducing EMI. ZnO's conductive nature allows it to reflect electromagnetic waves, preventing them from penetrating sensitive areas. The high surface area and nanoscale size scatter electromagnetic waves, diminishing their strength [46].

Metal oxide nanoparticles can be incorporated into paints and surface coatings to provide EMI shielding while maintaining transparency or desired aesthetics. ZnO nanoparticles can be embedded in polymer matrices to create lightweight, flexible, and effective EMI shielding materials. Conductive fabrics with ZnO nanoparticles are used in wearable electronics to protect against EMI. ZnO nanoparticles are used to design enclosures and housings for electronic components to prevent interference [47]. ZnO is relatively inexpensive and abundant compared to other shielding materials. ZnO nanoparticles are resistant to environmental degradation, ensuring long-term performance. It can be integrated into various forms, including coatings, films, and composites, to suit different applications [48].

It is ideal for applications where transparency is essential, such as in displays and windows, ensuring uniform dispersion of ZnO nanoparticles in composites and coatings, which can be challenging and critical for optimal performance. ZnO nanoparticles must be compatible with the base materials (e.g., polymers, resins) to achieve effective integration and performance. Advanced techniques may be required to produce and apply ZnO nanoparticles efficiently and uniformly. Zinc oxide (ZnO), Al_2O_3 [49], TiO_2, and SiO_2 nanoparticles are promising for EMI shielding applications. They offer a combination of cost-effectiveness, environmental stability, and versatile integration options. As research and development continue, ZnO nanoparticles will likely play an increasingly important role in protecting electronic devices and systems from electromagnetic interference. Some metal oxides, such as titanium dioxide (TiO_2) nanoparticles, have high dielectric constant and are used in composite material coatings and films for EMI shielding.

5.3.2.4 Thin-Film Conductive Oxides

Due to their unique properties, conductive oxides like indium tin oxide (ITO) are widely used in transparent shielding applications. Their excellent electrical conductivity makes it an efficient conductor in various electronic applications [50]. ITO coatings are durable and resistant to environmental factors like moisture and temperature variations. Indium tin oxide (ITO) plates are conductive due to their unique electronic structure and composition, which consists of a mixture of indium oxide (In_2O_3) and tin oxide (SnO_2). Composition and doping in indium oxide (In_2O_3) is a wide-bandgap semiconductor, which typically has poor electrical conductivity.

However, when it is doped with tin oxide, its electronic properties change. Tin oxide (SnO_2) acts as a dopant in the indium oxide lattice. When tin is added to indium oxide, some of the tin atoms replace indium atoms in the crystal lattice, and this substitution introduces free electrons into the material [51].

Carrier generation of substituting tin (Sn^{4+}) for indium (In^{3+}) in the lattice creates an excess of positive charges, which free electrons must balance. These free electrons are not bound to any atom and can move freely through the crystal lattice [52]. The movement of these electrons under an applied electric field gives rise to electrical conductivity. In the conduction band in ITO, free electrons occupy energy states near the bottom of the conduction band. The presence of these electrons in the conduction band allow ITO to conduct electricity [53]. Despite being a wide-bandgap material (around 3.5 to 4.3 eV), the high concentration of free electrons will enable ITO to exhibit metallic-like conductivity. One of the unique features of ITO is its combination of electrical conductivity and optical transparency in the visible spectrum [54].

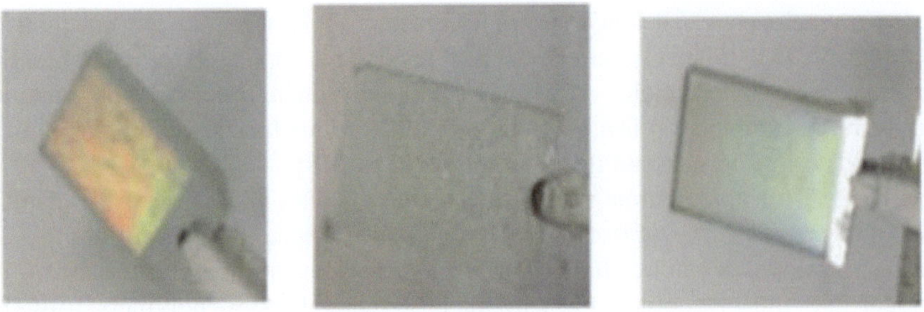

Fig. 5.13 Large plate original conductive glass [55]

The wide bandgap of ITO means it does not absorb visible light, allowing it to remain transparent while still conducting electricity due to the free electrons. ITO is used in the transparent electrodes of touchscreens, providing the necessary electrical pathways for detecting touch while maintaining screen visibility in touch-screen devices. It is also used in liquid crystal diodes (LCDs) and organic liquid crystal diodes (OLED) displays, and ITO layers help control pixel activation. ITO is employed in solar cells to create transparent conductive layers that allow light to pass through while conducting electricity. Transparent conductive oxides, like ITO, can shield electronic devices from electromagnetic interference without obstructing visibility. ITO-coated glass can be used in smart windows to adjust transparency in response to electrical signals, providing energy efficiency and comfort in buildings [54]. Figure 5.13 shows a large plate of original conductive glass [55]. ITO can be deposited on various substrates, including glass and plastics, to develop modern electronic and optoelectronic devices. However, indium is relatively rare and expensive, driving up the cost of ITO. ITO layers can be brittle, which might limit their use in flexible electronics [56, 57].

5.3.2.5 MXenes (Inorganic Layered Materials)

MXenes are a class of two-dimensional (2D) transition metal carbides, nitrides, and carbonitrides that have garnered significant attention for their exceptional electrical, mechanical, and thermal properties. These materials are up-and-coming for electromagnetic interference (EMI) shielding applications [58]. Figure 5.14 shows thermal reflection characteristics of composite inorganic coatings [59]. This thermal inorganic nano metal oxide nanomaterial will shield IR, UV, and other EM wave interference and thermal insulation properties.

MXenes exhibit high electrical conductivity, which is crucial for EMI shielding as it effectively reflects and partially absorbs electromagnetic waves. MXenes' layered structure provides a large surface area, which enhances their interaction with electromagnetic waves and improves their shielding effectiveness.

MXenes can be chemically modified to tailor their properties, such as electrical conductivity and surface functionality, making them versatile for different EMI

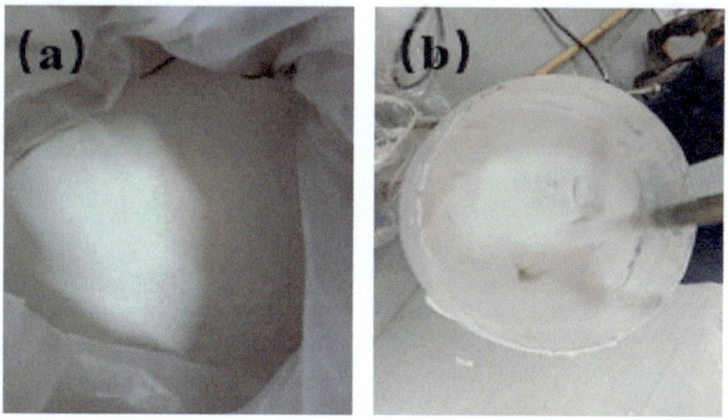

Fig. 5.14 Thermal reflection characteristics of composite inorganic coatings [59]

shielding requirements. MXenes possess high mechanical strength and flexibility, which are beneficial for developing robust and durable EMI shielding materials. Their excellent thermal stability ensures that MXenes can be used in environments with varying temperatures without degradation in performance [14].

MXenes' high electrical conductivity allows them to reflect incident electromagnetic waves, preventing them from passing through the material. Due to their large surface area and the presence of functional groups that can interact with the waves, MXenes can absorb electromagnetic waves, converting the electromagnetic energy into thermal energy. The layered structure of MXenes can cause multiple internal reflections of electromagnetic waves within the material, enhancing the absorption and dissipation of the waves. Figure 5.15 shows SEM images of MXene layered structures before and after acid leaching [60].

MXene applications in EMI Shielding MXenes can be used to create thin coatings and films that can be applied to electronic devices, enclosures, and other surfaces to provide effective EMI shielding. They can also be incorporated into polymer or ceramic matrices to develop composite materials with enhanced EMI shielding properties. These composites can be used in various industries, including aerospace, automotive, and telecommunications. MXenes can also be integrated into fabrics to produce lightweight, flexible EMI shielding textiles suitable for wearable electronics and protective clothing [12].

Advantages of MXenes for EMI Shielding MXenes provide high shielding effectiveness due to their excellent conductivity and layered structure. MXene-based materials can be lightweight, which is advantageous for applications where weight is critical. MXene materials are flexible and durable, making them suitable for various applications, including flexible electronics. Developing scalable and cost-effective methods for producing MXenes is essential for widespread adoption in EMI shielding applications [61]. Ensuring the long-term stability of MXene materials, particularly in harsh environments, is crucial for their practical use.

Integrating MXenes with other materials to create composites with synergistic properties requires further research and development. Due to their unique properties, MXenes represent a promising class of materials for EMI shielding. Ongoing research

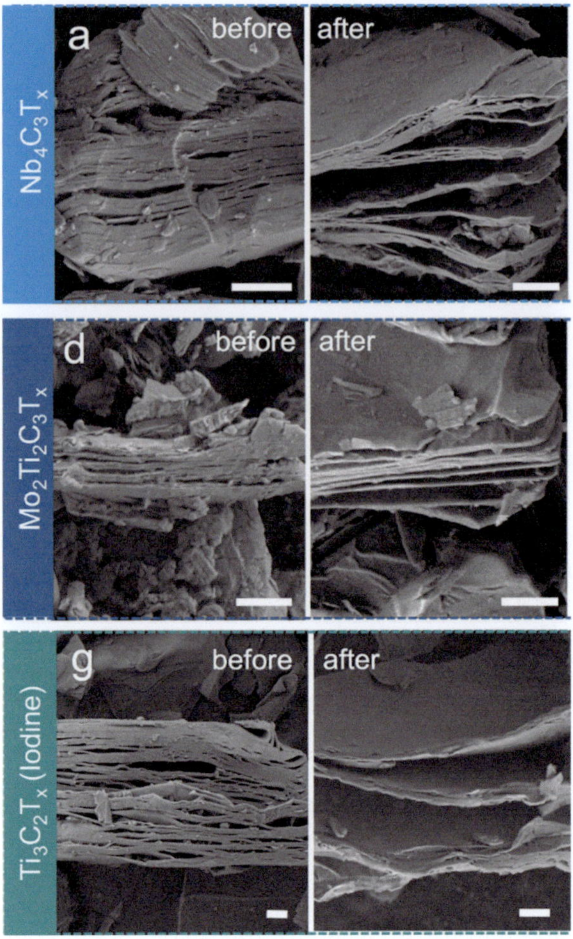

Fig. 5.15 SEM images of MXenes showing layered structures before and after acid leaching [60]

addresses the challenges associated with their production and integration to realize their full potential in various EMI shielding applications [14].

5.3.3 *Non-metal Oxide (Insulative, Ceramic)*

Non-metal oxide-based ceramics have the potential to be suitable candidates for various applications due to their ability to withstand high temperatures. Nevertheless, these materials are limited in their applicability due to their elevated density. On the contrary, a recommended approach would involve constructing structures using foam material. This method would effectively reduce the density of the structures while maintaining the desired level of electromagnetic interference (EMI) shielding properties [20].

Non-metal oxide (insulative, ceramic) materials are gaining attention for their use in electromagnetic interference (EMI) shielding due to their unique properties, such as high thermal stability, lightweight, corrosion resistance, and excellent insulating capabilities. These materials are often used in applications where traditional metal-based EMI shields are unsuitable, such as in high-temperature environments or lightweight, flexible electronic devices. Non-metallic oxides like ZrO_2, SiO_2, SiC–Al_2O_3 $MgAl_2O_4$, SiC, SiOC, Al_2O_3–ZrO_2, Al_2O_3, Si_3N_4, $MgAl_2O_4$, $ZrTiO_4$–TiO_2, Ti_2AlC, $MgAl_2O_4$, $CaSiO_3$ Si_2N_2O, $Ca_2ZnSi_2O_7$, Si_3N_4, Ti_3C_2Tx (MXenes), etc. can be used [62].

Ceramic materials have high strength, excellent wear resistance, antistatic properties, high fracture toughness, and high thermal and chemical stability. However, most of these materials do not exhibit satisfactory microwave-shielding properties. Thus, forming ceramic composites by combining a ceramic matrix phase with a conductive or EMI shielding material filler can ensure excellent shielding properties, even in challenging conditions [63].

Over the past few decades, researchers have extensively studied various carbon nanomaterials, including graphene, single- and multi-walled carbon nanotubes, and carbon fibers, to explore their potential as electromagnetic interference (EMI) shields. This study discusses the EM-shielding or absorbing properties of different ceramic composites, including graphene-ceramics, SiCN-based ceramic matrix composites, and ferrites. MXene ceramics, metal-organic frameworks, and sulfide ceramics are known for their exceptional electromagnetic interference (EMI) shielding properties [64].

5.3.3.1 Metal Nitrides and Carbides

Metal nitrides and carbides include titanium nitride (TiN) and silicon carbide (SiC), known for their hardness and conductive properties. Metal nitrides and carbides are gaining attention in electromagnetic interference (EMI) shielding applications due to their excellent electrical conductivity, thermal stability, and mechanical strength [65]. These materials exhibit high electrical conductivity, which is crucial for effective EMI shielding. They can withstand high temperatures, making them suitable for applications in harsh environments. Their robust mechanical properties ensure durability and resistance to wear and tear. They resist oxidation and corrosion, extending their operational life in various conditions [65].

Common Metal Nitrides and Carbides Used in EMI Shielding:

1. **Titanium Nitride (TiN)**: It has good electrical conductivity, high hardness, and durability. Because of its protective properties, it is often used in coatings.
2. **Tungsten Carbide (WC)** is extremely hard and durable. It has excellent conductivity and thermal resistance and is commonly used in wear-resistant coatings and cutting tools.

3. **Zirconium Carbide (ZrC)** has a high melting point and stability at elevated temperatures. It also has good electrical conductivity. Thus, it is utilized in high-temperature applications.
4. **Niobium Carbide (NbC)** has a high melting point, hardness, and good electrical conductivity. It is often used in metalworking and tool coatings [66].

With high thermal stability, metal nitrides and carbides offer solutions for extreme conditions. They can be applied as coatings on surfaces to provide effective EMI shielding while enhancing the durability of the coated material. They can be incorporated into composite materials to improve their shielding effectiveness and mechanical properties [67]. These are used to construct enclosures for electronic devices to prevent EMI from affecting the performance of sensitive components. It is utilized in critical components in aerospace and automotive industries where EMI shielding is essential, and materials must withstand harsh conditions [68].

The metal oxide, including metallic and non-metallic nanostructures, is incorporated in unique combinations of high-performance cement, which has high thermal stability, rigid waterproofing properties, and quick setting time. Ceramic cement for nuclear power plants shields them from harmful radiation [47, 48] and efficiently blocks electromagnetic interference. It is long-lasting and durable, reducing the need for frequent replacements. Some metal nitrides and carbides can be expensive to produce. The manufacturing and application processes can be complex and require specialized techniques. Overall, metal nitrides and carbides offer promising solutions for EMI shielding, combining excellent protective properties with durability and stability in demanding environments [68, 69].

Porous or foam ceramics are well-suited for harsh environments due to their exceptional heat resistance, thermal/chemical corrosion resistance, low density, and ability to withstand thermal shock. They can be used as EMI shielding material, outperforming metals or conductive polymer composites. Many efforts have been made to create ceramic foam composites that are both efficient and easy to process. These foams play a crucial role in EMI shielding, and various methods yield varying pore structures, which subsequently impact ceramic foams' mechanical and electromagnetic interference (EMI) properties. Researchers should carefully select appropriate methods and materials based on practical applications [69].

However, ceramic foam composites for EMI applications have significant progress to make. There is considerable potential for enhancing ceramic foam composites as electromagnetic interference (EMI) materials, as indicated in this section. Porous structures are essential for electromagnetic interference properties (EMI). They can enhance the transmission distance of electromagnetic waves and increase electromagnetic loss [70].

However, the impact of pore sizes on application performance remains uncertain. Further hypothetical deduction is crucial for a deeper understanding of the relationship between pore designs and performance. Additionally, it is essential to note that ceramic foams with low electrical conductivity cannot be directly utilized in EMI. The growth of high electrical conductivity nanomaterials like CNTs or graphene becomes necessary to address this limitation [71]. Controlling the uniformity of

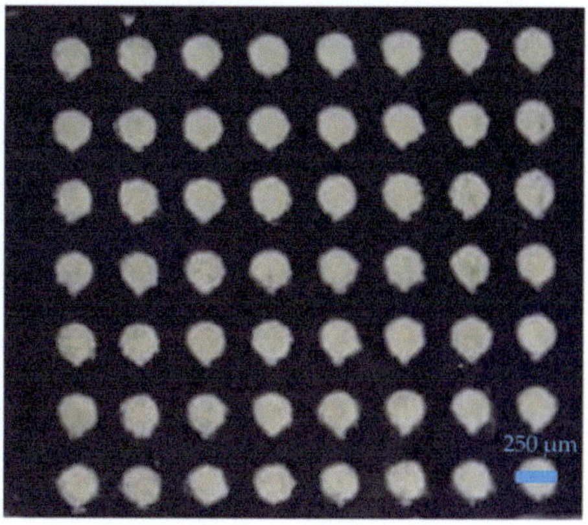

Fig. 5.16 Ferrite plate with periodical holes and introduced glass–ceramic composite in it [72]

nanomaterial growth while maintaining a specific porosity to enhance absorption and multireflection loss remains a significant challenge. Lastly, it is essential to consider the balance between pore structure, EMI performance, and mechanical properties. A ferrite plate with periodical holes and an introduced glass–ceramic composite is shown in Fig. 5.16 [72].

A work report by Sung-Hoon et al. [73] reported that while high conductivity is not needed for high-frequency EMI SE, conductive network conductivity dominates EMI shielding. For adequate EMI shielding, the composite conductivity should exceed 10 S/m. This study aimed to create an effective EMI shielding material. Thus, all CNT composites exceeded the volume-fraction percolation criterion. When CNT loading was decreased, the secondary filler's (excluded volume) effect on electrical conductivity grew, while at high CNT loading, it was modest. According to the power-law trend ($\sigma_{DC} \sim \sigma_0(p - \text{pc})\ \beta$) of fiber-type composites, low CNT loading is more sensitive than high CNT loading (where σ_{DC} is composite conductivity, σ_0 is a constant parameter, p is filler volume fraction, pc is percolation threshold, and β is critical exponent). Figure 5.17a–c shows absorption-dominated shielding. It is observed from Fig. 5.17 that SE(A) of CNT/silica/PDMS (6 wt% CNT) composite is higher than SE(R), showing microwave absorption is more essential in X-band EMI shielding. This pattern in the somewhat high conductivity composite instance matches our past studies. Multi-layer composites with varying compositions and contents (coupled layers) will be investigated to improve EMI SE and microwave absorption. Figure 5.17d shows that the CNT wt% was increased with a fixed micro-silica concentration to study the conductivity increment trend.

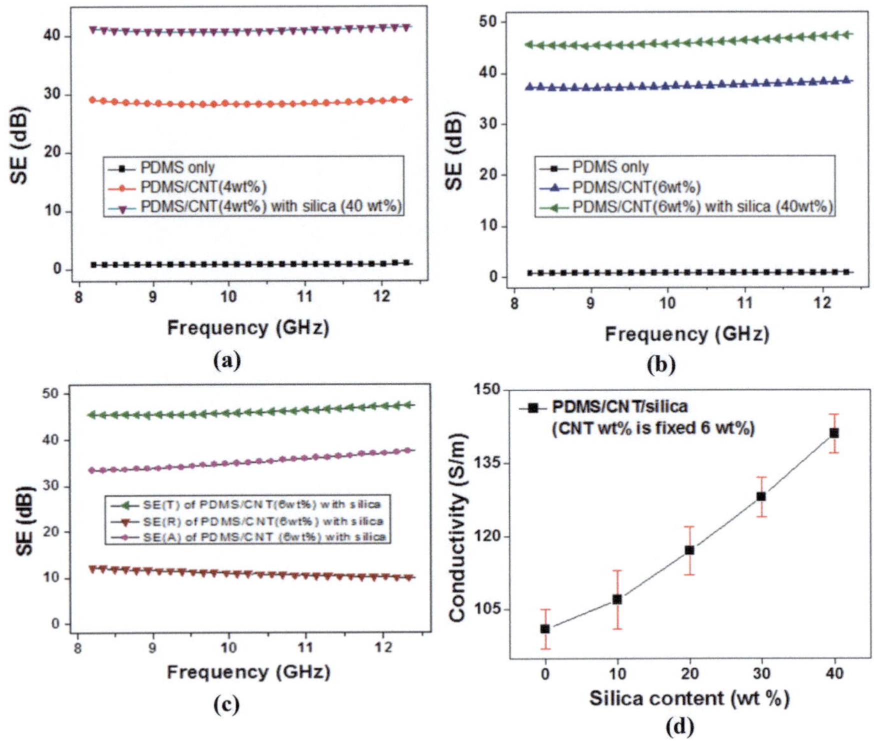

Fig. 5.17 **a–c** Comparison of CNT/micro-silica/PDMS and CNT/PDMS composites shows enhanced EMI shielding effectiveness (SE). **d** Conductivity of different loading of silica content [73]

5.3.4 Hybrid Nanocomposites

Combining metal nanoparticles with carbon-based nanomaterials or other inorganic nanomaterials creates synergistic effects, resulting in a high-performance EMI shielding composite material with tailored properties. Multiple properties cannot be achieved with a single material [74]. This hybrid material is the composite of two or more inorganic nanomaterials from insulators, semiconductors, and conductors that have enhanced interaction with electromagnetic waves [75].

The polymer could be insulative polymers like polyethylene, teflon, polypyrrole, polymethyl methacrylate (PMMA), acrylonitrile butyl styrene (ABS), or conductive polymers like polyaniline (PANi), Polypyrrole (PPy), and Polyacetylene (PAC) [76]. This polymer matrix was combined with inorganic nanomaterials like metal nanoparticles with good conductivity, including metal oxides, which are generally semiconductors. In some cases, highly insulative ceramic-natured nanoparticles are added proportionally with a critical concentration [42]. This will be in synergism to bring the desired EMI shielding ability, to achieve a hybrid material composed of inorganic

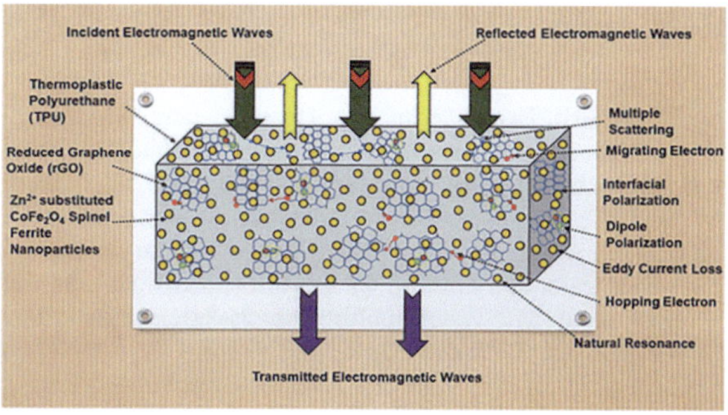

Fig. 5.18 Schematic illustration of the electromagnetic interference shielding mechanism in TPU-matrix-based nanocomposites with Zn^{2+}-substituted $CoFe_2O_4$ spinel ferrite nanoparticles and rGO as nanofillers [79]

and organic materials. Its electrical and magnetic properties can be tuned through size, shape, and composition [77]. It is ideal for applications where weight and flexibility are critical. It has the potential for large-scale production and integration into various substrates [78]. Figure 5.18 shows schematic illustration of the electromagnetic interference shielding mechanism in TPU-matrix-based nanocomposites with Zn^{2+}-substituted $CoFe_2O_4$ spinel ferrite nanoparticles and rGO as nanofillers [79].

Standard Components in Hybrid Nanocomposites are

1. **Metal Nanoparticles** silver, copper, and nickel provide high conductivity and adequate reflection and absorption of electromagnetic waves.
2. **Metal Oxides** zinc oxide (ZnO) and iron oxide (Fe_2O_3) offer additional dielectric properties and enhance EMI shielding through absorption and scattering.
3. **Carbon-Based Materials** carbon nanotubes (CNTs), graphene, and carbon black provide high surface area, conductivity, and additional EMI shielding through absorption and scattering.
4. **Polymers** polyimide, epoxy, and polystyrene are matrices that hold the nanomaterials together and provide flexibility and lightweight characteristics.

5.4 Conclusion

This chapter provides comprehensive overview of electromagnetic shielding technologies using inorganic materials. Additionally, it explores the role of nanotechnology in enhancing the performance of inorganic materials for shielding applications. Real-world examples showcased in various industries demonstrate the significant impact of innovative shielding solutions on improving overall system performance and reliability.

References

1. Woo, S.H., Lee, J., Kim, J.P., Hong, J., Jang, J.M., Ko, T.Y., Kang, Y.C., Kim, D.W., Kim, S.J.: Electrostatic assembly of charge-modulated MXene and graphene oxide into conductive porous nanohybrids via rapid annealing for EMI shielding. Appl. Mater. Today **39**, 102271 (2024)
2. Liu, J., Yang, Y., Zheng, X.: Chapter 2—The fundamentals of metal oxides for electrocatalytic water splitting. In: Qi, J., Korotcenkov, G. (eds.) Metal Oxides and Related Solids for Electrocatalytic Water Splitting, pp. 25–60. Elsevier (2022)
3. França, E.L.T., Santos, A.R., Assis, L.K.C.S., Castro-Lopes, S., Oliveira, D.M., Carvalho, A.S., Padrón Hernández, E.: Exploring electrodeposited iron and iron oxide nanostructures on porous alumina membrane for enhanced EMI shielding. J. Magn. Magn. Mater. **605**, 172310 (2024)
4. Choudhary, H.K., Kumar, R., Pawar, S.P., Sundararaj, U., Sahoo, B.: Effect of morphology and role of conductivity of embedded metallic nanoparticles on electromagnetic interference shielding of PVDF-carbonaceous-nanofiller composites. Carbon **164**, 357–368 (2020)
5. Qiu, L., Han, W., Yu, Q., Liu, Y., Chen, B., Yi, L.: Liquid metal induced segregated-like electromagnetic shielding composites with excellent photothermal conversion and strain sensing capacities. Compos. Sci. Technol. **257**, 110825 (2024)
6. Li, H., Li, J., Chu, W., Lin, J., He, P., Fan, W., Zhao, Q.: Ultrahigh absorption dominant EMI shielding polyimide composites with enhanced piezoelectric property. Compos. Sci. Technol. **257**, 110820 (2024)
7. Chaudhary, A., Gupta, V., Teotia, S., Nimanpure, S., Rajak, D.K.: Electromagnetic shielding capabilities of metal matrix composites. In: Brabazon, D. (ed.) Encyclopedia of Materials: Composites, pp. 428–441. Elsevier, Oxford (2021)
8. Katheria, A., Das, P., Singh, H., Nayak, J., Paul, S., Biswas, S., Das, N.C.: Highly flexible EMA/Fe_3O_4@g-C_3N_4 composite for thermal control and EMI shielding application. Colloids Surf. A **700**, 134756 (2024)
9. Yuan, C., Li, X., Huang, M., Li, F., Zhang, Z., Wang, C., Hu, W.: Multifunctional asymmetric foam with compression-enhanced EMI shielding effectiveness exhibits real-time tunability of reflection/absorption ratio and pressure sensing. Chem. Eng. J. **489**, 151359 (2024)
10. Dj, B.G., Wessley, G.J.J.: A study on EMI shielding in aircraft: introduction, methods and significance of using electrospun nanocomposites. J. Space Saf. Eng. **11**, 150–160 (2024)
11. Ni, L., Luo, Y., Shen, L., Qiu, C., Yan, L., Liang, M., Zhou, S., Zou, H.: Lightweight polyimide/silica aerogel composite foams and their derived carbon foams for excellent flame retardancy, thermal insulation and EMI shielding performance. Polymer **305**, 127158 (2024)
12. Katheria, A., Das, P., Nayak, J., Roy, B., Pal, A., Biswas, S., Das, N.C.: MXene and Fe_3O_4 decorated g-C_3N_4 incorporated high flexible hybrid polymer composite for enhanced electrical conductivity, EMI shielding and thermal conductivity. Next Mater. **6**, 100292 (2025)
13. Parmar, S., Ray, B., Ashok, A., Shilpa, U., Datar, S.: Highly tunable rGO composites for EMI shielding inks: a tailored approach to modulate broadband EMI shielding. Synth. Met. **301**, 117536 (2024)
14. Ouyang, W., Mei, L., Liu, Q., Ding, C., Liu, Y., Zhao, C., Xu, L., Lu, F., Luo, D., Miao, C., Bai, Y., Lu, Q., Luo, T., Wu, Z.: Ultrathin-flexible multifunctional MXene composite hydrogels with good mechanical properties-high strain sensitivity and ultra-broadband EMI shielding performances. Chem. Eng. J. **494**, 153068 (2024)
15. Rayar, A., Naveen, C.S., Onkarappa, H.S., Betageri, V.S., Prasanna, G.D.: EMI shielding applications of PANI-Ferrite nanocomposite materials: a review. Synth. Met. **295**, 117338 (2023)
16. Radoń, A., Hawełek, Ł., Łukowiec, D., Kubacki, J., Włodarczyk, P.: Dielectric and electromagnetic interference shielding properties of high entropy (Zn, Fe, Ni, Mg, Cd)Fe_2O_4 ferrite. Sci. Rep. **9**, 20078 (2019)

17. Shakir, M.F., Tariq, A., Rehan, Z.A., Nawab, Y., Abdul Rashid, I., Afzal, A., Hamid, U., Raza, F., Zubair, K., Rizwan, M.S., Riaz, S., Sultan, A., Muttaqi, M.: Effect of Nickel-spinal-Ferrites on EMI shielding properties of polystyrene/polyaniline blend. SN Appl. Sci. **2**, 706 (2020)
18. Kapte, I.S., Niyogi, S., Qureshi, P.J., Khade, V., Avanish Babu, T., Siva Kumar, K.V., Madhuri, W.: Microwave assisted cobalt nickel ferrite as exhaust thrust sensor and EMI shield. Ceram. Int. **50**, 30754–30762 (2024)
19. Mohapatra, P.P., Dobbidi, P.: Development of spinel ferrite-based composites for efficient EMI shielding. Mater. Chem. Phys. **301**, 127581 (2023)
20. Nasreen Taj, M., Daruka Prasad, B., Ramarao, N., Nagabhushana, H., Reddy, A., Ashwini, K.R., Vinuta, Dhanalakshmi, K.: Dielectric and structural properties of polyaniline-tungsten trioxide nanocomposites: for the packing of nano-electronic devices and EMI shielding. Nano-Struct. Nano-Objects **39**, 101219 (2024)
21. Suarez, A., Victoria, J., Torres, J., Martinez, P.A., Alcarria, A., Perez, J., Garcia-Olcina, R., Soret, J., Muetsch, S., Gerfer, A.: Performance study of split ferrite cores designed for EMI suppression on cables. Electronics **9**, 1992 (2020)
22. Zakaria, M.A., Menazea, A.A., Mostafa, A.M., Al-Ashkar, E.A.: Ultra-thin silver nanoparticles film prepared via pulsed laser deposition: synthesis, characterization, and its catalytic activity on reduction of 4-nitrophenol. Surf. Interfaces **19**, 100438 (2020)
23. Menamparambath, M.M., Muhammed Ajmal, C., Kim, K.H., Yang, D., Roh, J., Park, H.C., Kwak, C., Choi, J.-Y., Baik, S.: Silver nanowires decorated with silver nanoparticles for low-haze flexible transparent conductive films. Scient. Reports **5**, 16371 (2015)
24. Mahato, R., Masiul Islam, S., Singh, S.: Flexible piezo-resistive strain sensors based on silver nanowires and graphene nanoplatelets reinforced polydimethylsiloxane for human motion detection. Mater. Today Commun. **40**, 110056 (2024)
25. Zheng, B., Deng, Z., Chen, H., Chen, J., Meng, Z., Wang, C., Zheng, D., Su, C., Liang, X., Xie, W., Jiang, G., Liu, M.: Fabrication of high-reliability silver nanowire transparent film heaters and the associated electromigration failure mechanism. Thin Solid Films **800**, 140413 (2024)
26. Chen, C., Zhou, J., Li, H., Lin, H., Peng, Y., Wang, K., Li, X.: Research on femtosecond laser welding process of silver nanowire transparent electrodes. J. Market. Res. **26**, 9157–9168 (2023)
27. Kou, P., Yang, L., Chang, C., He, S.: Improved flexible transparent conductive electrodes based on silver nanowire networks by a simple sunlight illumination approach. Sci. Rep. **7**, 42052 (2017)
28. Park, S., Bang, J., Kim, B.S., Oh, S.J., Choi, J.H.: Metallic fusion of nanocrystal thin films for flexible and high-performance electromagnetic interference shielding materials. Mater. Today Adv. **12**, 100177 (2021)
29. Riaz, S., Naz, S., Younus, A., Javid, A., Akram, S., Nosheen, A., Ashraf, M.: Layer by layer deposition of PEDOT, silver and copper to develop durable, flexible, and EMI shielding and antibacterial textiles. Colloids Surf. A **650**, 129486 (2022)
30. Bobsin, A., Rodrigues, T.C., Fernandes, I.J., Ferreira, S.B., Peter, C.R., Hasenkamp, W., Moraes, C.A.M.: Copper and silver microparticles for high-performance conductive inks in electronic chip shielding. Mater. Chem. Phys. **315**, 129007 (2024)
31. Lee, S., Wern, C., Yi, S.: Novel fabrication of silver-coated copper nanowires with organic compound solution. Materials **15**, 1135 (2022)
32. Raagulan, K., Ghim, J.S., Braveenth, R., Jung, M.J., Lee, S.B., Chai, K.Y., Mi Kim, B., Lee, J.: EMI shielding of the hydrophobic, flexible, lightweight carbonless nano-plate composites. Nanomaterials **10**, 2086 (2020)
33. Pandey, A., Kumar, R., Mondal, D.P., Kumar, P., Singh, S.: Metal and metal matrix 2d nanomaterial composites: attractive alternatives for EMI shielding applications. In: Adapting 2d Nanomaterials for Advanced Applications, pp. 347–373. American Chemical Society (2020)
34. Wang, Y., Du, D., Zhou, Z., Xie, H., Li, J., Zhao, Y.: Reactive conductive ink capable of in situ and rapid synthesis of conductive patterns suitable for inkjet printing. Molecules **24**, 3548 (2019)

35. Mercante, L.A., Andre, R.S., Facure, M.H.M., Correa, D.S., Mattoso, L.H.C.: Recent progress in conductive electrospun materials for flexible electronics: energy, sensing, and electromagnetic shielding applications. Chem. Eng. J. **465**, 142847 (2023)
36. Wei, Z., Cheng, Y., Hu, X., Meng, Y., Zhan, Y., Li, Y., Xia, H., Jiang, X., Chen, Z.: Cellulose–derived carbon scaffolds with bidirectional gradient Fe_3O_4 distribution: integration of green EMI shielding and thermal management. Int. J. Biol. Macromol. **275**, 133724 (2024)
37. Jang, K.-P.: Development and experimental verification of inorganic electromagnetic pulse shielding paint for building interiors using carbon-based materials. Materials **17**, 2863 (2024)
38. Jiang, B., He, C., Zhao, N., Nash, P., Shi, C., Wang, Z.: Ultralight metal foams. Sci. Rep. **5**, 13825 (2015)
39. Yang, H., Li, Y., Ma, B., Zhu, Y.: Review and a theoretical approach on pressure drop correlations of flow through open-cell metal foam. Materials **14**, 3153 (2021)
40. Ali, A., Hussain, F., Kalsoom, A., Riaz, T., Zaman Khan, M., Zubair, Z., Shaker, K., Militky, J., Noman, M.T., Ashraf, M.: Multifunctional electrically conductive copper electroplated fabrics sensitizes by in-situ deposition of copper and silver nanoparticles. Nanomaterials **11**, 3097 (2021)
41. Wang, Z., Wang, D., Zhang, C., Chen, W., Meng, Q., Yuan, H., Yang, S.: A fluorinated polyimide based nano silver paste with high thermal resistance and outstanding thixotropic performance. Polymers **15**, 1150 (2023)
42. Nan, Z., Wei, W., Lin, Z., Chang, J., Hao, Y.: Flexible nanocomposite conductors for electromagnetic interference shielding. Nano-Micro Lett. **15**, 172 (2023)
43. Harshapriya, P., Kaur, P., Basandrai, D.: Synthesis and characterization of BFO/BTO/ZnO nanocomposites for promising EMI shielding applications. Inorg. Chem. Commun. **156**, 111186 (2023)
44. Abhilash, G.P., Sushmita, K., Bose, S., Shivakumara, C.: Functionalization of dielectric $BaTiO_3$ and semiconducting ZnO nanoparticles on rGO layers and their polymer composites: applications in EMI shielding. Synth. Met. **297**, 117387 (2023)
45. Thakur, N., Sharma, I., Thakur, P., Mahajan, M., Bhardwaj, S., Jasrotia, R., Dahshan, A., Elsaeedy, H.I., Kaur, S., Anand, G., Sharma, P., Batoo, K.M., Kumar, G.: Improvement in the structural, magnetic and electromagnetic behaviour of barium hexaferrites with yttrium doping for EMI shielding. J. Alloy. Compd. **976**, 173042 (2024)
46. Yadav, R.S., Kuřitka, I.: Recent advances on outstanding microwave absorption and electromagnetic interference shielding nanocomposites of ZnO semiconductor. Adv. Coll. Interface. Sci. **326**, 103137 (2024)
47. Tekin, İ, Pekgöz, M.: The effects of ZnO powder and different fiber types used in tuff/calcite-based geopolymers on electromagnetic shielding effectiveness. J. Build. Eng. **89**, 109343 (2024)
48. Zhu, H., Fu, K., Huang, T., Liu, H., Yang, B., Zhou, Y., Li, Y.: Highly conductive CFRP composite with Ag-coated T-ZnO interlayers for excellent lightning strike protection, EMI shielding and interlayer toughness. Compos. Part B Eng. **279**, 111448 (2024)
49. Zhang, Z., Huang, Y., Zhang, P., Wan, C.: Outstanding EMI shielding properties of Al_2O_3 reinforced with sandwiched graphene/CNTs. J. Eur. Ceram. Soc. **43**, 4082–4087 (2023)
50. Erdogan, N., Erden, F., Astarlioglu, A.T., Ozdemir, M., Ozbay, S., Aygun, G., Ozyuzer, L.: ITO/Au/ITO multilayer thin films on transparent polycarbonate with enhanced EMI shielding properties. Curr. Appl. Phys. **20**, 489–497 (2020)
51. Shahbaz Saleem, M., Bilal Hanif, M., Gregor, M., Motola, M., Faheem Khan, A.: Nanostructured multi-layer MgF2/ITO coatings prepared via e-beam evaporation for efficient electromagnetic interference shielding performance. Appl. Surf. Sci. **596**, 153584 (2022)
52. Bai, S., Guo, X., Zhang, X., Zhao, X., Yang, H.: $Ti_3C_2T_x$ MXene-AgNW composite flexible transparent conductive films for EMI shielding. Compos. A Appl. Sci. Manuf. **149**, 106545 (2021)
53. Kim, Y., Kim, N., Lee, S.H., Hyeong, S.-K., Lee, J.-H., Lee, J., Bae, J.S., Cho, I.S., Choi, J.-Y., Kim, S.Y., Yu, H.K.: Enhanced ultra high frequency EMI shielding with controlled ITO nano-branch width via different tin material types. Nanoscale **15**, 13635–13644 (2023)

54. Khan, A.F., Abadeen, Z.u., Hanif, M.B., Saleem, M.S., Naveed, K., Bashir, K., Ahmad, I.: Multi-layer MgF_2/ITO coatings for electromagnetic interference shielding. Mater. Chem. Phys. **272**, 125009 (2021)
55. Wen, H., Weng, B., Wang, B., Xiao, W., Liu, X., Wang, Y., Zhang, M., Huang, H.: Advancements in transparent conductive oxides for photoelectrochemical applications. Nanomaterials **14**, 591 (2024)
56. Alam, M.J., Cameron, D.C.: Optical and electrical properties of transparent conductive ITO thin films deposited by sol–gel process. Thin Solid Films **377–378**, 455–459 (2000)
57. Koseoglu, H., Turkoglu, F., Kurt, M., Yaman, M.D., Akca, F.G., Aygun, G., Ozyuzer, L.: Improvement of optical and electrical properties of ITO thin films by electro-annealing. Vacuum **120**, 8–13 (2015)
58. Kumar, S., Taunk, M.: 2D MXenes for flexible device applications. Materials Today Physics **46**, 101483 (2024)
59. Shang, J., Wang, M., Wang, P., Li, G., Yang, M., Li, Y.: Study on thermal reflection characteristics of composite inorganic coatings. Appl. Sci. **14**, 6898 (2024)
60. Zhang, T., Chang, L., Zhang, X., Wan, H., Liu, N., Zhou, L., Xiao, X.: Simultaneously tuning interlayer spacing and termination of MXenes by Lewis-basic halides. Nat. Commun. **13**, 6731 (2022)
61. Yang, D., Zhou, M., Fu, H.: Highly conductive, breathable and magnetic MXene/Fe_3O_4/PEDOT:PSS coated carbon cloth for high-efficiency electromagnetic interference shielding. Prog. Org. Coat. **192**, 108518 (2024)
62. Gao, Q., Wang, X., Schubert, D.W., Liu, X.: Review on polymer/MXene composites for electromagnetic interference shielding applications. Adv. Nanocompos. **1**, 52–76 (2024)
63. Zhang, Q., Wang, Q., Cui, J., Zhao, S., Zhang, G., Gao, A., Yan, Y.: Structural design and preparation of $Ti_3C_2T_x$ MXene/polymer composites for absorption-dominated electromagnetic interference shielding. Nanoscale Adv. **5**, 3549–3574 (2023)
64. He, P., Liu, Z.-Y., Mao, G.-B., Liu, Q., Zheng, M.-J., Zuo, R.-Z., Cao, W.-Q., Hou, Z.-L., Yuan, J., Cao, M.-S.: MXene films: toward high-performance electromagnetic interference shielding and supercapacitor electrode. Compos. A Appl. Sci. Manuf. **157**, 106935 (2022)
65. Guo, D., Tan, B., Jiang, X., Gao, G., Lin, Y.: Multifunctional polyimide/boron nitride nanosheet/$Ti_3C_2T_x$ MXene composite film with three-dimensional conductive network for integrated thermal conductive, electromagnetic interference shielding, and Joule heating performances. Compos. Sci. Technol. **254**, 110690 (2024)
66. Wypych, G.: 34—Fillers. EMI shielding. In: Wypych, G. (ed.) Encyclopedia of Polymer and Rubber Additives, pp. 207–213. ChemTec Publishing (2024)
67. Yang, Y., Bi, R., Ren, W., Sun, Y., Zhao, H., Duan, H.: Oriented rGO framework induced hBN confined network in PDMS for absorption enhanced EMI shielding and through-plane thermal conductivity. Compos. Sci. Technol. **243**, 110259 (2023)
68. He, Y., Liu, Z., Zhang, B., Wu, H., Liu, H., Nishimura, K., Xu, Y., Huang, Z.: Dual sinter-aids to fabricate high-entropy transition metal carbide ceramic and their properties. Ceram. Int. **50**, 28899–28906 (2024)
69. Deeraj, B.D.S., Shebin, K.J., Bose, S., Sampath, S., Joseph, K.: Electrospun carbon fibers embedded with core–shell TiC@TiO_2 nanostructures and their epoxy composites for potential EMI shielding in the Ku band. Nano-Struct. Nano-Objects **32**, 100912 (2022)
70. Wu, Q., Zeng, Z., Xiao, L.: From 2D graphene and MXene nanolayers to 3D biomimetic porous composite aerogels for electromagnetic interference shielding. Compos. A Appl. Sci. Manuf. **177**, 107939 (2024)
71. Yang, Y., Liu, S., Jin, L., Li, G., Quan, B., Chen, Q., Tang, X., Ji, X., Wu, H., Lu, X., Qu, J.: A novel multisource energy harvester with enhanced thermal conductivity for efficient energy harvesting and superior EMI shielding. Compos. A Appl. Sci. Manuf. **175**, 107803 (2023)
72. Tumarkin, A., Tyurnina, N., Tyurnina, Z., Sinelshchikova, O., Drozdovsky, A., Gagarin, A., Sapego, E., Karamov, A., Bogdan, A.: Composite metamaterial: ferrite matrix with ferroelectric inclusions. Coatings **13**, 117 (2023)

73. Park, S.-H., Ha, J.-H.: Improved electromagnetic interference shielding properties through the use of segregate carbon nanotube networks. Materials **12**, 1395 (2019)
74. Zeng, Y., Luo, X., Yu, K., Qian, K.: EMI shielding performance of phenolic-based carbon foam modified with GO/SiO_2 hybrid nanomaterials. Chem. Phys. Lett. **715**, 166–172 (2019)
75. Patadia, M., Quinn, A., Tank, M., Jolowsky, C., Luiz, L., Psulkowski, A., Kurilich, M., De Leon, A., Liang, R., Sweat, R.: Enhanced multifunctionality in carbon fiber/carbon nanotube reinforced PEEK hybrid composites: Superior combination of mechanical properties, electrical conductivity, and EMI shielding. Compos. B Eng. **284**, 111674 (2024)
76. Jagadeesh Chandra, R.B., Shivamurthy, B., Kulkarni, S.D., Kumar, M.S.: Hybrid polymer composites for EMI shielding application—a review. Mater. Res. Exp. **6**, 082008 (2019)
77. Ardiansyah, A., Heryanto, H., Tahir, D.: Textile shields: current bibliometric analysis of fabric integration for EMI shielding and prospects for future research. Radiat. Phys. Chem. **225**, 112112 (2024)
78. Omana, L., Chandran, A., John, R.E., Wilson, R., George, K.C., Unnikrishnan, N.V., Varghese, S.S., George, G., Simon, S.M., Paul, I.: Recent advances in polymer nanocomposites for electromagnetic interference shielding: a review. ACS Omega **7**, 25921–25947 (2022)
79. Anju, Yadav, R.S., Pötschke, P., Pionteck, J., Krause, B., Kuřitka, I., Vilcakova, J., Skoda, D., Urbánek, P., Machovsky, M., Masař, M., Urbánek, M., Jurca, M., Kalina, L., Havlica, J.: High-performance, lightweight, and flexible thermoplastic polyurethane nanocomposites with Zn^{2+}-substituted $CoFe_2O_4$ nanoparticles and reduced graphene oxide as shielding materials against electromagnetic pollution. ACS Omega. **6**, 28098–28118 (2021)

Chapter 6
Carbonaceous Nanofillers in Polymer Matrix

Abstract Polymer is a versatile material known for its low cost, light weight, flexibility, ease of processing, and dimensional properties. However, the mechanical characteristics of polymeric materials often need to be improved for engineering applications. As a result, research has focused on developing new and enhanced polymeric materials with EMI shielding properties. Polymer blending has been a common approach to creating materials with unique characteristics. Yet, like pure polymers, these blends exhibit insufficient physical properties for engineering use. Over the past few decades, studies have explored the incorporation of various fillers and nanofillers into polymer matrices. The inclusion of nanofillers has resulted in materials that are cost-effective, lightweight, easy to process, and possess desirable mechanical properties. This chapter discusses carbon-based nanofillers, which can replace the conventional metal-based fillers frequently used in EMI shielding applications.

6.1 Introduction

Carbon nanomaterials have unique properties, which exhibit unique advantages in technological applications. No other materials can replace the superiority and uniqueness of carbon-based materials. The multifunctional capabilities of carbon-reinforced polymers (CRPs) are often limited by their restricted multifunctional properties, which hinder their broader application [1]. Researchers have highlighted carbon-based nanostructured materials to tackle the increase in EMI shielding problems. The impressive mechanical and electrical properties of carbon nanofibers, graphene, carbon nanodiamonds, fullerenes, carbon dots, and carbon nanotubes (CNTs) make them unique nanomaterials [2]. Incorporating nanofillers enhances thermoplastic-based composites' properties, leading to the development of reinforced hybrid composites [2]. Carbon-based materials such as carbon black, carbon fiber, and carbon nanotubes (CNT) are under investigation as lightweight alternatives for electromagnetic interference (EMI) shielding [3]. Carbon nanostructures, such as carbon nanotubes and graphene, are increasingly significant in this context. Carbon

S. K. Koul et al., *Microwave Absorbers*, Engineering Materials,
https://doi.org/10.1007/978-981-96-9596-6_6

nanotubes are frequently employed to enhance the properties of these composites, resulting in substantial improvements in the mechanical, thermal, and electrical performance of the resultant nanocomposites [4].

The increasing demand for compact and lightweight electronics and communication systems has heightened concerns regarding electromagnetic interference (EMI) [5]. This has heightened the demand for lightweight, flexible materials that offer effective EMI shielding. Polymers, by themselves, are insufficient for achieving high electrical conductivity. Conducting polymers, in contrast, exhibit brittleness, possess low mechanical stability, and undergo decomposition at elevated temperatures.

Insulative polymers and semi-conductive intrinsic polymers are made as composites to improve electrical conductivity, mechanical strength, flexibility, lightweight, moldability, ease of processing, and cost-effectiveness [6]. Carbon-based materials, such as graphene, nanofibers, and carbon nanotubes (CNTs), are added to these polymer composites to improve their electrical conductivity. These materials are particularly advantageous for EMI shielding [7].

However, excessive amounts of carbonaceous filler can cause manufacturing challenges, such as aggregation, which leads to poor dispersion. To address this issue, carbon bucky papers (BPs) are placed between polymer sheets to enhance carbon content and improve electrical conductivity. More research is needed on the EMI shielding effectiveness of nanocarbon [6]. Different allotropes of nanocarbon are frequently used in various applications to prepare composites for EMI shielding, including CNTs and graphene, as shown in Fig. 6.1. Nano graphene particles, nanodiamonds, fullerene, and carbon dots-like materials exist in different forms, but all comprise carbon [1].

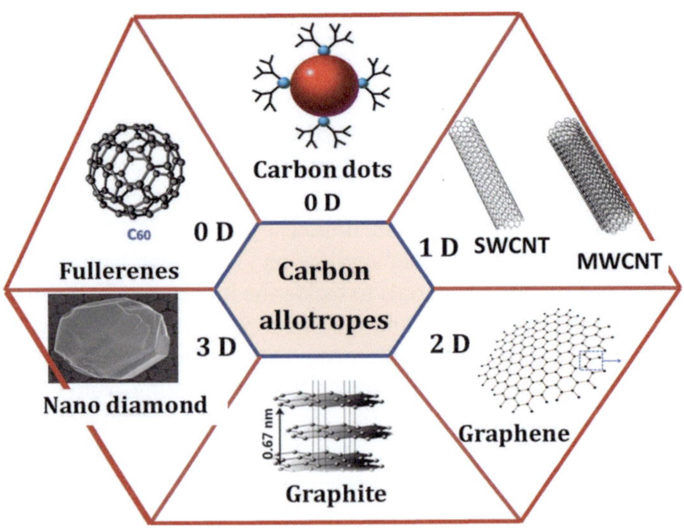

Fig. 6.1 Carbon-based nanomaterials

6.2 Graphene

Graphene is a material of significant interest that has garnered extensive research attention in recent years due to its unique properties. Graphene is utilized in polymer composites, improving electromagnetic interference (EMI), electromagnetic compatibility (EMC), and electrical, mechanical, and thermal properties. Graphene exhibits the lowest electrical percolation threshold compared to other conductive fillers. Due to its high specific surface area, tunable electrical properties, and potential applications in electrical conducting materials, graphene offers significant shielding efficiency, characterized by its light weight and technical properties. Traditional EMI shielding materials, such as metals, tend to be bulky and heavy, which is problematic in applications where weight is a concern, such as parts for aerospace or portable electronics.

Graphene offers a superior shielding-to-weight ratio, making it ideal for lightweight applications. For electromagnetic compatibility (EMC) applications, graphene helps to minimize electromagnetic interference between different components within a device. Its use in coatings and composites reduces unwanted emissions while improving signal integrity, which is essential for electronic systems operating without external interference.

Figure 6.2 shows SEM image of (a–d) graphene and (b) the electrical conductivity of the graphene–polydimethylsiloxane (G-PDMS) composite. The SEM images of graphene reveal its sheet-like morphology. At the same time, the conductivity graph indicates that an increase in the volume percentage of graphene enhances the electrical conductivity of the polymer composite [8].

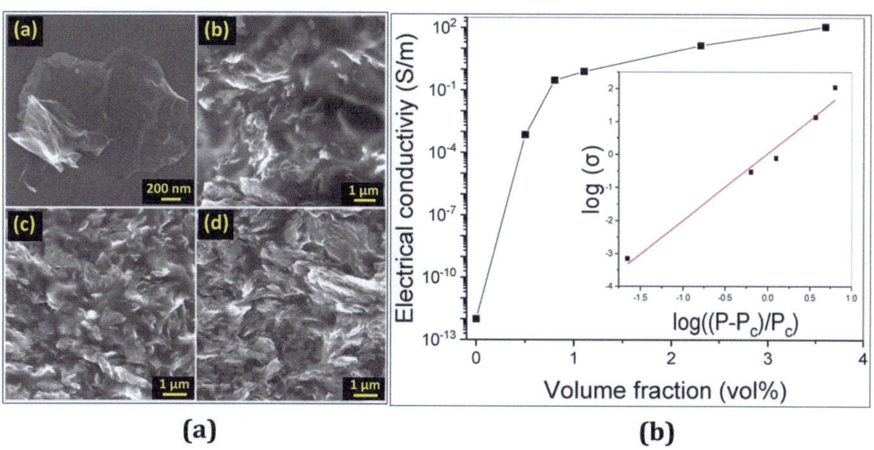

Fig. 6.2 SEM image of **a** graphene. **b** Electrical conductivity of the graphene/PDMS composite [8]

Reduced graphene oxide (rGO), a derivative of graphene, is primarily utilized in composite development and is considered a promising candidate for electromagnetic interference (EMI) and electromagnetic compatibility (EMC) applications. This potential is attributed to its controllable dielectric loss, which results from residual functional groups and defects [9]. These characteristics increase dielectric loss via defect polarization and dipole relaxation mechanisms. The functional groups on its surface enable the anchoring of other dielectric and magnetic materials, enhancing the material's electromagnetic loss capability and facilitating tailored impedance matching.

Integrating graphene or its derivatives with polymers can result in substantial electromagnetic interference shielding effectiveness through mechanisms of electric and dielectric loss [10]. The effectiveness of EMI shielding in composites containing conductive fillers, whether in film or solid-state form, is primarily determined by their electrical conductivity [11]. To address this issue, materials that mainly absorb electromagnetic waves are gaining popularity as electromagnetic interference (EMI) shields, with the expectation that the absorption/reflection (A/R) ratio will surpass unity [12]. A higher A/R ratio signifies reduced reflection, indicating that the material predominantly attenuates incident waves via dielectric or magnetic loss [13, 14].

Initial research has focused on developing EMI shielding materials by incorporating graphene as a filler via conventional polymer processing techniques, including extrusion and injection. Previous studies have identified multiple problems associated with the processing methods mentioned above. Achieving effective EMI shielding effectiveness (SE) necessitates a high graphene content, which increases production costs [15].

Secondly, an increased quantity of graphene elevates the product's density, which opposes the objective of weight reduction. Furthermore, addressing the agglomeration of graphene resulting from robust Van der Waals forces presents significant challenges. In recent years, there has been an increase in the development of graphene/polymer-based porous materials for electromagnetic interference (EMI) shielding, which has addressed these challenges [16]. Porous structures present distinct advantages over traditional compact shielding materials. They increase the flexibility of the shielding materials, reduce their density, improve the dispersibility of graphene during construction, and enhance the materials' capacity to attenuate electromagnetic (EM) waves via multiple reflections within the matrix.

Careful design of the microstructure of shielding materials offers the potential for enhanced EMI shielding performance [17]. Numerous assembly techniques have emerged to produce porous EMI shielding materials. Figure 6.3 presents field emission scanning electron microscopy (FE-SEM) images of (a) graphene oxide (GO) and (b) reduced graphene oxide (RGO) [18]. This study demonstrates that reduced graphene coated with Fe_3O_4 nanoparticles and blended with natural rubber exhibits unique properties, including enhanced processability, improved flexibility, and high electromagnetic interference (EMI) shielding effectiveness in the X-band. This composite material is applicable in strain sensor-like applications.

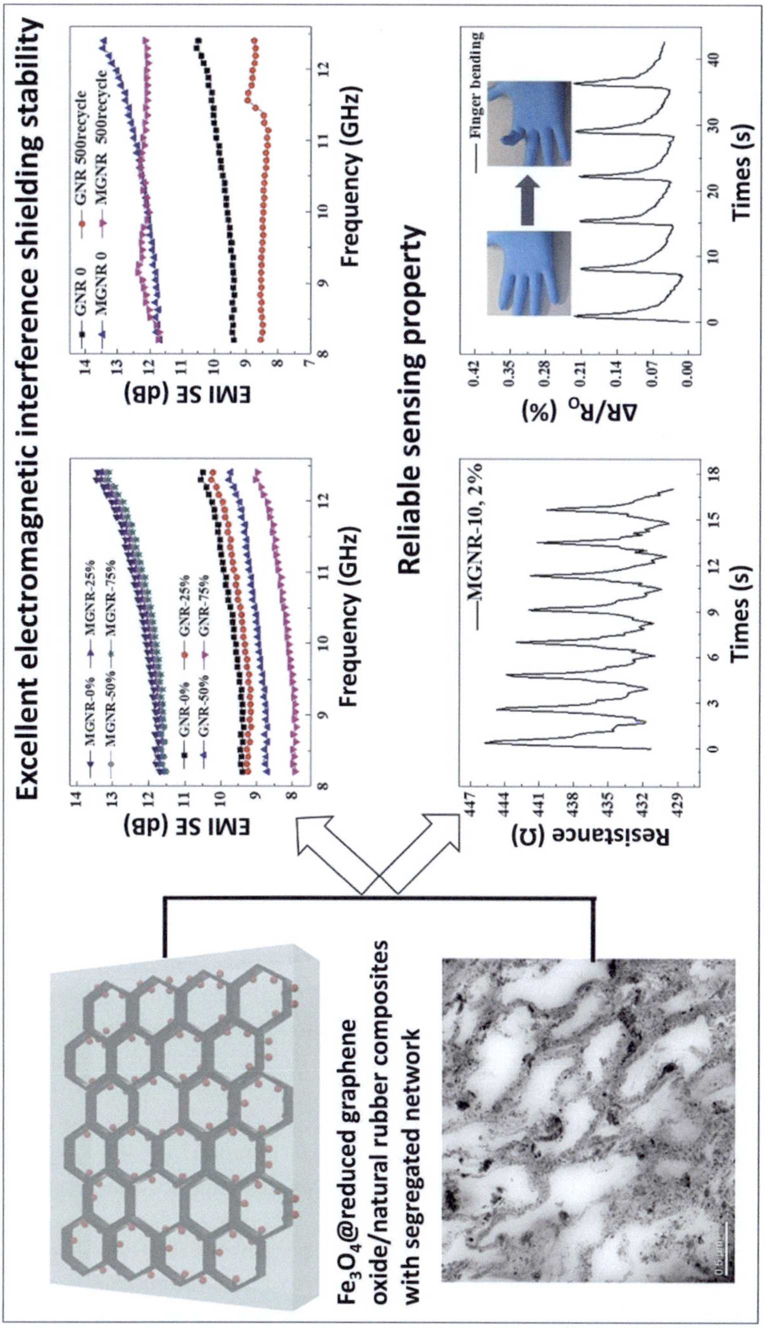

Fig. 6.3 EMI shielding segregated magnetic graphene network in rubber composites [18]

6.3 Nanotubes

CNTs are cylindrical structures formed from rolled-up sheets of graphene. The recent literature indicates a significant increase in carbon nanotube (CNT) utilization due to its superior characteristics and properties [19]. The nanoscale characteristics show extremely low thermal expansion coefficient (CTE), exceptional mechanical strength, large specific surface area (SSA), and high thermal conductivity [20]. Carbon nanotubes (CNTs) are regarded as optimal nanofillers for enhancing the physical properties of fiber-reinforced polymer (FRP) materials across a range of advanced applications, including microelectronics, chemical sensors, electromagnetic interference shielding (EMI), microwave absorption, hydrogen storage, and aerospace sectors [21]. Carbon nanotubes (CNTs) represent an important category within the carbon family [22]. They exhibit several beneficial properties, such as a high specific surface area, remarkable flexibility, significant mechanical stiffness, low mass density, and a high aspect ratio (length-to-diameter ratio). They are characterized as a 1D structure with a frictionless surface and effective π–π stacking interactions with aromatic compounds [23].

Following the discovery of carbon nanotubes (CNTs) by Iijima in 1991, they have attracted considerable attention, and their various applications continue to be explored in research. Carbon nanotubes (CNTs) are elongated macromolecules characterized by a radius of a few nanometers and lengths that can reach several micrometers [24]. Half-fullerene-like structures frequently cap the ends of these tubes, with their walls formed by a hexagonal lattice of carbon atoms originating from graphene sheets. Carbon nanotubes (CNTs) can be classified into three categories: single-walled carbon nanotubes (SWCNTs), double-walled carbon nanotubes (DWCNTs), and multi-walled carbon nanotubes (MWCNTs) [25].

Single-walled carbon nanotubes (SWCNTs) comprise a singular concentric graphene cylinder. Double-walled carbon nanotubes (DWCNTs) consist of two concentric cylinders. Various studies indicate that single-walled carbon nanotubes (SWCNT) demonstrate superior electromagnetic interference (EMI) shielding compared to multi-walled carbon nanotubes (MWCNT), as SWCNT exhibit lower resistance to electrical charges than MWCNT [24]. Another critical characteristic of SWCNT is its composition, which consists of a single layer of graphene wall. This results in a higher aspect ratio than MWCNT and fewer adjuvant walls, facilitating the accumulation of free electrical charges. Multi-walled carbon nanotubes consist of several cylindrical graphene layers, with an interlayer spacing of approximately 0.34 nm, sustained by Van der Waals forces [26]. Current synthesis methods for carbon nanotubes (CNTs), including electric laser ablation, arc discharge, and chemical vapor deposition (CVD), are utilized in commercial applications to produce significant amounts of CNTs.

One potential method for manufacturing hybrid composites involves dispersing carbon nanotubes (CNTs) into the polymer matrix to prepare nanocomposite for electrical applications. Adding a small number of nanotubes to polymers improves composite properties; adding 1% CNTs can increase tensile strength by 25% and stiffness by 36–42% [27].

Improper dispersion of carbon nanotubes (CNTs) can lead to accumulation, reducing composite properties. Adding polar bonds through functionalization enhances dispersion by treating CNT surfaces with chemicals like $-CH_3$, –OH, $-NH_3$, –SH, and –COOH. However, this method is limited in incorporating high CNT concentrations. A new CNT sheet technique has been developed for composite manufacturing to overcome this issue [28]. Among the various methods for producing CNT films, floating catalyst chemical vapor deposition (FCCVD) stands out as simple, environmentally friendly, and suitable for large-scale production. A carbon-containing gas and a catalyst, usually iron-based, are introduced into a heated chamber, producing various carbon nanotubes [29].

Carbon shells enhance carbon nanotube formation on an iron catalyst, producing a CNT sock network. Layering solvent creates dense CNT sheets. Due to sheet thickness, resin starvation occurs in CNT sheet/carbon fiber prepreg composites [30]. The resin must penetrate further into the CNT sheet to ensure proper adhesion between the carbon fiber prepreg and CNT layers. The decreased thickness of the CNT sock allows for better resin flow in the through-thickness direction, improving infiltration and creating reliable EMI shielding materials [31].

Praveen et al. [32, 33] prepared a series of LDPE/MWCNT nanocomposites with MWCNT content ranging from 0.05 to 5 weight (wt) % using a melt mixing method, followed by hydraulic pressing. The prepared nanocomposites were analyzed using differential scanning calorimetry (DSC), revealing no interaction between MWCNT and LDPE. The studies indicated an improvement in crystallinity by 5 wt%.

The incorporation of MWCNT enhanced the roughness of the fracture surface, as demonstrated by SEM analysis. MWCNT exhibited inhomogeneous distribution at lower weight percentages; however, a transition to homogeneous dispersion was noted at higher weights, with minimal accumulation observed. The mechanical properties of tensile strength, Young's modulus, and yield strength of nanocomposites were significantly enhanced compared to pure ultra-high-density polyethylene (UHDPE). The roughness profile of the nanocomposite increases with higher MWCNT content. Strong adhesion or coupling between the filler and polymer is essential for the polymer nanocomposite to demonstrate superior properties to the pristine polymer. The functionalization of carbon nanotubes is vital to achieving strong adhesion.

Composites prepared from CNTs have no alteration in the crystalline structure of the polymer (polyethylene) due to the addition of fillers. When high aspect ratio CNTs are used as the filler, the EMI shielding is believed to have improved, and the electrical percolation threshold is lower. However, using CNT in large quantities will increase the production cost of shielding material. Since the operation cost for CVD from which nanotubes will be prepared is a higher process, LDPE/CNT nanocomposite was prepared by melt mixing under ambient conditions [33]. The SEM images of nanocomposites prepared from isotactic polypropylene (iPP) and TPU with varying concentrations of MWCNT are shown in Fig. 6.4. Figure 6.5 shows the EMI shielding performance of those polymer nanocomposites. From this figure, we can observe that

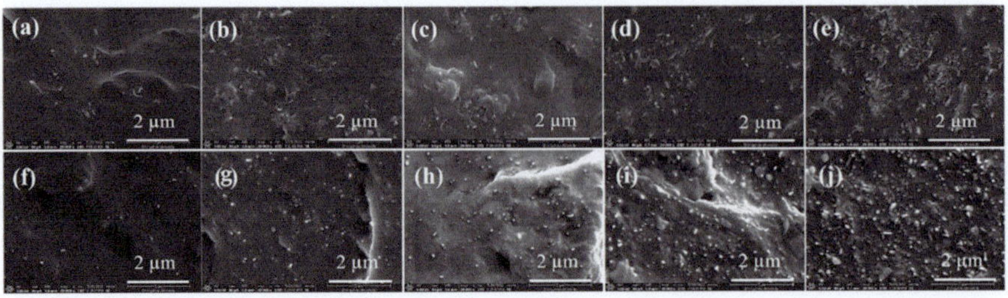

Fig. 6.4 SEM images of iPP/MWCNT and TPU/MWCNT composites at different wt% levels [34]

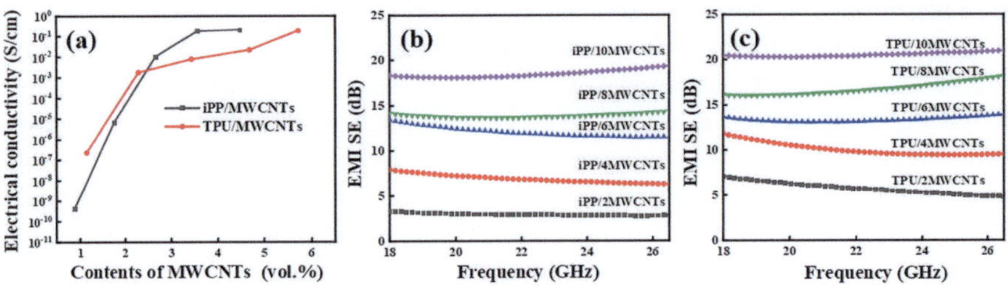

Fig. 6.5 Volume conductivity variation, EMI shielding [34]

as the concentration of CNT increases, the conductivity also increases. In SE graphs, we can observe an optimal concentration of CNT addition shielding effectiveness of 20 dB, the maximum value for the entire TPU series. The polymer composite with isotactic polypropylene (iPP) yielded slightly less shielding than TPU at a particular wt% of CNT observed. This is due to the better electrical conductivity of the TPU composite compared to the iPP system, as seen in Fig. 6.4 [34].

6.3.1 SWCNT

Single-walled carbon nanotubes (SWCNTs) consist of a single hollow cylinder; due to their tiny diameter, high aspect ratio, lightweight nature, exceptional mechanical strength, electrical conductivity, and ease of percolation, SWCNTs surpass traditional carbon fillers [35]. A recent study suggests that SWCNTs/polymer nanocomposites offer a promising approach to introducing new electrical properties. When SWCNTs are combined with polymers, achieving higher electromagnetic interference (EMI) shielding efficiency (SE) with a low percentage of SWCNTs by weight is possible. The morphology and composition of CNTs play a crucial role in determining shielding effectiveness (SE) [36, 37]. SEM image of single-walled carbon nanotubes [38] is shown in Fig. 6.6. These single-walled carbon nanotubes need precise environmental reaction conditions for synthesizing in a CVD chamber.

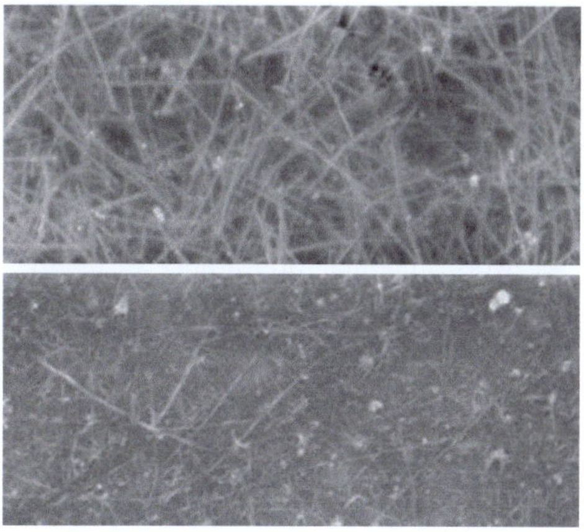

Fig. 6.6 SEM image of single-walled carbon nanotubes [38]

SWCNT has superior electrical conductivity as it has only one graphene-layered wall, hence less resistance to the movement of electrical charges. In contrast, MWCNT has more graphene walls, so there will be relatively higher resistance than SWCNT.

6.3.2 MWCNT

The choice of material is also an essential part of the design of composites since, in most industries, the material must be quickly and readily available, cheap, and easy to use [39]. The choice of material also depends on its requirement for being biodegradable. The elemental material carbon provides all these desired properties; therefore, many companies prefer carbon-based composite material designs.

In this respect, a rolled graphene sheet is considered a carbonaceous cylindrical nanomaterial in CNTs [40]. The SWCNTs consist of a single-layer graphene sheet. In contrast, MWCNTs consist of multiple layers of graphene sheets in concentric circles [41]. From a mechanical point of view, MWCNTs are more appropriate for reinforcing polymer composites, including carbon-reinforced epoxy resins. MWCNTs have high tensile strength and elastic modulus, which can significantly improve the mechanical properties of polymer composites. For example, MWCNTs are stronger than steel in tension, with a tensile strength of 10–50 GPa compared to 0.5 GPa for mild steel. When subjected to heavy loads, MWCNTs help distribute stress and transfer loads throughout the composite matrix. The MWCNT production cost is lower than that of SWCNT, and the electrical conductivity of MWCNT is slightly lower than that of SWCNT. However, MWCNT has better mechanical properties than SWCNT. So MWCNT, if used in polymer composite preparation, has some advantages such as cost-effectiveness, good mechanical properties, and stability; with slightly less electrical conductivity, it will give good shielding effectiveness.

So MWCNT incorporated composites have been practically implicated in EM attenuation applications with cost-effectiveness, making them viable materials for civil, military, and aviation applications.

However, a new type of MWCNT, helical MWCNTs, has recently been applied, and CNT is preferred as a filler in polymer composite [42]. In Fig. 6.7, SEM images of the MWCNTs at different magnifications [43] are shown, where we can observe the "worm, noodle" appearance of synthesized MWCNT from the CVD process. Each nanotube has a 2–100 nm diameter and a length range of several micrometers to over 30 μm.

Figure 6.8 shows SE graphs obtained from polymer composites fabricated from two different fillers. Different microfibers are carbon fiber-reinforced polymer (CFRP) and glass fiber-reinforced polymer (GFRP). In cases where the impact of microfiber on CFRP-based samples is shown for EMI shielding capability, the EMI shielding effectiveness exceeded 15 dB.

Here, the role of MWCNT is to increase electrical conductivity. Furthermore, intrinsic conductivity associated with CF makes it slightly more conductive than

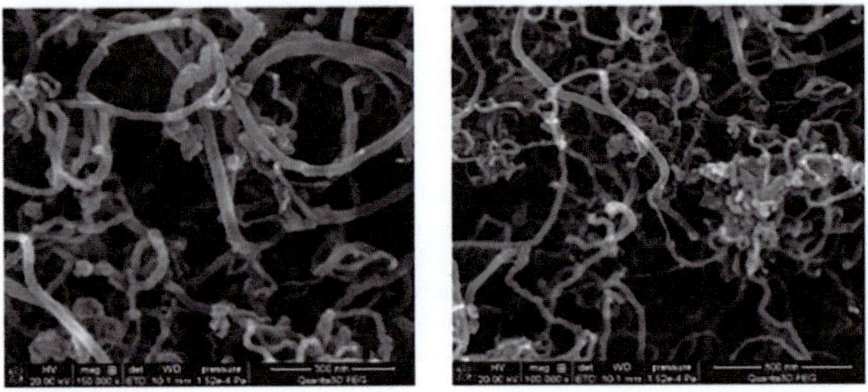

Fig. 6.7 SEM images of the MWCNTs at different magnifications [43]

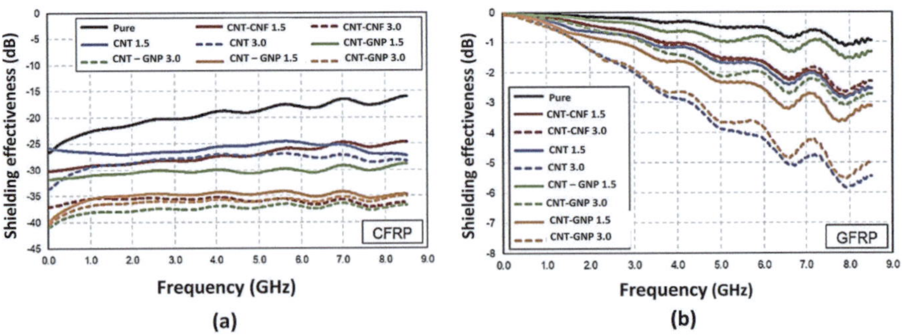

Fig. 6.8 EMI shielding effectiveness of samples fabricated with **a** CFRP and **b** GFRP matrix [44]

ceramic, non-conductive-natured GF composite. So, MWCNT, in any case, improves the SE capability with several dielectric losses. This suggests that carbon fiber composites possessing strong electrical conductivity could result in significant EMI shielding effectiveness. The combined impact of carbon nanomaterial (CNMs) embedded in carbon fiber-based composites can enhance EMI shielding effectiveness. The samples made of GFRP showed an EMI shielding performance of around 6 dB at most, which is less than that of CFRP composites, with the MWCNT-CF 3% sample showing 37 dB SE.

6.4 Carbon Nanofibers (CNFs)

Carbon nanofibers exhibit superior mechanical qualities, elevated electrical conductivity, and high thermal conductivity, which can be integrated into various matrices such as thermoplastics, thermosets, elastomers, ceramics, and metals. CNFs are cylindrical or conical nanostructures that range in diameter from a few nanometers to hundreds of nanometers, with lengths varying from under a micron to millimeters. CNFs have an electronic structure like graphite and conductive properties. The mesoporous nanofibers exhibit excellent thermal stability up to 900 °C in air. The synthesis is simple, eco-friendly, and more cost-effective than carbon nanotubes. It must be noted that CNFs exhibit lesser surface area, more defects, and more significant structural lengths than CNTs [45]. Various synthesis methods for CNFs include electrospinning, vapor phase, and solvothermal growth. Electro-spun CNFs are frequently prepared by carbonizing prepared polyacrylonitrile (PAN) nanofibers under inert gas pressure [46].

Carbon nanofibers (CNFs), vapor-grown carbon fibers (VGCFs), and vapor-grown carbon nanofibers (VGCNFs) are examples of cylindrical nanostructures. They consist of graphene layers arranged in stacked cones, cups, or plates. Carbon nanotubes consist of graphene layers rolled up into flawless cylinders. Figure 6.9 displays SEM pictures representing electro-spun nanofibers: (a) pre-carbonization and (b) post-carbonization [47].

In CFs, carbon is versatile in forming chemical bonds due to its flexibility, which allows it to create many stable organic and inorganic molecules. Elemental carbon exists as several allotropes (i.e., variants), including diamond, graphite, and fullerenes. These forms consist of elemental carbon but vary primarily in their properties. Besides mechanical properties, CNFs are used for their improved properties, such as thermal, electrical, and electromagnetic shielding [48]. As carbon is a cheap and readily available material, CNFs are extensively used as additives in composite materials. These are extremely small, with dimensions in the nanometer scale. Thus, specialized microscopic techniques such as Scanning Tunneling Microscopy and Atomic Force Microscopy are necessary to characterize the properties of CNFs [49].

Scientists worldwide are using carbon nanofibers to create Lithium-ion battery electrodes that promise four times the storage capacity of today's batteries [50]. Researchers also use nanofibers to develop sensors that change color as they absorb

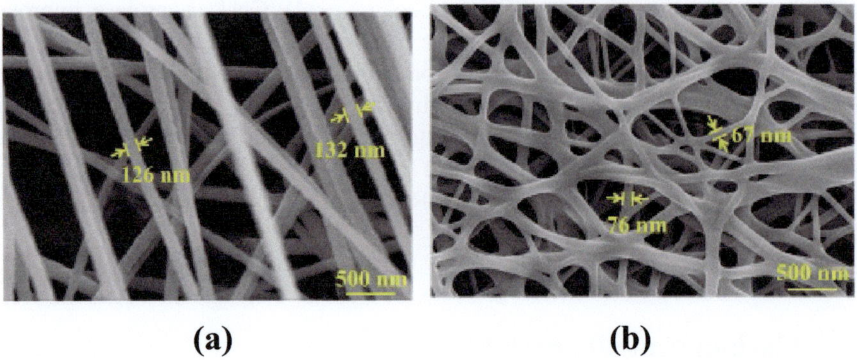

Fig. 6.9 SEM images of electro-spun nanofibers: **a** pre-carbonization, **b** post-carbonization [47]

chemical vapors. This porous carbon nanofiber's unique structure results in good EMI shielding performance when used as a non-conductive filler, as well as high reflectivity and chemical and mechanical stability [51].

Electron field emission is the emission of electrons induced by an electrostatic field. The most common context is field emission from a solid surface into a vacuum [52]. Field emission can, however, occur from a solid or liquid surface into a vacuum, air, a fluid, or any dielectric, which might be non-conducting or only weakly conducting. The other phenomenon associated with field emission is the promotion of electrons from the valence to the conduction band of semiconductors induced by the field, which also helps with EMC and EMI applications [48]. CNFs possess a surface area, accumulating electrical charges on the surface. Incorporating this into any non-conducting or conducting polymer will result in a conducting or semiconducting composite with a regular structure, making it suitable for EMI shielding applications. Recent studies have shown that carbon nanofibers exhibit biocompatibility, making them ideal for applications in space jackets and personal garments that require radiation shielding [53].

Additional carbon-based materials utilized for EMI shielding comprise carbon nanotubes (CNTs), graphene, carbon aerogels, and carbon black. Carbon nanofibers (CNFs) are anticipated to have a crucial impact on the future of electromagnetic interference (EMI) shielding owing to their numerous benefits. Carbon nanofibers are lightweight, vital for managing electromagnetic pollution in mobile devices, and protect human society and surrounding devices. Carbon nanofibers exhibit outstanding electrical conductivity, a crucial factor for electromagnetic interference shielding. Carbon nanofibers exhibit significant flexibility, making them advantageous for materials used in electromagnetic interference shielding. Cost-efficiency is essential, as CNFs provide an economical solution for EMI shielding materials. Carbon nanofibers can be prepared relatively quickly, presenting a substantial benefit for materials used in electromagnetic interference shielding. Carbon nanofibers are readily available, delivering a considerable benefit for materials used in electromagnetic interference shielding. The superiority and importance of one such case was

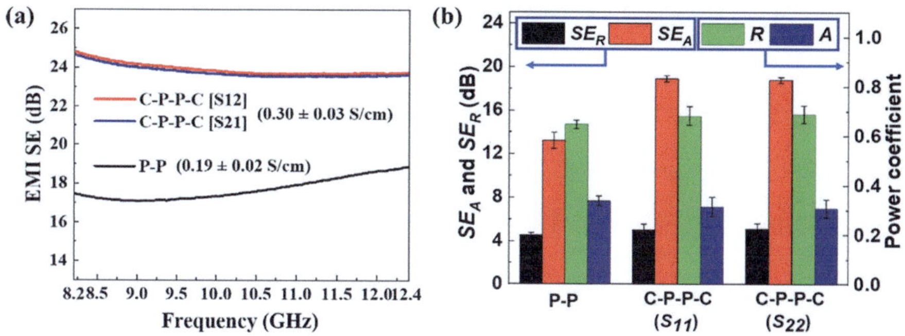

Fig. 6.10 **a** EMI SE in terms of S_{12} and S_{21} parameters; **b** SE_A, SE_R, and power coefficient of nanofiber polymer composites calculated using S_{11} and S_{22} parameters [47]

studied by Ma et al. [47]. Using carbon nanofibers, EMI SE, and electrical conductivities, polymer composites in terms of S_{12} and S_{21} values having SE of 25 dB were obtained, and SE_A, SE_R, and power coefficient of nanofiber polymer composites were calculated using S_{11} and S_{22} which are also discussed here [47]. These results are depicted in Fig. 6.10.

6.5 Carbon-Based Quantum Dots

Nanometer-sized 0D materials made of carbon, known as quantum dots, have attracted significant attention from researchers in various disciplines [54]. Carbon nanomaterials with zero dimensions are categorized as carbon quantum dots (CQDs) and graphene quantum dots (GQDs). Both have *sp*2 hybridization features, such as minimal toxicity, effective photoluminescence, easy and inexpensive synthesis, solubility in water, photostability, and consistent emission without blinking [55, 56].

GQDs are graphene sheets with a diameter and thickness of under 100 nm and consist of 10 graphene layers. GQDs possess a smaller bandgap than graphene, leading to their utilization in optical and electronic applications [57, 58]. Nonetheless, it is possible to utilize graphene in those scenarios by decreasing its horizontal dimensions and transforming it into GQDs [59].

The lateral size of GQDs can be controlled to modulate their bandgap up to the different electrical conductivity regions, as well as the number of layers, type of edge, and doping level; therefore, GQDs have great promise for EMI shielding applications [56, 60]. Either top-down or bottom-up methodologies could produce graphene quantum dots (GQDs). This involves the synthesis and fabrication of 0D graphene quantum dots (GQDs) from specific carbon source materials [carbon fiber, graphene oxide, graphene nanoribbons (GNRs)] using various methods. Methods encompass electron beam lithography, chemical oxidation, solvothermal processes, microwave-assisted hydrothermal techniques, micro fluidization, electrochemical approaches,

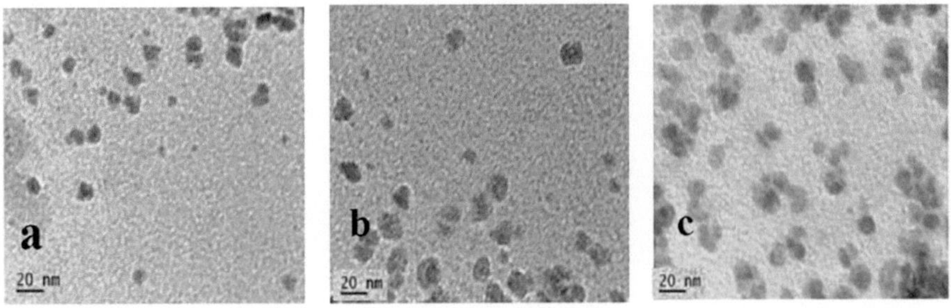

Fig. 6.11 TEM images of carbon quantum dots [63]

pyrolysis, and oxidation [61, 62]. From the TEM images of carbon quantum dots [63] shown in Fig. 6.11, we can observe that the tiny dot-like structured nano quantum has a size less than a few nm.

6.6 Nanodiamonds

Nanodiamonds are tiny carbon particles shaped like truncated octahedra (that measure approximately 2–8 nm) in diameter. The smallest monocrystalline nanodiamonds, known as detonated nanodiamonds, are created through the detonation of carbon-based explosives, followed by a purification process. They have attractive features of diamond nanoparticles, including their surface properties and non-toxic nature.

The structure of a nanodiamond can be separated into three parts: the outer surface made up of carbon atoms with various functional groups, the center region containing sp^3 hybridized carbon atoms, and the inner diamond core. The external coating consists of a hydrogenated surface, including H and OH groups, and measures 0.6 nm in diameter, making up approximately 61% of the overall carbon content. The inner core has a potential diameter of 2.4 nm, with about 40% of the carbon in the core and its inner layer [64].

Various techniques such as CVD, plasma-assisted CVD, laser ablation, ball milling, autoclave, and irradiation can all be used to create nanodiamond particles [65]. An explosive combination containing oxygen-deficient 2-methyl-1,3,5-trinitrobenzene is the preferred method for creating nanodiamonds through detonation. When detonated in a non-reactive environment, explosives create nanodiamond particles in temperatures as high as around 1200 °C [66]. This process typically produces particles ranging in size from 4 to 5 nm [67]. Furthermore, bigger aggregates of nanodiamonds ranging from 100 to 200 nm have been reported, and even greater masses can be formed during the detonation process. Moreover, the technique may result in the creation of nano-onion-like carbon nanoparticles [66].

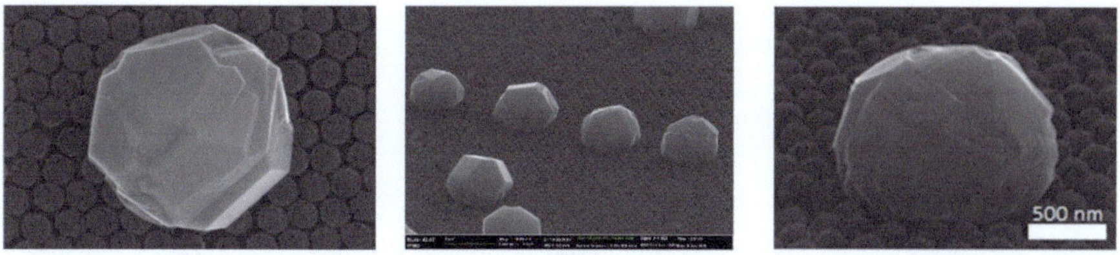

Fig. 6.12 SEM images of the HFCVD diamond nanoparticles grown on SiO_2 [69]

The nanocomposites are impressive for being strong, stable at high temperatures, good at conducting heat, and flame-retardant [67]. Polymer/nanodiamond nanocomposites are helpful in energy, electronics, and space because they are small, have a lot of surface area, and have special qualities. Researchers have shown keen interest in studying nanodiamonds' structural properties and dispersion within polymer matrices [68]. SEM images of the HFCVD diamond nanoparticles grown on SiO_2 are shown in Fig. 6.12 [69]. Nanodimond resembles real diamonds in morphology and shape.

6.7 g-C_3N_4

g-C_3N_4 is a polymer created from common elements that can efficiently soak visible light due to its bulk bandgap energy of around 2.7 eV [70]. This n-type semiconductor is known for its unique characteristics, including physicochemical stability, non-toxicity, and straightforward synthesis, making it a promising metal-free photocatalyst. Following the initial study on g-C_3N_4's photocatalytic activity for hydrogen production in 2009, many related publications have surged, revealing various aspects of g-C_3N_4 [71]. Like other organic semiconductors, g-C_3N_4 generates Frenkel excitons that may result in quick recombination of charge carriers.

Additionally, g-C_3N_4 faces challenges such as a low surface area, insufficient active sites for redox reactions, and limited sunlight absorption [72]. Several strategies can address these issues to enhance g-C_3N_4-based photocatalysts, including morphological design, doping, and creating g-C3N4 hybrids and nanocomposites. Utilizing 2D g-C_3N_4 nanosheets can significantly increase its surface area [73]. Moreover, carbon nanosheets (CNNs) exhibit a wider bandgap, improved electron transport in the in-plane direction, and maintained charge carrier recombination due to the quantum confinement effect, in contrast to their bulk form. The higher specific surface area of the CNNs indicates more active sites on the surface, leading to a higher rate of EMI shielding [74] in Fig. 6.13. SEM images of all g-C_3N_4 samples [75] are shown.

Fig. 6.13 SEM images of all g-C_3N_4 samples [75]

6.8 Fullerene

Fullerene molecules, discovered in 1985, have symmetrical nanocarbon structures. They comprise polygons, specifically pentagons and hexagons [76]. These molecules are called after the number of carbon atoms they contain. One of the most well-known fullerene compounds is C60, often known as Buckminster-fullerene [77].

The resulting nanocomposites are notable for their strength, thermal stability, thermal conductivity, and flame resistance [78]. Polymer/nanodiamond nanocomposites are helpful in the energy, electronics, and space industries due to their small size, large surface area, and unique properties [79]. Researchers have shown a strong interest in exploring nanodiamond structural characteristics and dispersion in polymer matrices. Figure 6.14 shows fullerene analogs [80].

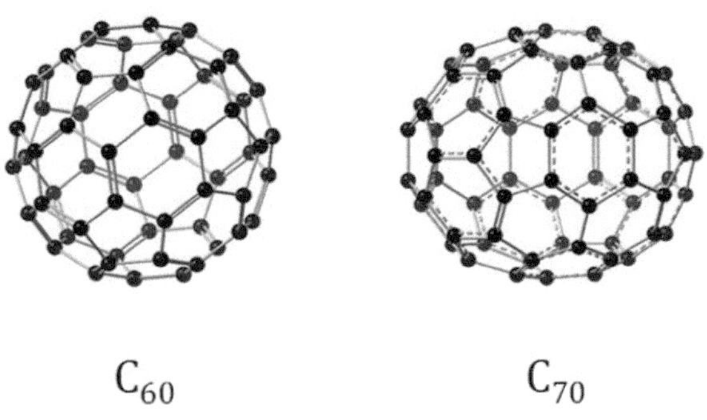

Fig. 6.14 Fullerene analogs [80]

6.9 Carbon Black

Carbon black is an essential kind of nanocarbon. It is almost entirely composed of elemental carbon and is utilized as a nanofiller in polymer matrices. Small carbon black nanoparticles can be smaller than 100 nm. Carbon black's microstructure is typically identified as quasi-graphitic. These tiny nanoparticles prefer to combine, generating chain formations that can measure up to 500 nm. The weak Van der Waals forces between carbon black particles cause them to aggregate [81].

These collections can create bunches resembling grapes, and with more gathering, they can form massive carbon black masses [50]. The interconnected chain-like formation of carbon black particles indicates a significant electron conduction event. A tiny amount of carbon black particles can exceed the percolation threshold and create a conductive network. Polymer/carbon black nanocomposites have practical applications in aerospace, such as antistatic, electrostatic dissipation, and radiation shielding [82]. SEM images of composites filled with black carbon (a) and (b), frequency dependences of real ε' and imaginary ε'' parts of complex permittivity, (c) and reflection loss of composites filled with carbon black (d) are shown in Fig. 6.15 [83]. Carbon black incorporated in any polymer matrix gives stability, flexibility, temperature stability, and electrical conductivity. Carbon black loading increases conductivity, and hence EMI shielding will be increased using reflection loss as reported in Kruzelak et al. [83].

6.10 Polymeric Nanocomposites

The above-discussed carbonaceous nanomaterials can be applied as powder, coatings, deposition, paste, and paint on the desired surfaces. These materials are futuristic, practical, efficient shielding materials if converted into polymer composite. When one or more carbonaceous materials are used in a single polymer to form a hybrid-natured composite, nanocomposites embody a substantial group of composite materials that integrate nanomaterial reinforcements, acquiring all the beneficial attributes of fillers and additional advantages above singular fillers [84]. The functional properties of nanocomposites are improved by adding nanofillers or nano-reinforcements to polymer matrices [85]. These nanoparticles have a high surface area, allowing effective interaction with the matrix. Improvements in the properties of the polymer matrix have been noted even with low additions of nanofillers [86]. Nanofillers like carbon nanotubes and nanofibers can significantly improve the mechanical characteristics of polymers. One potential improvement in the composite's strength could come from nanocarbon fillers [87].

Polymer composites have significant stiffness, toughness, impact resistance, and fatigue resistance characteristics. Furthermore, nanocarbons serve as excellent additives that improve the heat resistance of materials [88]. Nano-clays, metal nanoparticles, and nanocarbons have proven to be efficient nanofillers for polymers. The

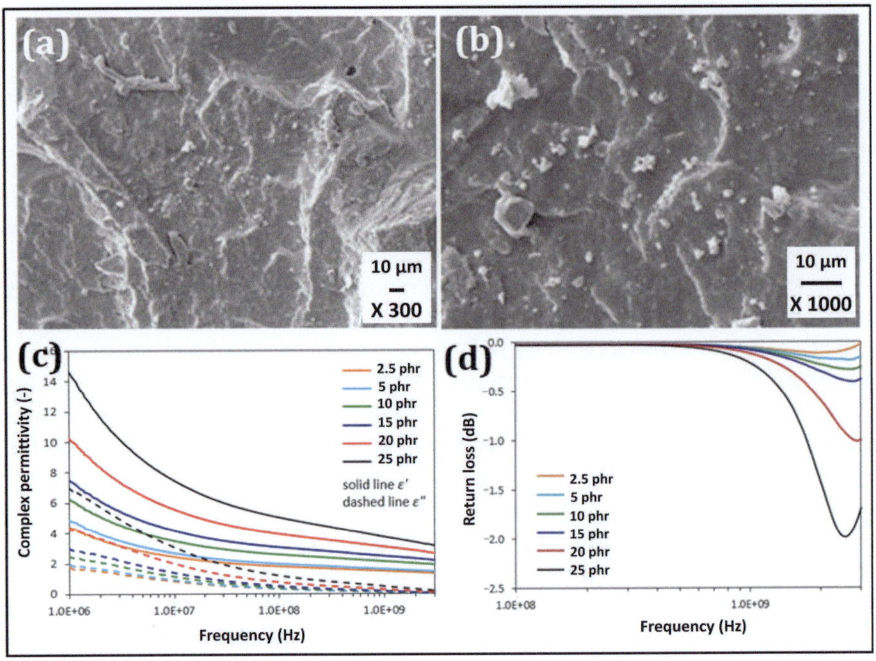

Fig. 6.15 SEM images, frequency dependences, and reflection loss [83]

properties of nanocomposites, such as electron transport, thermal conductivity, flame resistance, radiation shielding, and sensing capabilities, are greatly enhanced by including these nanofillers [89]. Different methods have been used to create nanocomposites, such as solution casting, melt extrusion, and in situ polymerization [90]. Nanoparticles have been successfully added to polymers through stirring, ultrasonication, and mechanical mixing [91]. Polymeric nanocomposites can be created in different formats, such as films, coatings, laminates, and fibers, with nanocomposite fibers generally produced through electrospinning [92].

The primary obstacle in developing nanocomposites lies in attaining effective dispersion of nanofiller particles, which may differ in size and include impurities. Several methods have been explored to manipulate the distribution of nanoparticles and enhance their physical properties [93]. However, the clustering of nanoparticles due to Van der Waals forces or electrostatic interactions can result in insufficient bonding between the matrix and the nanofiller. This gathering limits the best use of nano-enhancements [94]. To tackle this issue, the nanoparticles undergo a range of post-treatment techniques to isolate the individual particles. The shape and dimensionality of nanofillers affect the dispersion characteristics of nanocarbon nanoparticles. For example, 1D carbon nanotubes display distinct dispersion patterns compared to 0D spherical particles [95]. Carbon nanotubes frequently create entangled bundles, presenting challenges for their dispersion. Consequently, various strategies have been devised for functionalizing or modifying carbon nanotubes. These

Fig. 6.16 Cross-sectional SEM of cryogenically fractured LDPE-based composite [33]. Reprinted with permission from Elsevier

approaches have been implemented on single-walled carbon nanotubes (SWCNT), employing noncovalent and covalent functionalization techniques [96].

Surface defects in functional carbon nanotubes allow them to form micromechanical interlocking with the matrix, induce weak Van der Waals interactions, or form covalent bonds [97]. These covalent bonds can significantly improve the interfacial interaction between the matrix and the nanofillers. Robust covalent bonding facilitates enhanced stress transfer across the nanocomposite material, positively impacting its mechanical properties [98]. Carbon nanotubes that incorporate endohedral fullerenes exhibit enhanced characteristics in their interactions with macromolecular chains. Direct covalent sidewall functionalization markedly enhances the chemical reactivity of carbon nanotubes with polymers [99]. Different methods for covalent functionalization of carbon nanotubes encompass hydrogenation, chlorination, bromination, and the Diels–Alder reaction. A common technique for covalent modification involves using acid treatment to chemically oxidize the nanotube surface, adding carboxylic acid groups and different oxygen-containing functionalities [100]. This functionalization improves the compatibility between the matrix and the nanotubes, while the modification techniques can ensure consistent surface functionalities on the nanotube surface [101] as shown in Fig. 6.16. Cross-sectional SEM of cryogenically fractured LDPE-based composite is shown in this Figure [33].

6.11 Case Study 1 (LPDE:MWCNT:GNP:$LaFeO_3$)

Examining lanthanum ferrite with MWCNT and graphene in a polyethylene polymer matrix for X-band frequency examination.

6.11.1 LDPE Composite Fabrication

A blend of $LaFeO_3$, MWCNT, and GNP in THF was stirred for 15 min and evaporated in an oven. The mixture was then added to the melt mixing process with the "Brabender®-Plastic Order" tool for blending at 130 °C. It was molded using hot compression equipment at 110 bar and 120 °C for 1 h, resulting in a 3.5 mm thick rectangular mold measuring 20 mm by 30 mm. Fine-tuning the polymer filler is essential to maintaining the composite's mechanical strength [33, 102, 103]. The polymer nanocomposite was developed using various filler material combinations and labeled according to Table 6.1.

Figure 6.17 shows images of $LaFeO_3$ captured by an SEM at different magnification levels (a and b) and EDS of $LaFeO_3$ nanoparticles shown in panels (c and d). Figure 6.18a, b displays scanning electron micrographs of GNP and MWCNTs. As shown in Fig. 6.18a, the surface of graphene comprises several sheets of loosely bonded graphene produced via chemical exfoliation and microwave irradiation [105]. Figure 6.18b displays the MWCNT picture, which depicts the carbon nanotubes looking like worms. The CVD technique yields MWCNTs with a diameter of 40 nm and lengths ranging from 200 to 300 nm [106].

Electromagnetic waves can be absorbed, transmitted, or reflected depending on the medium. In this case, the impedance mismatch causes the incident electromagnetic wave to be reflected at the air–material boundary; the leftover energy is then transmitted into the material. The material then undergoes magnetic losses and attenuation due to the dielectric. Finally, the material–air interface transfers the residual incident energy to the air [107].

Reflectivity quantifies the power reflected, transmissivity quantifies the power transmitted, and absorptivity quantifies the energy absorbed. The polymer composite's reflectivity is determined using the S-parameter acquired from (6.1)–(6.2), which details the composite's proposed transmissivity, while (6.3) illustrates its absorptivity:

$$R = |S_{11}|^2 \tag{6.1}$$

$$T = |S_{21}|^2 \tag{6.2}$$

Table 6.1 Sample combination [104]

S. No.	Polymer (In wt.%)	Conductive filler (In wt.%)		Magnetic filler (In wt.%)
	(LDPE)	MWCNT	GNP	($LaFeO_3$)
1	50	5	45	0
2	50	5	40	5
3	50	5	35	10
4	50	5	25	20
5	50	5	15	30
6	50	5	5	40
7	50	0	0	50
8	50	0	50	0

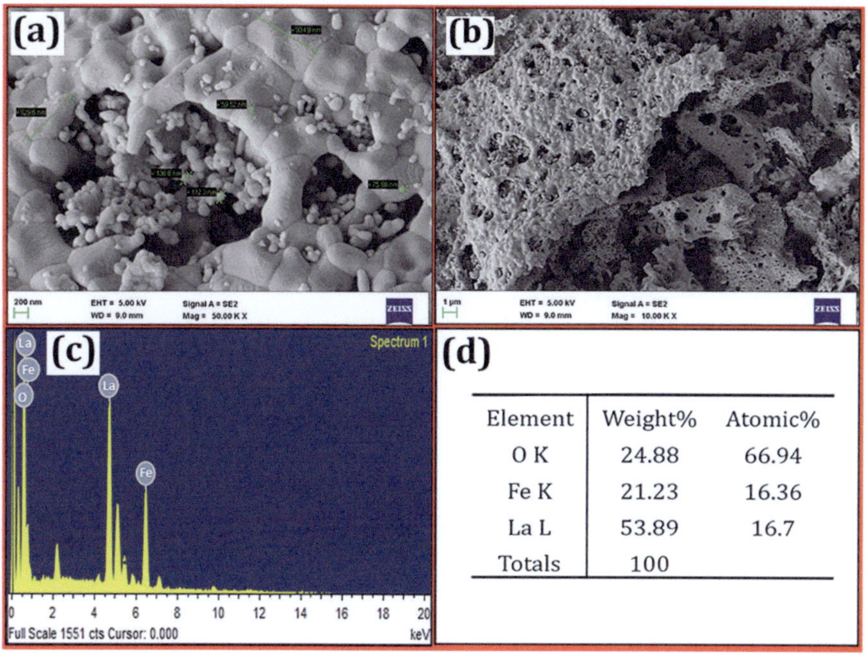

Element	Weight%	Atomic%
O K	24.88	66.94
Fe K	21.23	16.36
La L	53.89	16.7
Totals	100	

Fig. 6.17 Images of $LaFeO_3$ captured by an SEM at different magnification levels (**a** and **b**), EDS of $LaFeO_3$ nanoparticles shown in panels (**c** and **d**) [104]. Reprinted with permission from Elsevier

Fig. 6.18 **a**, **b** show the SEM micrographs of the GNP and MWCNTs [104]

$$A = 1 - R - T \tag{6.3}$$

The *S*-parameters assess the proposed material's reflection, transmission, and absorption characteristics. The *S*-parameter measurement assesses the electromagnetic shielding of composite materials. Impedance mismatches cause power to rebound into the air, with available power entering the material ($P_{av} = P_{inc} - P_{ref}$). Equation (6.4) calculates reflection loss, while (6.6) determines shielding effectiveness [108]:

$$SE_R = 10\log_{10}\left(\frac{P_{inc}}{P_{av}}\right) = -10\log_{10}(1-R) \tag{6.4}$$

$$SE_A = 10\log_{10}\left(\frac{P_{av}}{P_t}\right) = -10\log_{10}\left(\frac{T}{1-R}\right) \tag{6.5}$$

$$SE_T = SE_R + SE_A \tag{6.6}$$

SE_T indicates overall shielding efficacy, while SE_R shows incident power ratio. This suggests that reflection causes the incident wave to travel less into the substance [42]. A material's SE determines its ability to reduce electromagnetic waves. The available power to transmitted power ratio shows effectiveness. Equations (6.1)–(6.7) helps one to calculate shielding parameters.

6.11.2 EMI Characterization

Electromagnetic shielding mitigates dangers posed by a wide variety of electromagnetic radiation. The reflection and absorption of electromagnetic waves by materials are part of it. The decibel scale depicts the proportion of the electromagnetic wave's incident power (P_i) to its transmitted power (P_t). Reflection, absorption, and SE, along with multiple reflections, comprise the shielding:

$$SE_T = -10\log(P_t/P_i) = SE_R + SE_A + SE_{MR} \tag{6.7}$$

Mobile charges, such as holes and free electrons, contribute to impedance mismatching, enabling electromagnetic waves to reflect between air and absorber materials [109]. The polymer composite exhibited numerous reflections attributable to its inhomogeneous medium dispersion and an extensive interfacial area. Upon the absorption of the electromagnetic pulse, the magnetic fillers experienced a reduction in thermal energy and exhibited vibrational activity [110]. The behavior of electromagnetic waves as they travel through the medium is affected by the thickness of the material and the presence of dielectric, ohmic, and magnetic losses—electrical charge conduction, tunneling, and hopping cause more significant ohmic loss by dissipating energy [102]. The energy required to dissipate to counteract the effects of filler dipole reorientation leads to polarization losses [111]. Material dipole polarization becomes plausible when it effectively addresses defects, functional groups, and interfaces to a greater extent [112].

Impedance mismatching causes EM waves to reflect between air and absorber material [109]. The polymer composite had many reflections due to its inhomogeneous medium dispersion and larger interfacial area. After absorbing the EM pulse, the magnetic fillers lost heat and vibrated [110]. Material thickness, dielectric, ohmic, and magnetic losses influence EM waves traveling through a medium. Conduction, tunneling, and hopping increase ohmic loss [102]. When a material has fillers (like particles in a composite), these fillers may introduce additional dipole reorientation, which causes energy dissipation. This reorientation can result in increased dielectric losses. Counteracting this requires additional energy input [113]. Material dipole polarization is conceivable if it handles defects, functional groups, and interfaces more [112].

However, magnetic loss is influenced by a variety of material processes. Magnetic filters exhibit exchange resonance, hysteresis loss, eddy current loss, and natural resonance [111]. As a result, magnetic filler $LaFeO_3$ is required for optimal shielding [37]. SE_{MR} can be ignored if material shielding is over 10 dB and thickness exceeds skin depth. Effective SE_T is $SE_A + SE_R$ [33].

Various combinations of LDPE-doped SE with MWCNT and GNP fillers and $LaFeO_3$ were tested from 8.2 to 12.4 GHz using waveguide analysis. The nanocomposite material's SE_T was evaluated, with sample 50:5:40:5 showing the highest SE of 33.57 dB at 10.38 GHz. Sample 50:5:45:0 had a peak SE of 32.64 dB, 50:5:35:10 had 29.65 dB, 50:5:25:20 had 25.26 dB, and 50:5:15:30 had 23.33 dB. The results are plotted in Fig. 6.19a. Unmodified LDPE lacked protection. Some samples lacked shielding due to poor coordination between fillers, networks, and routes, reducing dielectric and magnetic properties [112].

SE_A graphs in Fig. 6.19b exhibit a similar design, aligning with SE_T trends and showcasing SE_A's significant impact. The suggested LDPE:MWCNT:GNP:$LaFeO_3$ mixture for absorption protection has been successful, with sample 50:5:40:5 achieving the highest SE_A of 44.76 dB at 10.4 GHz. Other samples, such as 50:5:45:0, 50:5:35:10, 50:5:25:20, and 50:5:15:30, also show varying levels of SE_A at 10.4 GHz. Samples lacking electromagnetic wave shielding through absorption include 50:5:5:40, 50:0:0:50, 50:5:0:45, and pure LDPE [114].

SE_R charts in Fig. 6.19c show different distributions of reflection evaluation samples. A ratio of 50:5:40:5 has the highest SE_R, while 50:5:45:0 reaches peak SE_R of 3.3 dB at 9.8 GHz. Sample 50:5:15:30 peaks at 3.16 dB at 10 GHz, and 50:5:25:20 and 50:5:35:10 show 3.1 dB at 10 GHz. Other samples have lower SE_R values, with pure LDPE at 1 dB, indicating it is electromagnetically neutral. Graph 6.19d emphasizes SE_A and SE_R effects on SE_T.

SE graphs revealed that both SE_A and SE_R contribute to SE_T. This will demonstrate the benefits of graphene-MWCNT conductive fillers [115]. High-coercivity $LaFeO_3$ nanoparticles produce magnetic loss. At a weight ratio of 50:5:40:5, LDPE: MWCNT: GNP: $LaFeO_3$ offers the optimum protection. Dipole, interfacial polarization, and reflections from graphene/MWCNT fillers shield with $LaFeO_3$ inducing heat release. Maximum shielding needs 5% ferrite, unlike 0% sample [116]. This demonstrates that polymer composites require magnetic and conductive materials for effective shielding [102].

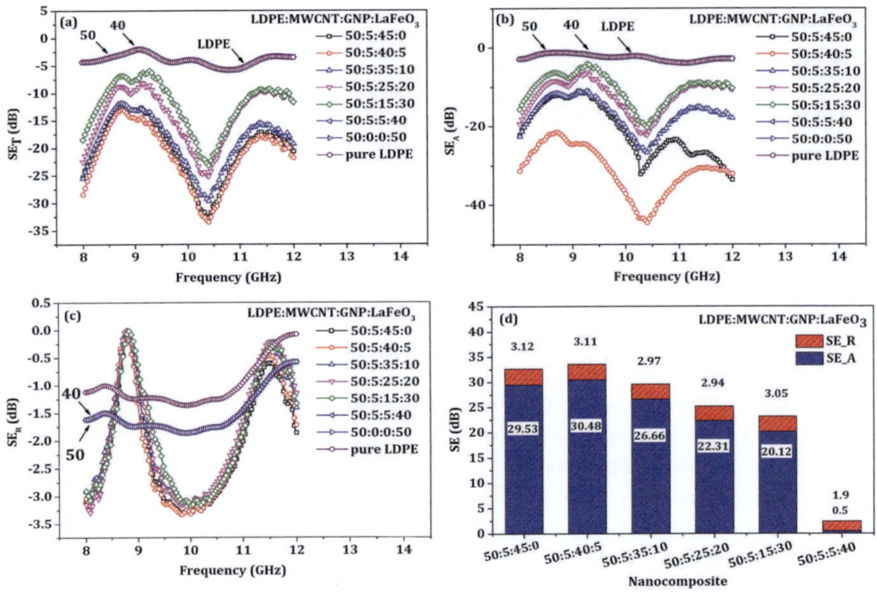

Fig. 6.19 A nanocomposite comprising multi-walled carbon nanotubes, illustrated through a bar column that delineates the coefficients of absorption, reflection, and transmission [104]. Reprinted with permission from Elsevier

Nanoparticles of $LaFeO_3$ absorb the electromagnetic pulse and vibrate at their natural frequency, generating molecular heat. It will spread through networks of graphene and multi-walled carbon nanotubes [102]. Polyethylene matrix combined with MWCNT, and graphene conduct networks and pathways. Electric charges on the surface will shift as conductive additives absorb electromagnetic waves. Absorption gives graphene and MWCNT access to unbound electrical charges. Charge accumulation at interfaces causes interfacial polarization to occur [114]. The impedance mismatch of EM wave moving between air and material, encountering fillers in porous polyethylene, gives rise to reflections. Electrical conductivity will increase when conductive paths form at a particular concentration within the sample [114, 115]. The material containing $LaFeO_3$, MWCNT, and GNP with magnetic and dielectric losses reduces the energy of electromagnetic waves and shielding effectively in the frequency band 8.2–12.4 GHz [33]. Table 6.2 demonstrates that other researchers used similar materials and shielding settings to support our findings.

The research presented a polymer nanocomposite with a base of low-density polyethylene, including conductive fillers like multi-walled carbon nanotubes and graphene and magnetic fillers in the shape of $LaFeO_3$ nanoparticles. The combustion process yields $LaFeO_3$ nanoparticles. Its high saturation magnetization and coercivity distinguish $LaFeO_3$. The MWCNT exhibits a width of 30–40 nm and a tubular structure, in contrast to graphene's layered, sheet-like configuration. In the frequency range of 8.2–12.4 GHz, the concentration of 50:5:40:5 was a significant

Table 6.2 Literature review of recent research using comparable materials [104]

S. No.	Material	Max. $\|SE_T\|$ dB	Frequency range	Reference
1	PEI:CNT	32.3	8.2–12.4 GHz	[23]
2	LDPE:Graphene: MWCNT:$CoFe_2O_4$	50	8.2–12.4 GHz	[33]
3	HDPE:CNT	21.8	8.2–12.4 GHz	[117]
4	Co:MWCNT:polyethylene	55	12.8–17.8 GHz	[118]
5	PET:PANI	23.95	8.2–12.4 GHz	[119]
6	UHMWPE:LLDPE:biochar	48.7	300–1500 MHz	[120]
7	PE:K-CB:FCNF	33	8.2–12.4 GHz	[121]
8	CF:PC film	38.6	30–1500 MHz	[122]
9	PEDOT:PSS	29.8	8.2–12.4 GHz	[123]
10	Graphite nanobelts:UHMWPE	35	8.2–12.4 GHz	[124]
11	PE:PP:carbon	36	25.5–37.5 GHz	[125]
12	LDPE:Graphene:MWCNT:α-Fe_2O_3	40	8.2–12.4 GHz	[126]
13	LDPE:Graphene:MWCNT:$LaFe_2O_4$	33.57	8.2–12.4 GHz	[104]

factor with a standard error of 33.57 dB. MWCNT and GNP contribute to various losses, while the magnetic properties of $LaFeO_3$ result in different losses. Impedance disparity between air and polymer causes reflection, and conduction channels at critical concentration allow the movement of electrons, leading to a practical SE value.

6.12 Case Study 2 (LPDE:MWCNT:GNP:$ZrFe_2O_4$)

6.12.1 Introduction

In a case study, the work by Praveen et al. [32] reveals the importance of carbon-based nanomaterials, namely MWCNT and graphene, which were incorporated along with ferrites in polyethylene polymer composite. Considering prior investigations conducted by various organizations, we noted constraints in implementing a distinctive and significant amalgamation of conductive and magnetic fillers. As a result, we studied the critical concentration of MWCNT, magnetic nanoparticles, and graphene when incorporated into polyethylene polymer and their synergistic effects. The process of melt mixing facilitated the creation of nanocomposites, which underwent evaluation for their efficacy in electromagnetic field protection. This study explores the magnetic properties of $ZrFe_2O_4$ nanoparticles for their potential use in absorbing microwaves.

Nanomaterials like carbon nanofiber, MWCNT, graphene, MXenes, and metallic nanoparticles are being studied for their potential as lightweight shield materials due to their strong conductivity and low density. MWCNT and graphene are particularly promising for their excellent 2D conductivity and versatility in functionalization. There is growing interest in creating synthetic polymer-based materials for EMI shielding. Society is at risk from environmental pollution and the energy crisis.

Zirconium ferrite ($ZrFe_2O_4$) is a ceramic made of zirconium and iron oxides, known for its magnetic and electrical properties. Due to its stability, thermal shock resistance, and condition tolerance, it is used in inductors, transformers, and for high-temperature applications. It is also used in high-temperature ceramics, refractories, and medical applications like MRI and drug delivery using nanoparticles.

Biswas et al. [127] and team studied combining PVDF and SAN with multi-walled carbon nanotubes in cobalt ferrite via pyrene derivative, giving − 44.8 dB SE. Pawar et al. [128] investigated MWNT-grafted Fe_3O_4 nanoparticles in PC/SAN blends, reaching − 32.5 dB SE in X-band and Ku-band. Harish et al. [127] studied coral-shaped YIG particles in polyaniline–wax composite, reaching − 44.8 dB SE. Sujatha et al. [103] studied iron nanoparticles coated on graphite in polyethylene for EMI shielding, achieving 49 dB SE at X-band with $ZrFe_2O_4$-reduced graphene oxide nanocomposites.

The study highlights the challenges of using conductive and magnetic fillers like MWCNT, magnetic nanoparticles, and graphene in polyethylene. Nanocomposites were produced using melt mixing and tested for electromagnetic field protection, investigating $ZrFe_2O_4$ nanoparticles' magnetic properties for microwave absorber applications.

LDPE powder and fillers like MWCNT, GNP, and $ZrFe_2O_4$ were mixed in THF for 50 min to achieve a uniform dispersion. The solvent was removed in a hot oven at 66 °C for 60 min. Using a Brabender® plastic order instrument at 115 °C, the materials were melted for 10 min at 40 rpm. Sentech® press molds material at 120 °C with 80N force, using a 10 × 10 cm die post extraction [126].

6.12.2 XRD Studies

The X-ray diffraction (XRD) pattern for $ZrFe_2O_4$ is presented in Fig. 6.20. The standard database includes zirconium iron oxide, COD 916457 of Fe_2O_3, and PDF card 00-002-0821, classified under the Hexagonal Crystal system. The observed peaks for $ZrFe_2O_4$ exhibit a strong correlation with the reference file. $ZrFe_2O_4$ nanoparticles exhibit a hexagonal structure, classified under Space Group: R-3c (167). Peaks at angles (in degrees) 24.4, 33.7, 35.8, 41.1, 50, 54, 62, and 64 match specific crystal planes. The characteristic peaks of zirconium are identified at angles 31.7, 34.5, and 49.7. Validation of $ZrFe_2O_4$ formation in the material is required [25, 35].

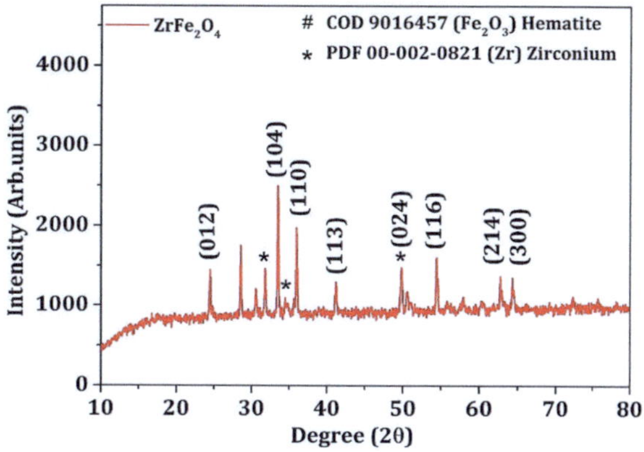

Fig. 6.20 XRD of $ZrFe_2O_4$ synthesized via solution combustion method [32]. Reprinted with permission from Elsevier

6.12.3 SEM/EDS Analysis

SEM images of SCS-generated $ZrFe_2O_4$ nanoparticles at 100 and 50 kx are shown in Fig. 6.21a–d. Polygonal nanoparticles have low regional aggregation. Many smaller particles, ranging from 31 to 90 nm, are found. The material's elemental abundance was shown in the EDS spectrum and a percentage table, and there were no additional pollutants (Fig. 6.21c, d). Elemental mapping (Fig. 6.22) reveals nanoparticle distribution and aqueous combustion-derived nanoparticles' abundance. Crystalline nanoparticle shape, appearance, and dispersion are shown in electron micrographs. EDS spectrum depicts Zr, Fe, and oxygen in $ZrFe_2O_4$ nanoparticles.

Figure 6.23 illustrates the morphology and filler dispersion of LDPE composites containing 5 wt% $ZrFe_2O_4$, which has undergone cryogenic cracking [38]. The cross-sectional area was subjected to color mapping and SEM analysis. The study sample exhibited a higher concentration of 40 wt%, resulting in a predominant distribution of GNP as observed through scanning electron microscopy. Additionally, $ZrFe_2O_4$ porous agglomerated clusters are distributed throughout the matrix. Figure 6.24 illustrates the elemental analysis results conducted through EDS color mapping. Color mapping photographs demonstrate the uniform distribution of carbon-based and magnetic $ZrFe_2O_4$ across all locations, confirming the fillers' composite nature and consistent dispersion (Fig. 6.22 is an illustration of polymer composite sample undergoing cryogenic fracturing). The photographs were captured using a 1 kx system, employing appropriate gold sputtering techniques. The LDPE polymer matrix incorporates MWCNT clusters, GNP, and $ZrFe_2O_4$ crystals. The Brabender® plastic order instrument employs a uniform melt mixing method, resulting in the elemental mapping of the polymer composite exhibiting a consistent distribution of elements.

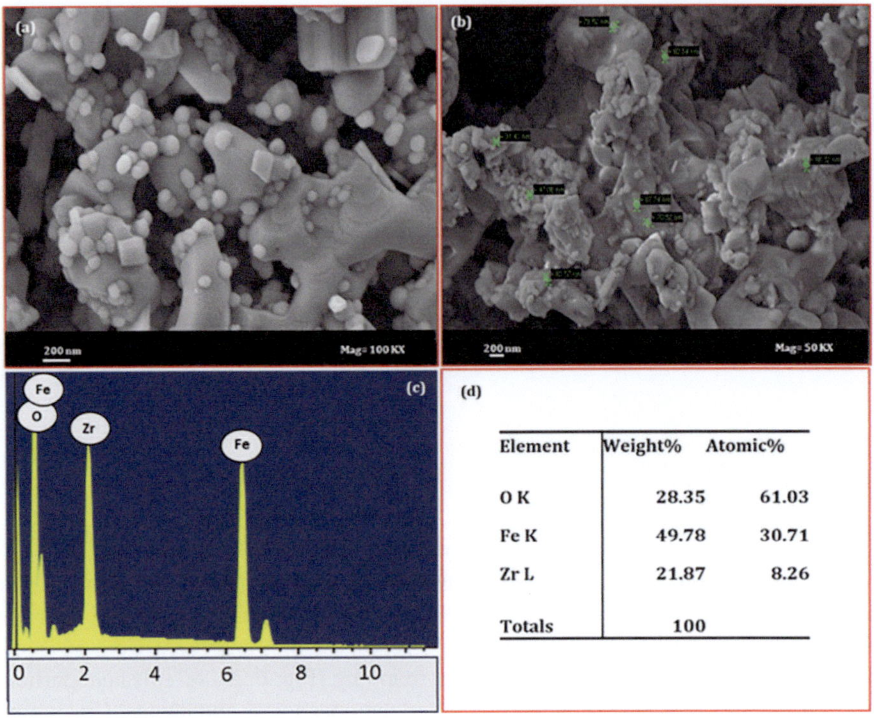

Element	Weight%	Atomic%
O K	28.35	61.03
Fe K	49.78	30.71
Zr L	21.87	8.26
Totals	100	

Fig. 6.21 FESEM of $ZrFe_2O_4$ at different magnifications (**a** and **b**). EDS results of $ZrFe_2O_4$ (**c** and **d**) [32]. Reprinted with permission from Elsevier

6.12.4 Mechanical Test

Composite samples were tested with a tensiometer to determine mechanical properties. Graph data (Fig. 6.25) shows that the 50:5:45:0 composition has the weakest strength among samples, with results on ultimate strength, slope, elongation, and work done [42]. Samples with more graphene were weaker. Unfilled LDPE had strong load-bearing capacity; engineering slope tests showed modulus variations. Pure LDPE had the lowest and graphene highest load-bearing capacity. Unexpectedly, polymer composites containing higher amounts of graphene exhibit reduced stress resistance. As the initial stress increases, the resistance to deformation also rises, leading to a stiffening of the material. A trade-off must be considered based on the composite material. As the concentration of the composite increases, GNPs are integrated within the polymer chains. The addition of GNP and other fillers to polymers results in the migration of polymer chains, leading to a reduction in stickiness and adhesiveness. An increase in GNP and MWCNT concentration causes micro-cracks in polymer nanocomposite.

Fig. 6.22 Elemental color mapping images of $ZrFe_2O_4$ [32]. Reprinted with permission from Elsevier

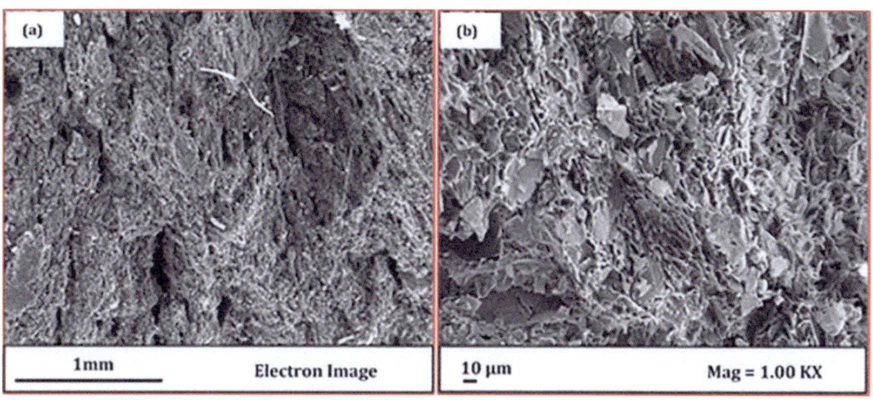

Fig. 6.23 Cross-sectional SEM of cryogenically fractured LDPE-based composite [32]. Reprinted with permission from Elsevier

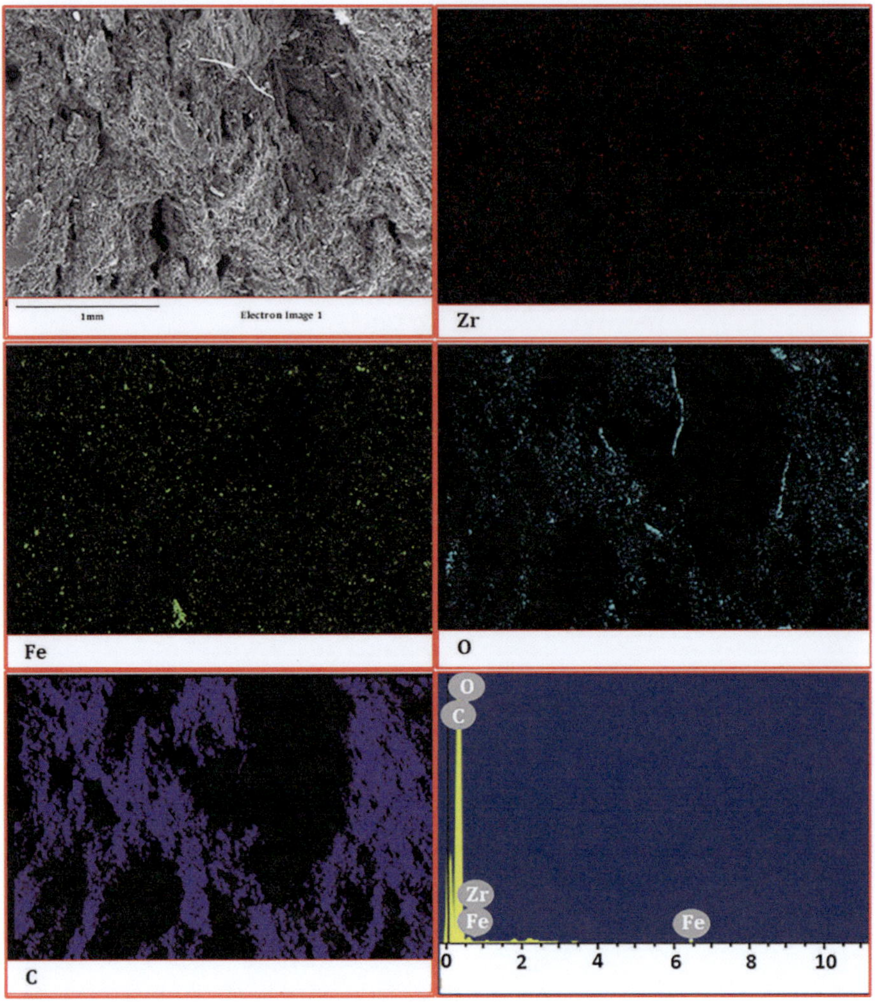

Fig. 6.24 Elemental mapping shows a composite sample surface with a 50:5:40:5 composition ratio [32]. Reprinted with permission from Elsevier

Ferrite nanoparticles require a smaller volume in the hybrid composite than GNP and MWCNT. The composite exhibits enhanced mechanical strength as the concentration of GNP diminishes and the amount of ferrite rises. Pure LDPE contains no fillers within its polymer chains. Consequently, the adhesiveness between polymers will be enhanced. All the polymer composite samples exhibited high filler loading, significant mechanical strength, and favorable X-band EMI SE results.

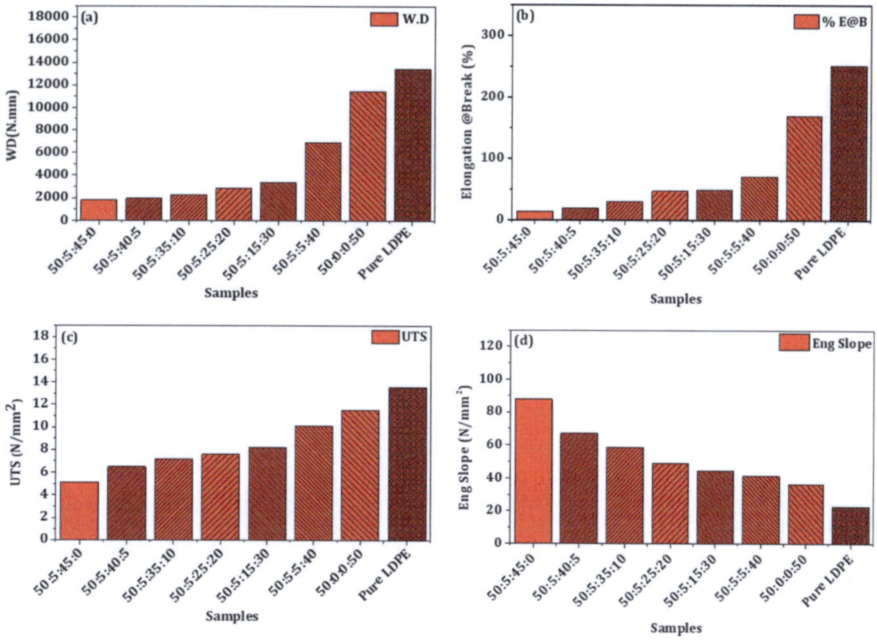

Fig. 6.25 Polymer composite properties include work done, elongation, and strength [32]. Reprinted with permission from Elsevier

6.12.5 Evaluating the Efficiency of EMI

Raising conductivity enhances SE value: $ZrFe_2O_4$, LDPE, graphene, and MWCNT blended in Brabender Plasticorder®. Adjusting graphene and ferrite ratios tested LDPE-MWCNT compatibility above 5% weight ratio. It shields electric and magnetic elements in EM wave directions perpendicular to each other. The efficiency of EM wave protection can be increased by adding X-band fillers with magnetic and dielectric losses. This study investigates the total shielding effectiveness (SE_T) in the X-band frequency for LDPE: MWCNT: GNP: $ZrFe_2O_4$ at various levels of "x" nanocomposites (0, 5, 10, 20, 30, 40, and 50 wt%) with a lateral thickness of 3.5 mm.

The SE_T graphs (Fig. 6.26) represent the cumulative SE_R and SE_A, as defined by (6.3) and (6.4). Figure 6.27 shows that the 50:5:40:5 sample has better shielding at 10.3 GHz and reaches 17.45 dB at 8.6 GHz. The specimen with a ratio of 50:5:45:0 shows a lowest SE of 9.65 dB and a highest SE of 21.41 dB at a frequency of 10.36 GHz. The setup with a ratio of 50:5:35:10 produces a top SE of 18.36 dB and a bottom of 7.4 dB. The sample 50:5:25:20 shows a minimum of 10.3- and 14.7-dB SE, while the remaining samples, 50:5:15:30 and 50:5:5:40, display maximum values of 6.5 and 5.09 dB, respectively.

In Fig. 6.27, SE_A plots are displayed. The plots matched with SE_A, indicating that SE_A has a key role in offering protection, whereas SE_R has a limited impact.

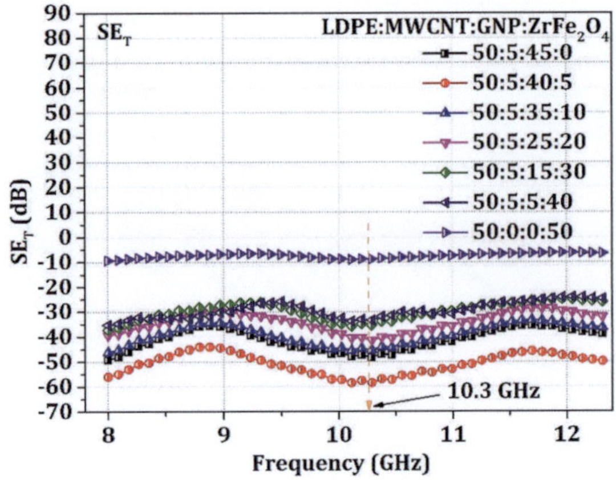

Fig. 6.26 The nanocomposite of LDPE, MWCNT, GNP, and $ZrFe_2O_4$ is a total shielding effectiveness (SE_T) [32]. Reprinted with permission from Elsevier

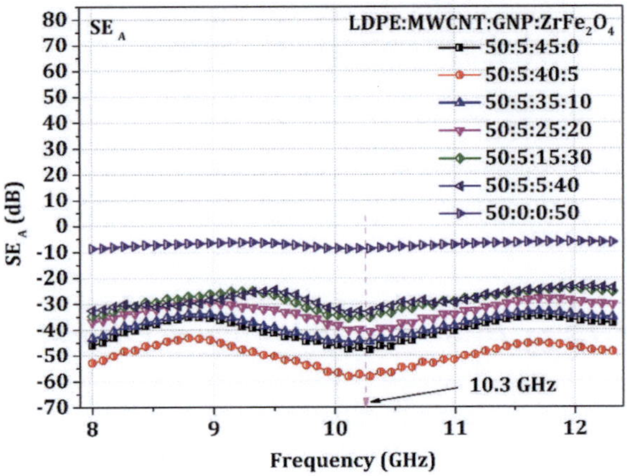

Fig. 6.27 Shielding effectiveness due to absorption (SE_A) of the LDPE: MWCNT: GNP: $ZrFe_2O_4$ nanocomposite [32]. Reprinted with permission from Elsevier

Absorption, rather than reflection, is the material's primary method for electromagnetic wave shielding. The sample containing 50% of one component, 5% of another, 40% of a third, and 5% of a fourth shows the lowest SE_A of 15.6 dB and the highest SE_A of 31.1 dB. The highest SE_A for sample 50:5:45:0 is 20.5 dB, while the lowest is 8.05 dB. Sample with 50:5:35:10 composition has max 17.77 dB SE_A, min 6.5 dB. The signal-to-noise ratio varies from 2.14 to 13.95 dB across the 50:5:25:20 sample. The sample with a proportion of 50:5:15:30 shows the lowest value of 0.71 dB and a highest of 7.57 dB. The sample containing 50% *x*, 5% *y*, 5% *z*, and 40% *w* varies from 0.5 to 3 decibels.

Figure 6.28 presents examples of SE_R graph polymer composites. SE_T will experience minimal impact from SE_R. The 50:5:40:5 sample exhibits a minimum SE_R of 0.42 dB and a maximum of 3.09 dB. The maximum SE_R for sample 50:5:45:0 is 3.09 dB, while the minimum is 0.42 dB. The SE_R values for the samples with ratios

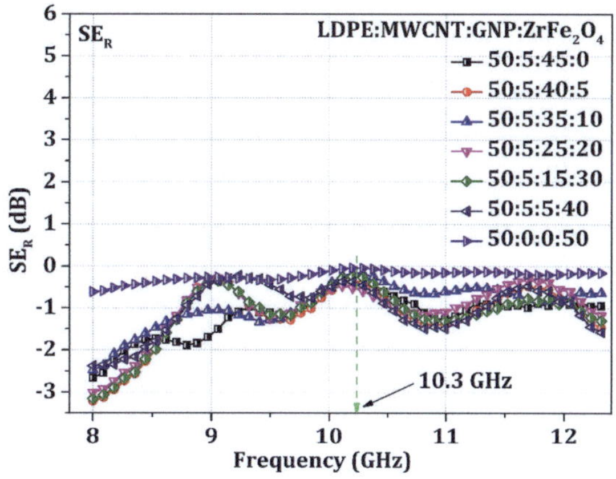

Fig. 6.28 Shielding effectiveness (SE_R) due to reflection of the LDPE: MWCNT: GNP: $ZrFe_2O_4$ nanocomposite [32]. Reprinted with permission from Elsevier

of 50:5:35:10, 50:5:25:20, 50:5:15:30, and 50:5:5:40 range from 0.24 to 3 dB, with varying minimum and maximum values.

Figure 6.29 displays a bar-and-column graph illustrating the Significance of Effectiveness (SE) in terms of both percentage and value. The graph shows the contributions of the Shielding Effectiveness Ratio (SE_R) and Shielding Effectiveness Attenuation (SE_A) to the overall SE_T. The sample with the highest SE_T value, 50:5:40:5, indicates that SE_A is more important than SE_R. A recent study found that a polymer blended with 5% MWCNT, GNP, and varying amounts of $ZrFe_2O_4$ nanoparticles can attenuate X-band electromagnetic waves. Sample 50:5:40:5 with conductive fillers shows MWCNT and GNP absorb EM waves efficiently. On the other hand, the second sample, including both conductive and magnetic fillers, shows improved performance due to the combined action of 5% ferrite nanoparticles, MWCNT, and GNP. Other composite samples have lower SE values, indicating ineffective filler synergy. Table 6.3 lists the frequency range, SE_T value, and standards for EMI shielding composites, aiding material design.

6.13 Conclusions

In this present study, LDPE: MWCNT: GNP: $ZrFe_2O_4$ nanocomposite material was improved and showed 40 dB or 99.997% shielding efficacy. Multiple losses from electromagnetic radiation contact with materials create the shielding effect. The composite's magnetic filler causes loss of magnetic, natural resonance, and eddy current. Conductive infill causes ohmic losses, dipole polarization, and dielectric relaxation. The fillers' interfacial polarization polarizes the polymer composite. Many conducting routes result from a capacitor effect and increased AC conductivity. The substance shields better when coupled synergistically at key.

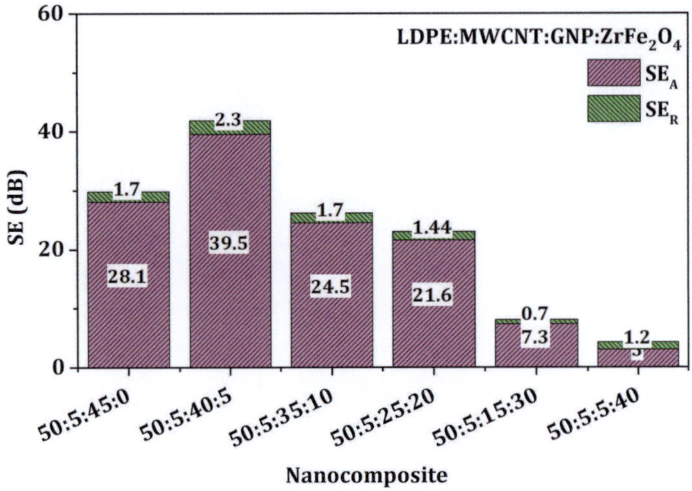

Fig. 6.29 The contribution of absorption and reflection to shielding effectiveness is shown in dB and the percentage for the LDPE nanocomposite [32]. Reprinted with permission from Elsevier

Table 6.3 Comparison of EMI shielding values when carbonaceous filler is used in preparing polymer composites [32]

S. No.	Materials	Frequency range GHz	SE_T value in dB	Reference
1	PVDF/$CoFe_2O_4$@CNTs	8.2–12.4	26.09	[26]
2	$CoFe_2O_4$-MWCNT-GNP-LDPE	8.2–12.4	49	[33]
3	rGO-$BaTiO_3$-ABS: rGO-ZnO-ABS	8.2–12.4	59.8/45.1	[100]
4	FeO@Graphite-MWCNT-GNP-LDPE	8.2–12.4	49	[103]
5	$LaFe_2O_4$-MWCNT-GNP-LDPE	8.2–12.4	40	[104]
6	α-Fe_2O_3-MWCNT-GNP-LDPE	8.2–12.4	40	[126]
7	$CoFe_2O_4$: Graphene: Epoxy (GrEpCoF)	8–18	22	[129]
8	$ZnFe_2O_4$: MWCNT	0.45–1.42	40.65	[130]
9	Gd2O3: Fe_3O_4-PC-MWCNTs	8–18	28/33	[131]
10	$MnFe_2O_4$-MWCNT-RGO-PVDF	8–18	38	[132]
11	$ZrFe_2O_4$-MWCNT-GNP-LDPE	8.2–12.4	40	[32]

Reprinted with permission from Elsevier

References

1. Kumar, R., Sahoo, S., Joanni, E., Singh, R.K., Tan, W.K., Kar, K.K., Matsuda, A.: Recent progress on carbon-based composite materials for microwave electromagnetic interference shielding. Carbon **177**, 304–331 (2021)

2. Liu, H., Yang, Y., Tian, N., You, C., Yang, Y.: Foam-structured carbon materials and composites for electromagnetic interference shielding: design principles and structural evolution. Carbon **217**, 118608 (2024)
3. Ali, A., Rahimian Koloor, S.S., Alshehri, A.H., Arockiarajan, A.: Carbon nanotube characteristics and enhancement effects on the mechanical features of polymer-based materials and structures—a review. J. Mater. Res. Technol. **24**, 6495–6521 (2023)
4. Fenta, E.W., Mebratie, B.A.: Advancements in carbon nanotube-polymer composites: Enhancing properties and applications through advanced manufacturing techniques. Heliyon **10**(16), e36490 (2024)
5. Liu, J.-T., Zheng, Y.-C., Hou, X., Feng, X.-R., Jiang, K., Wang, M.: Structured carbon for electromagnetic shielding and microwave absorption from carbonization of waste polymer: a review. Chem. Eng. J. **496**, 154013 (2024)
6. Sharma, S., Verma, A., Rangappa, S.M., Siengchin, S., Ogata, S.: Recent progressive developments in conductive-fillers based polymer nanocomposites (CFPNC's) and conducting polymeric nanocomposites (CPNC's) for multifaceted sensing applications. J. Market. Res. **26**, 5921–5974 (2023)
7. Wang, C., Murugadoss, V., Kong, J., He, Z., Mai, X., Shao, Q., Chen, Y., Guo, L., Liu, C., Angaiah, S., Guo, Z.: Overview of carbon nanostructures and nanocomposites for electromagnetic wave shielding. Carbon **140**, 696–733 (2018)
8. Ha, J.-H., Hong, S.-K., Ryu, J.-K., Bae, J., Park, S.-H.: Development of multi-functional graphene polymer composites having electromagnetic interference shielding and de-icing properties. Polymers **11**(12), 2101 (2019)
9. Kausar, A., Ahmad, I., Zhao, T., Aldaghri, O., Ibnaouf, K.H., Eisa, M.H., Lam, T.D.: Graphene nanocomposites for electromagnetic interference shielding—trends and advancements. J. Compos. Sci. **7**(9), 384 (2023)
10. Liu, H., Wu, S., You, C., Tian, N., Li, Y., Chopra, N.: Recent progress in morphological engineering of carbon materials for electromagnetic interference shielding. Carbon **172**, 569–596 (2021)
11. Das, P., Deoghare, A.B., Ranjan Maity, S.: Exploring the potential of graphene as an EMI shielding material—an overview. Mater. Today: Proc. **22**, 1737–1744 (2020)
12. Jovanović, S., Huskić, M., Kepić, D., Yasir, M., Haddadi, K.: A review on graphene and graphene composites for application in electromagnetic shielding. Graphene 2D Mater. **8**(3), 59–80 (2023)
13. Chen, J., Liu, Y.-L., Sun, D.-X., Qi, X.-D., Yang, J.-H., Wang, Y.: Recent progress in structural design of graphene/polymer porous composites toward electromagnetic interference shielding application. Chem. Eng. J. **495**, 153586 (2024)
14. Ata, S., Bukhari, S.A., Bibi, I., Mohsin, I.-U., Shoaib, M., Majid, F., Iqbal, M., Alshammari, F.H., Alfryyan, N., Alwadai, N.: Magnetite/graphene oxide/Prussian blue composite with robust effectiveness for electromagnetic interference shielding. Ceram. Int. **48**(2), 1690–1698 (2022)
15. Zhang, H., Zheng, X., Jiang, R., Liu, Z., Li, W., Zhou, X.: Research progress of functional composite electromagnetic shielding materials. Eur. Polymer J. **185**, 111825 (2023)
16. Cheng, Z., Wang, R., Wang, Y., Cao, Y., Shen, Y., Huang, Y., Chen, Y.: Recent advances in graphene aerogels as absorption-dominated electromagnetic interference shielding materials. Carbon **205**, 112–137 (2023)
17. Pavlou, C., Pastore Carbone, M.G., Manikas, A.C., Trakakis, G., Koral, C., Papari, G., Andreone, A., Galiotis, C.: Effective EMI shielding behaviour of thin graphene/PMMA nanolaminates in the THz range. Nat. Commun. **12**(1), 4655 (2021)
18. Wang, J., Liu, B., Cheng, Y., Ma, Z., Zhan, Y., Xia, H.: Constructing a segregated magnetic graphene network in rubber composites for integrating electromagnetic interference shielding stability and multi-sensing performance. Polymers **13**(19), 3277 (2021)
19. Wu, N., Hu, Q., Wei, R., Mai, X., Naik, N., Pan, D., Guo, Z., Shi, Z.: Review on the electromagnetic interference shielding properties of carbon based materials and their novel composites: recent progress, challenges and prospects. Carbon **176**, 88–105 (2021)

20. Huang, W., Mei, H., Yan, Y., Xia, J., Cheng, L.: Regulating synergistic micro-nano pore structure of CNT sponge to control EMI shielding performance. J. Market. Res. **29**, 1155–1164 (2024)
21. Al-Saleh, M.H., El-Methaly, H.M.: Influence of PLA/HDPE ratio and CNT content on the morphology, electrical, and EMI shielding of CNT-filled PLA/HDPE blends. Synth. Met. **304**, 117592 (2024)
22. Mei, H., Zhao, X., Gui, X., Lu, D., Han, D., Xiao, S., Cheng, L.: SiC encapsulated Fe@CNT ultra-high absorptive shielding material for high temperature resistant EMI shielding. Ceram. Int. **45**(14), 17144–17151 (2019)
23. Feng, D., Liu, P., Wang, Q.: Exploiting the piezoresistivity and EMI shielding of polyetherimide/carbon nanotube foams by tailoring their porous morphology and segregated CNT networks. Compos. A Appl. Sci. Manuf. **124**, 105463 (2019)
24. Sushmita, K., Formanek, P., Krause, B., Pötschke, P., Bose, S.: Distribution of carbon nanotubes in polycarbonate-based blends for electromagnetic interference shielding. ACS Appl. Nano Mater. **5**(1), 662–677 (2022)
25. Huang, B., Yue, J., Fan, B., Tang, X.-Z., Liu, Y., Huang, X.: Constructing hierarchical structure via in situ growth of CNT in SiO_2-coated carbon foam for high-performance EMI shielding application. Compos. Sci. Technol. **222**, 109372 (2022)
26. Kazmi, S.J., Nadeem, M., Warsi, M.A., Manzoor, S., Shabbir, B., Hussain, S.: PVDF/CFO-anchored CNTs ternary composite system with enhanced EMI shielding and EMW absorption properties. J. Alloy. Compd. **903**, 163938 (2022)
27. Javaria Kazmi, S., Nadeem, M., Younis, A., Loomba, S., Shabbir, B., Manzoor, S., Hussain, S.: PANI/CFO@CNTs ternary composite system for EMI shielding applications. J. Magn. Magn. Mater. **563**, 170037 (2022)
28. Zhang, Z., Huang, Y., Zhang, P., Wan, C.: Outstanding EMI shielding properties of Al_2O_3 reinforced with sandwiched graphene/CNTs. J. Eur. Ceram. Soc. **43**(9), 4082–4087 (2023)
29. Huang, B., Wang, Z., Tu, J., Liu, C., Xu, P., Ding, Y.: Interfacial distribution and compatibilization of imidazolium functionalized CNTs in poly(lactic acid)/polycaprolactone composites with excellent EMI shielding and mechanical properties. Int. J. Biol. Macromol. **227**, 1182–1190 (2023)
30. Fu, H., Bai, Y., Duan, S., Zhou, H., Gong, W.: Structure design of multi-layered ABS/CNTs composite foams for EMI shielding application with low reflection and high absorption characteristics. Appl. Surf. Sci. **624**, 157168 (2023)
31. Zhu, Y., Wang, Q., Han, Y., Li, L., Cao, M.: Constructing WSe2@CNTs heterojunction to tune attenuation capability for efficient microwave absorbing and green EMI shielding. Appl. Surf. Sci. **592**, 153253 (2022)
32. Praveen, M., Harichandra, B.P., Krishna, R.H., Kumar, M., Karthikeya, G.S., Swamy, H.R., Koul, S., Nagabhushana, B.M.: Enhanced EM shielding: synergetic effect of zirconium ferrite, MWCNT, and graphene in LDPE polymer composite to counteract electromagnetic radiation in the X-band range. Colloids Surf. A **699**, 134535 (2024)
33. Praveen, M., Karthikeya, G.S., Krishna, R.H., Mamatha, G.M., Manjunatha, C., Khosla, A., Nagabhushana, B.M.: The role of magnetic nano $CoFe_2O_4$ and conductive MWCNT/graphene in LDPE-based composites for electromagnetic interference shielding in X-band. Diam. Relat. Mater. **130**, 109501 (2022)
34. Li, Y., Yu, W., Ruan, Q., Li, K., Guo, X., Bai, Z., Chen, J.: Enhanced high-performance iPP/TPU/MWCNT nanocomposite for electromagnetic interference shielding. Polymers **16**(13), 1837 (2024)
35. Siva Kumar, O.V.P.R., Sundaramoorthy, A., Padmapriya, V.S., Raman, D.N.: Preparation of freestanding films from SWCNT/PANI nanocomposites using different blending techniques and characterization of their EMI shielding effectiveness in X-band. Phosphorus, Sulfur Silicon Relat. Elem. **197**(3), 209–217 (2021)
36. Ivanova, N.V., Vershinina, A.I., Lomakin, M.V., Russakov, D.M., Gordaya, O.R., Chirkova, I.M., Kosobutsky, A.V., Krasnikov, D.V., Chetyrkina, M.R., Nasibulin, A.G., Shandakov, S.D.: SWCNT fibers decorated with Au nanoparticles as voltammetric sensors for arsenic (III) determination. Microchem. J. **206**, 111482 (2024)

37. Cao, J., Zhang, Z., Ding, Y., Fan, J., Yu, Y., Dong, H., Yan, J., Li, H., Wang, Y., Liao, Y.: One-step aerosol synthesis of iron nanoparticles coated single-walled carbon nanotubes (Fe@SWCNT) for multifunctional composite textiles with electromagnetic interference shielding properties. J. Alloy. Compd. **986**, 174139 (2024)
38. Loghin, F., Rivadeneyra, A., Becherer, M., Lugli, P., Bobinger, M.: A facile and efficient protocol for preparing residual-free single-walled carbon nanotube films for stable sensing applications. Nanomaterials **9**(3), 471 (2019)
39. Yang, X., Luo, J., Xue, Y., Ren, H., Yang, Z., Yang, C., Zhang, J., Cai, Z., Liu, Y., Zhang, H., Yu, J.: MWCNT functionalized CF/PEKK composites with optimized EMI shielding and mechanical properties. Compos. Commun. **42**, 101679 (2023)
40. Petrychuk, M.V., Oliynyk, V.V., Zagorodnii, V.V., Ogurtsov, N.A., Pud, A.A.: PVDF/poly(3-methylthiophene)/MWCNT nanocomposites for EMI shielding in the microwave range. Heliyon **9**(12), e23101 (2023)
41. Jia, F., Lu, Z., Li, S., Zhang, J., Liu, Y., Wang, H., Xu, X., Du, A., Guo, D., Yan, N.: Asymmetric c-MWCNT/AgNWs/PANFs hybrid film constructed by tailoring conductive-blocks strategy for efficient EMI shielding. Carbon **217**, 118600 (2024)
42. Pal, R., Goyal, S.L., Rawal, I., Gupta, A.K.: Tailoring of EMI shielding properties of polyaniline with MWCNTs embedment in X-band (8.2–12.4 GHz). J. Phys. Chem. Solids **169**, 110867 (2022)
43. Pączkowski, P., Sigareva, N.V., Gorelov, B.M., Terets, M.I., Sementsov, Y.I., Kartel, M.T., Gawdzik, B.: The influence of carbon nanotubes on the physical and chemical properties of nanocomposites based on unsaturated polyester resin. Nanomaterials **13**(23), 2981 (2023)
44. Jang, D., Kim, B.-J., Nam, I.-W.: A comprehensive study on EMI shielding performance of carbon nanomaterials-embedded CFRP or GFRP composites. Polymers **14**(23), 5224 (2022)
45. Vidyashree, M.P., Sushmita, K., Nagarajan, P., Kokila, M.K., Bose, S.: Mimicking 'sea-urchin' like heirarchical carbon structures self-assembled from carbon fibers for green EMI shielding. Chem. Eng. J. Adv. **13**, 100430 (2023)
46. Patle, V.K., Mehta, Y., Kumar, R.: Nickel and iron nanoparticles decorated carbon fibers reinforced phenolic resin-based carbon composites foam for excellent electromagnetic interference shielding. Diam. Relat. Mater. **145**, 111069 (2024)
47. Ma, Y., Zhuang, Y., Li, C., Shen, X., Zhang, L.: Improving electromagnetic interference shielding while retaining mechanical properties of carbon fiber-based composites by introducing carbon nanofiber sheet into laminate structure. Polymers **14**(9), 1658 (2022)
48. Damodaren, P., Rengaraj, R., Rohit, D., Anand, H., Venkatakrishnan, G.R.: Performance comparison of copper and carbon nano fibre braiding in cable data transfer application. Mater. Today: Proc. **72**, 3012–3016 (2023)
49. Hu, Y., Jiang, M., Cong, X., Liu, G., Yi, X., Liu, X.: Lightweight, multifunctional recycled carbon fibre/MXene/PEDOT: PSS nonwoven veils with double-layered structure for excellent electromagnetic interference shielding. Chem. Eng. J. **489**, 151122 (2024)
50. Sharma, G.K., James, N.R.: Carbon black incorporated carbon nanofiber based polydimethylsiloxane composite for electromagnetic interference shielding. Carbon Trends **8**, 100177 (2022)
51. Kanwal, R., Maqsood, M.F., Raza, M.A., Inam, A., Waris, M., Rehman, Z.U., Mehdi, S.M.Z., Abbas, N., Lee, N.: Polypyrrole coated carbon fiber/magnetite/graphene oxide reinforced hybrid epoxy composites for high strength and electromagnetic interference shielding. Mater. Today Commun. **38**, 107684 (2024)
52. Li, X., Xu, T., Cao, W., Wang, M., Chen, F., Jin, L., Song, N., Sun, S., Ding, P.: Graphene/carbon fiber network constructed by co-carbonization strategy for functional integrated polyimide composites with enhanced electromagnetic shielding and thermal conductive properties. Chem. Eng. J. **464**, 142595 (2023)
53. Jiang, D., Tian, S., Li, H., Du, Z., Liu, T., Yan, D., Zhou, L., Bai, S., Qiang, X.: Lightweight HfC nanowire-carbon fiber/graphene aerogel composites for high-efficiency electromagnetic interference shielding. Carbon **219**, 118788 (2024)

54. Kausar, A.: Chapter 5—Graphene quantum dots, graphene nanoplatelets, and graphene nanoribbons with polymers. In: Kausar, A. (ed.) Graphene to Polymer/Graphene Nanocomposites, pp. 91–116 (2022)
55. Rani, P., Ahamed, B., Deshmukh, K.: Electromagnetic interference shielding properties of graphene quantum-dots reinforced poly(vinyl alcohol)/polypyrrole blend nanocomposites. J. Appl. Polym. Sci. **137**(45), 49392 (2020)
56. El-Shamy, A.G.: Polyvinyl alcohol and silver decorated carbon quantum-dots for new nanocomposites with application electromagnetic interface (EMI) shielding. Prog. Org. Coat. **146**, 105747 (2020)
57. Qi, H., Teng, M., Liu, M., Liu, S., Li, J., Yu, H., Teng, C., Huang, Z., Liu, H., Shao, Q., Umar, A., Ding, T., Gao, Q., Guo, Z.: Biomass-derived nitrogen-doped carbon quantum dots: highly selective fluorescent probe for detecting Fe^{3+} ions and tetracyclines. J. Colloid Interface Sci. **539**, 332–341 (2019)
58. Al Naim, A.F., Ibrahim, S.S., El-Shamy, A.G.: A new class of electromagnetic shields based on carbon dots adorning Te nanorods embedded into PEDOT:PSS for protection from electromagnetic (EM) pollutions. Prog. Org. Coat. **161**, 106509 (2021)
59. El-Shamy, A.G.: Novel conducting PVA/Carbon quantum dots (CQDs) nanocomposite for high anti-electromagnetic wave performance. J. Alloy. Compd. **810**, 151940 (2019)
60. Ramana, L.N., Dinh, L.N.M., Agarwal, V.: Influence of surface charge of graphene quantum dots on their uptake and clearance in melanoma cells††Electronic supplementary information (ESI) available. Nanoscale Adv. **3**(12), 3513–3521 (2021). https://doi.org/10.1039/d0na00935k
61. Alshahrani, H., Arun Prakash, V.R.: Development of highly flexible electromagnetic interference shielding composites for electronic applications using Cobalt/Hevea brasiliensis seed husk carbon dots/Bamboo microfibre-polyvinyl alcohol. Ind. Crops Prod. **191**, 115967 (2023)
62. Radha Krushna, B.R., Sharma, S.C., Srinivasan, A.R., Sahu, S., Ponnazhagan, K., George, A., Govardhan Rathla, K.S., Manjula, M.V., Shivakumar, V., Devaraja, S., Veera Vanitha, D., Nagabhushana, H.: Carbon dots as a distinctive platform fabricated through a sustainable approach for versatile applications. Colloids Surf. A **703**, 135135 (2024)
63. Anpalagan, K., Yin, H., Cole, I., Zhang, T., Lai, D.T.H.: Quantum yield enhancement of carbon quantum dots using chemical-free precursors for sensing Cr (VI) Ions. Inorganics **12**(4), 96 (2024)
64. Bahramian, R, Nezafati, M.B., Hamed Aboutalebi, S.: Recent progress and prospects of two-dimensional materials for electromagnetic interference shielding. FlatChem **47**, 100722 (2024)
65. Paramanik, B., Das, D.: The effect of CO_2 addition to the ($C_2H_2 + H_2$) gas system on the low-temperature growth of diamond-like carbon (DLC) with prominent nano-diamond phase. Mater. Today: Proc. **62**, 5057–5060 (2022)
66. Fanetti, S., Romi, S., Crichton, W., Rosenthal, A., Scelta, D., Alabarse, F., Bini, R., Santoro, M.: Quasi-isotropic high pressure, large volume synthesis of a polymeric composite incorporating diamond-like carbon nano-threads. Diam. Relat. Mater. **136**, 109912 (2023)
67. Liu, X.-S., Fu, Q.-G., Wang, H., Song, Q.: Improvement of mechanical property and ablation resistance of carbon/carbon composites by adding nano-diamond particles. Diam. Relat. Mater. **104**, 107751 (2020)
68. Su, Q., Shi, W., Li, D., Lai, J., Jiang, L., Wang, L., Ding, W., Xia, Y.: Effects of carbon concentration on properties of nano-diamond films. Appl. Surf. Sci. **258**(10), 4645–4648 (2012)
69. Grudinkin, S.A., Bogdanov, K.V., Tolmachev, V.A., Baranov, M.A., Kaliya, I.E., Golubev, V.G., Baranov, A.V.: Multifunctional core/shell diamond nanoparticles combining unique thermal and light properties for future biological applications. Nanomaterials **13**(24), 3124 (2023)
70. Katheria, A., Das, P., Nayak, J., Roy, B., Pal, A., Biswas, S., Das, N.C.: MXene and Fe_3O_4 decorated g-C_3N_4 incorporated high flexible hybrid polymer composite for enhanced electrical conductivity, EMI shielding and thermal conductivity. Next Mater. **6**, 100292 (2025)

71. Hussain, I., Yesmin, S.: Enhanced microwave absorption of g-C_3N_4/poly(vinylidene difluoride)/carbon black composites. Mater. Chem. Phys. **309**, 128337 (2023)
72. Anand, S., Muniyappan, S., Racik, K.M., Manikandan, A., Mani, D., Nandhini, S., Karuppasamy, P., Pandian, M.S., Ramasamy, P., Chandar, N.K.: Fabrication of binary to quaternary PVDF based flexible composite films and ultrathin sandwich structured quaternary PVDF/CB/g-C_3N_4/$BaFe_{11}$.5Al0.5O19 composite films for efficient EMI shielding performance. Synth. Met. **291**, 117199 (2022)
73. Zhou, Q., Qian, K., Fang, J., Miao, M., Cao, S., Feng, X.: UV-light modulated Ti_3C_2Tx MXene/g-C_3N_4 heterojunction film for electromagnetic interference shielding. Compos. A Appl. Sci. Manuf. **134**, 105899 (2020)
74. Katheria, A., Das, P., Bhagat, A., Nayak, J., Paul, S., Das, N.C.: Fe_3O_4@g-C_3N_4 and MWCNT embedded highly flexible polymeric hybrid composite for simultaneous thermal control and suppressing microwave radiation. J. Alloy. Compd. **988**, 174287 (2024)
75. Li, K., Chen, M., Chen, L., Zhao, S., Xue, W., Han, Z., Han, Y.: Synthesis of g-C_3N_4 derived from different precursors for photodegradation of sulfamethazine under visible light. Processes **11**(2), 528 (2023)
76. Dhall, S., Nathawat, R., Sood, K.: Chapter 1—carbon-based nanomaterials. In: Dhall, S. (ed.) Carbon Nanomaterials and Their Nanocomposite-Based Chemiresistive Gas Sensors, pp. 3–39: Elsevier (2023)
77. Jayalakshmy, M.S., Mishra, R.K.: Chapter fourteen—applications of carbon-based nanofiller-incorporated rubber composites in the fields of tire engineering, flexible electronics and EMI shielding. In: Yaragalla, S., Mishra, R.K., Thomas, S., Kalarikkal, N., Maria, H.J. (eds.) Carbon-Based Nanofillers and Their Rubber Nanocomposites, pp. 441–472. Elsevier (2019)
78. Kausar, A.: Potential of polymer/fullerene nanocomposites for anticorrosion applications in the biomedical field. J. Compos. Sci. **6**(12), 394 (2022)
79. Ford, W.T.: Fullerene-containing polymers. In: Mark, J.E. (ed.) Polymer Data Handbook: Second Edition. Oxford University Press (2009)
80. Rodrigues, D.J.L., Santos, L.M.N.B.F., Melo, A., Lima, C.F.R.A.C.: Theoretical study on the diels-alder reaction of fullerenes: analysis of isomerism, aromaticity, and solvation. Organics **3**(4), 364–379 (2022)
81. Govindasamy, T., Mathew, N.K., Asapu, V.K., Subramanian, V., Subramanian, B.: Investigation on evaluation of Fe_3S_4–Carbon black nanohybrids for EMI shield in X-band region. Diam. Relat. Mater. **131**, 109608 (2023)
82. Habibi, N., Pourjavadi, A.: Superhydrophobic and thermally conductive carbon black/hexagonal boron nitride@Fe_3O_4/cellulose composite paper for electromagnetic interference shielding. Synth. Met. **285**, 117008 (2022)
83. Kruželák, J., Kvasničáková, A., Hložeková, K., Dosoudil, R., Gořalík, M., Hudec, I.: Electromagnetic interference shielding and physical-mechanical characteristics of rubber composites filled with manganese-zinc ferrite and carbon black. Polymers **13**(4), 616 (2021)
84. Lee, J., Cho, S.-B., Dimitrov, K.C., Lee, Y., Khang, D.-Y.: Conducting polymer hollow nanostructures by surfactant-free Ouzo emulsion for an exceptional EMI shielding performance. Eur. Polymer J. **206**, 112771 (2024)
85. Rashid, A.B., Haque, M., Islam, S.M.M., Uddin Labib, K.M.R.: Nanotechnology-enhanced fiber-reinforced polymer composites: recent advancements on processing techniques and applications. Heliyon **10**(2), e24692 (2024)
86. Chhetri, S., Kuila, T.: Polymer composites with 3D graphene architectures as high-performance EMI shielding materials: a review. RSC Appl. Polym. **2**(4), 507–533 (2024)
87. Díez-Pascual, A.M., Naffakh, M., Marco, C., Ellis, G.: Mechanical and electrical properties of carbon nanotube/poly(phenylene sulphide) composites incorporating polyetherimide and inorganic fullerene-like nanoparticles. Compos. A Appl. Sci. Manuf. **43**(4), 603–612 (2012)
88. Chakraborty, T., Saha, S., Gupta, K., Dutta, S., Mahapatra, A.S., Mondal, D., Pradhan, A.K., Chakraborty, C., Das, S., Sutradhar, S.: An effective microwave absorber using multi-layer design of carbon allotrope based Co2Z hexaferrite-polymer nanocomposite film for EMI shielding applications. Chem. Eng. J. **487**, 150323 (2024)

89. Akbar, M.I., Armynah, B., Tahir, D.: Sandwich composite structure (SCS): review of wood-based materials with polymers and metals as a filler for electromagnetic interference (EMI) shielding. Ind. Crops Prod. **215**, 118619 (2024)
90. Kumari, S., Dalal, J., Kumar, A., Pal, R., Chahal, R., Ohlan, A.: Enhanced microwave absorption properties of conducting polymer@graphene composite to counteract electromagnetic radiation pollution: green EMI shielding††Electronic supplementary information (ESI) available. RSC Adv. **14**(1), 662–676 (2024). https://doi.org/10.1039/d3ra07245b
91. Choudhury, S.N., Das, P., Bhawal, P., Pal, A., Banerji, P., Das, N.C.: Double percolation behavior through the preferential distribution of conductive black in polymer blends to boost electrical properties and EMI shielding effectiveness. Mater. Today Commun. **35**, 106109 (2023)
92. Shakir, H.M.F., Zhao, T., Zubair, K., Sarwar, M.K., Rehan, Z.A., Fatima, K., Aziz, H.R.: Fabrication and EMI (electromagnetic interference) shielding performance of polyester fabrics coated with polyaniline via in-situ polymerization. Mater. Today Commun. **37**, 106971 (2023)
93. Calheiros Souto, L.F., Soares, B.G.: Electromagnetic wave absorption, EMI shielding effectiveness and electrical properties of ethylene—vinyl Acetate (EVA)/Polyaniline (PAni) blends prepared by in situ polymerization. Synth. Met. **298**, 117441 (2023)
94. Chauhan, S., Nikhil Mohan, P., Raju, K.C.J., Ghotia, S., Dwivedi, N., Dhand, C., Singh, S., Kumar, P.: Free-standing polymer/multiwalled carbon nanotubes composite thin films for high thermal conductivity and prominent EMI shielding. Colloids Surf. A **673**, 131811 (2023)
95. Bai, Y., Zhang, B., Fei, G., Ma, Z.: Composite polymeric film for stretchable, self-healing, recyclable EMI shielding and Joule heating. Chem. Eng. J. **478**, 147382 (2023)
96. Al-Saleh, M.H., Al-Sharman, M.M.: Influence of carbon nanofiller geometry on EMI shielding and electrical percolation behaviors of polymer composites. Synth. Met. **294**, 117314 (2023)
97. Nimra, S.S., Rehan, Z.A., Ali, S.H., Atir, S., Fatima, K., Shahzadi, F., Shakir, H.M.F., Alamir, M.A., Ali El-Bagory, T.M.A., Shahid, I.: Electrically conductive fibers fabrication and characterization via in-situ polymerization of aniline for the protection against EMI and thermal imaging signals. J. Mater. Res. Technol. **23**, 2399–2409 (2023)
98. Ma, L., Hamidinejad, M., Wei, L., Zhao, B., Park, C.B.: Absorption-dominant EMI shielding polymer composite foams: microstructure and geometry optimization. Mater. Today Phys. **30**, 100940 (2023)
99. Hong, J., Kwon, J., Im, D., Ko, J., Nam, C.Y., Yang, H.G., Shin, S.H., Hong, S.M., Hwang, S.S., Yoon, H.G., Lee, A.S.: Best practices for correlating electrical conductivity with broadband EMI shielding in binary filler-based conducting polymer composites. Chem. Eng. J. **455**, 140528 (2023)
100. Abhilash, G.P., Sushmita, K., Bose, S., Shivakumara, C.: Functionalization of dielectric $BaTiO_3$ and semiconducting ZnO nanoparticles on rGO layers and their polymer composites: applications in EMI shielding. Synth. Met. **297**, 117387 (2023)
101. Madhusudhan, C.K., Faisal, M., Maruthi, N., Raghavendra, N., Mahendra, K., Abdul Kadar, C.H.: 24—functionalized conducting polymer nanocomposites for EMI shielding applications. In: Patel, G., Deshmukh, K., Hussain, C.M. (eds.) Advances in Functionalized Polymer Nanocomposites, pp. 959–1000: Woodhead Publishing (2024)
102. Choudhary, H.K., Pawar, S.P., Kumar, R., Anupama, A.V., Bose, S., Sahoo, B.: Mechanistic insight into the critical concentration of barium hexaferrite and the conductive polymeric phase with respect to synergistically electromagnetic interference (EMI) shielding. **2**(2), 830–841 (2017)
103. Basa Reddy, S., Praveen, M., Basavarajappa, T., Bapu Tandale, V., Pachipulusu, R., Hegde, S.: EMI shielding performance of electroless coated iron nanoparticles on graphite in low-density polyethylene composite for X-band applications. J. Indian Chem. Soc. **100**(4), 100962 (2023)
104. Madhu Kumar, T.M., Kini, H.J., Praveen, M., Kumar, M.: Electromagnetic interference shielding performance of lanthanum ferrite with MWCNT and graphene in the polyethylene polymer matrix in X-band frequency. Diam. Relat. Mater. **141**, 110701 (2024)
105. Ravindran, A.R., Feng, C., Huang, S., Wang, Y., Zhao, Z., Yang, J.: Effects of graphene nanoplatelet size and surface area on the AC electrical conductivity and dielectric constant of epoxy nanocomposites. **10**(5), 477 (2018)

106. Karimi Bakhtiyar, N., Javadpour, R., Zeinali Heris, S., Mohammadpourfard, M.: Improving the thermal characteristics of a cooling tower by replacing the operating fluid with functionalized and non-functionalized aqueous MWCNT nanofluids. Case Stud. Therm. Eng. **39**, 102422 (2022)
107. Yao, L., Cao, W., Zhao, J., Zheng, Q., Wang, Y., Jiang, S., Pan, Q., Song, J., Zhu, Y., Cao, M.: Regulating bifunctional flower-like $NiFe_2O_4$/graphene for green EMI shielding and lithium ion storage. J. Mater. Sci. Technol. **127**, 48–60 (2022)
108. Rengaswamy, K., Asapu, V.K., Sundara, R., Venkatachalam, S.: Effective attenuation of electromagnetic waves by Ag adorned MWCNT-polybenzoxazine composites for EMI shielding application. Compos. Sci. Technol. **223**, 109411 (2022)
109. Fu, H., Chen, L., Liu, D., Zhang, Y., Cao, Y., Wu, C., Yong, Z., Yu, Y., Li, Q.: Multifunctional NiCo@RGO/SWNTs foam with oriented pore structure for excellent electromagnetic interference shielding. Chem. Eng. J. **454**, 140324 (2023)
110. Mani, D., Vu, M.C., Lim, C.-S., Kim, J.-B., Jeong, T.-H., Kim, H.J., Islam, M.A., Lim, J.-H., Kim, K.-M., Kim, S.-R.: Stretching induced alignment of graphene nanoplatelets in polyurethane films for superior in-plane thermal conductivity and electromagnetic interference shielding. Carbon **201**, 568–576 (2023)
111. Choudhary, H.K., Kumar, R., Pawar, S.P., Sundararaj, U., Sahoo, B.: Effect of morphology and role of conductivity of embedded metallic nanoparticles on electromagnetic interference shielding of PVDF-carbonaceous-nanofiller composites. Carbon **164**, 357–368 (2020)
112. Suthar, M., Roy, P.K.: Structural, electromagnetic, and Ku-band absorption characterization of La-Mg substituted Y-type barium hexaferrite for EMI shielding application. Mater. Sci. Eng. B **283**, 115801 (2022)
113. Ryu, S.H., Han, Y.K., Kwon, S.J., Kim, T., Jung, B.M., Lee, S.-B., Park, B.: Absorption-dominant, low reflection EMI shielding materials with integrated metal mesh/TPU/CIP composite. Chem. Eng. J. **428**, 131167 (2022)
114. Yadav, R.S., Kuřitka, I., Vilcakova, J., Machovsky, M., Skoda, D., Urbánek, P., Masař, M., Jurča, M., Urbánek, M., Kalina, L., Havlica, J.: $NiFe_2O_4$ nanoparticles synthesized by dextrin from corn-mediated sol-gel combustion method and its polypropylene nanocomposites engineered with reduced graphene oxide for the reduction of electromagnetic pollution. ACS Omega **4**(26), 22069–22081 (2019)
115. Rohini, R., Bose, S.: Extraordinary improvement in mechanical properties and absorption-driven microwave shielding through epoxy-grafted graphene "Interconnects." ACS Omega **3**(3), 3200–3210 (2018)
116. Menon, A.V., Madras, G., Bose, S.: Mussel-inspired self-healing polyurethane with "Flower-like" magnetic MoS_2 as efficient microwave absorbers. ACS Appl. Polym. Mater. **1**(9), 2417–2429 (2019)
117. Moazen, S., Sahebian, S., Haddad-Sabzevar, M.: Low percolation behavior of HDPE/CNT nanocomposites for EMI shielding application: random distribution to segregated structure. Synth. Met. **281**, 116900 (2021)
118. Kazakova, M.A., Semikolenova, N.V., Korovin, E.Y., Zhuravlev, V.A., Selyutin, A.G., Velikanov, D.A., Moseenkov, S.I., Andreev, A.S., Lapina, O.B., Suslyaev, V.I., Matsko, M.A., Zakharov, V.A., Lacaillerie, J.-B.E.: Co/multi-walled carbon nanotubes/polyethylene composites for microwave absorption: tuning the effectiveness of electromagnetic shielding by varying the components ratio. Compos. Sci. Technol. **207**, 108731 (2021)
119. Zhang, Y., Pan, T., Yang, Z.: Flexible polyethylene terephthalate/polyaniline composite paper with bending durability and effective electromagnetic shielding performance. Chem. Eng. J. **389**, 124433 (2020)
120. Li, S., Huang, A., Chen, Y.-J., Li, D., Turng, L.-S.: Highly filled biochar/ultra-high molecular weight polyethylene/linear low density polyethylene composites for high-performance electromagnetic interference shielding. Compos. B Eng. **153**, 277–284 (2018)
121. Mondal, S., Ravindren, R., Bhawal, P., Shin, B., Ganguly, S., Nah, C., Das, N.C.: Combination effect of carbon nanofiber and ketjen carbon black hybrid nanofillers on mechanical, electrical, and electromagnetic interference shielding properties of chlorinated polyethylene nanocomposites. Compos. B Eng. **197**, 108071 (2020)

122. Tang, W., Lu, L., Xing, D., Fang, H., Liu, Q., Teh, K.S.: A carbon-fabric/polycarbonate sandwiched film with high tensile and EMI shielding comprehensive properties: an experimental study. Compos. B Eng. **152**, 8–16 (2018)
123. Jin, X.-Z., Yang, Z.-Y., Huang, C.-H., Yang, J.-H., Wang, Y.: PEDOT:PSS/MXene/PEG composites with remarkable thermal management performance and excellent HF-band & X-band electromagnetic interference shielding efficiency for electronic packaging. Chem. Eng. J. **448**, 137599 (2022)
124. Alaferdov, A.V., Lebedev, O.V., Roggero, U.F.S., Hernandez-Figueroa, H.E., Nista, S.V.G., Trindade, G.M., Danilov, Y.A., Ozerin, A.N., Moshkalev, S.A.: Highly conductive nanographite/ultra-high-molecular-weight polyethylene composite. Results Mater. **15**, 100298 (2022)
125. Mamunya, Y., Matzui, L., Vovchenko, L., Maruzhenko, O., Oliynyk, V., Pusz, S., Kumanek, B., Szeluga, U.: Influence of conductive nano- and microfiller distribution on electrical conductivity and EMI shielding properties of polymer/carbon composites. Compos. Sci. Technol. **170**, 51–59 (2019)
126. Manjappa, P., Rajan, H.K., Mahesh, M.G., Sadananda, K.G., Channegowda, M., Shivashankar, G.K., Mutt, N.B.: Effective attenuation of electromagnetic waves by synergetic effect of & alpha;-Fe_2O_3 and MWCNT/graphene in LDPE-based composites for EMI applications. **15**(24), 9006 (2022)
127. Biswas, S., Kar, G.P., Bose, S.: Microwave absorbers designed from PVDF/SAN blends containing multiwall carbon nanotubes anchored cobalt ferrite via a pyrene derivative. J. Mater. Chem. A **3**(23), 12413–12426 (2015)
128. Pawar, S.P., Marathe, D.A., Pattabhi, K., Bose, S.: Electromagnetic interference shielding through MWNT grafted Fe_3O_4 nanoparticles in PC/SAN blends. J. Mater. Chem. A **3**(2), 656–669 (2015)
129. Manobalan, S., Bose, S., Sumangala, T.P.: Effect of thickness on the EMI shielding effectiveness of epoxy composites with cobalt ferrite and graphene. Mater. Today: Proc. **90**, 128–132 (2023)
130. Tang, Y., Yin, P., Zhang, L., Wang, J., Feng, X., Wang, K., Dai, J.: Novel carbon encapsulated zinc ferrite/MWCNTs composite: preparation and low-frequency microwave absorption investigation. Ceram. Int. **46**(18, Part A), 28250–28261 (2020)
131. Sushmita, K., Menon, A.V., Sharma, S., Abhyankar, A.C., Madras, G., Bose, S.: Mechanistic insight into the nature of dopants in graphene derivatives influencing electromagnetic interference shielding properties in hybrid polymer nanocomposites. J. Phys. Chem. C **123**(4), 2579–2590 (2019)
132. Srivastava, R.K., Xavier, P., Gupta, S.N., Kar, G.P., Bose, S., Sood, A.K.: Excellent electromagnetic interference shielding by graphene- $MnFe_2O_4$-multiwalled carbon nanotube hybrids at very low weight percentage in polymer matrix. **1**(18), 5995–6003 (2016)

Chapter 7
Polymers for Electromagnetic Shielding

Abstract The high shielding efficiency (SE) of high electrical conductivity metals and their alloys makes them appropriate as shielding materials. Their weight, stiffness, and corrodibility have driven studies on substitutes for metals. Therefore, EMI shielding materials must be lightweight, thin, and fabrication-friendly. The material's biodegradability and environmental friendliness will also help develop sustainable engineering products. Polymeric materials and composites possess light weight and corrosion resistance, other favorable properties that aid EMI shielding. This chapter offers a comprehensive review of polymeric and composite shielding materials. Nanofiller-infused polymers for the design of novel EMI shielding material are also explored in this chapter.

7.1 Introduction

Our modern lifestyles demand more electromagnetic (EM) energy to support complex and compact systems across various industrial, military, aerospace, and commercial sectors [1]. The rise in electromagnetic pollution can significantly disrupt electrical devices and present hazards to nearby humans. High-quality shielding materials are essential to creating efficient electrical systems. Addressing EMI issues involves optimizing lossy dielectric and magnetic materials to effectively absorb or reflect EM radiation from electronic devices [2]. Shielding materials focusing on absorption is generally more beneficial for devices than those that reflect, as reflected waves can interfere with other nearby devices [3, 4].

Absorbing materials must be thin, lightweight, of low-density and, wide bandwidth, and durable [5]. An EMI shield is designed with a polymer base that can effectively absorb EM energy. Recent research indicates a rising demand for cost-effective microwave absorbent materials (MAMs), driven by the increased use of microwave-frequency devices and systems in military applications [6].

Many studies have highlighted the development of lightweight, thin, non-corrosive, and efficient absorbers suitable for a wide range of microwave frequencies. Research shows that effective EMI shielding materials can be created using metals,

S. K. Koul et al., *Microwave Absorbers*, Engineering Materials,
https://doi.org/10.1007/978-981-96-9596-6_7

metal oxides, and conducting polymer-based composites [7]. Significant findings regarding the contributions of polyaniline (PANI) composites, polymer blends with different metals and metal oxides, polypyrrole (PPy) composites, and polythiophene (PTh) composites are found in the literature.

Insulative polymers are also essential in this field since they have good thermal stability, mechanical properties, remolding ability, corrosion resistance, and conductivity. Insulative polymers will be tuned into conductive nanocomposites by incorporating some conducting nanofillers. This chapter comprehensively explores various composites for electromagnetic interference (EMI) shielding [8]. Not only conducting polymers but many lossy materials incorporated in the polymers have an advantage over electrically conductive fillers. Fillers are involved in many losses like dielectric loss, ohmic loss, magnetic loss, heat loss, and eddy current loss. Usually, semiconducting and ceramic nanomaterials are involved in this type of EM wave attenuation.

Metal-based shields are modified into various structural forms to overcome these issues. However, there is a growing need for flexible, adjustable, non-corrosive, and cost-effective shields made from conducting polymers [9]. Compared to various composites and polymer composite materials designed for EMI shielding, conducting polymer-based composites show considerable potential. The increasing demand for these composites in EMI shielding and microwave attenuation across different industries highlights the need for research to explore recent advancements in this area [10].

7.2 Classification of Polymers Based on Electrical Conductivity

Polymers can be classified using various criteria. This classification is essential to ensuring that each polymer is placed in the correct category based on its suitability for different industrial applications. Polymers are generally divided into two main types: natural and synthetic. They can also be further classified as biodegradable or nonbiodegradable [11]. Additionally, polymers can be categorized as organic or inorganic, and they may be classified as either linear chain or branched chain polymers. Lastly, they can be identified as thermoplastic or thermoset plastics. Ionic conductive polymers are essential for ion conduction in electrolytic solutions and are commonly used in battery membranes and conductivity cells. The ability of polymers to transfer electrical charge makes them functional polymers [12]. In Fig. 7.1 classification of conductive polymers is presented [13, 14].

Polymers are the pure medium of hydrocarbons, which have some unique properties; however, polymers do not apply to EMI shielding. Industrially, relevant nanomaterials would be added as fillers to design a polymer nanocomposite, which is functional for EMI shielding.

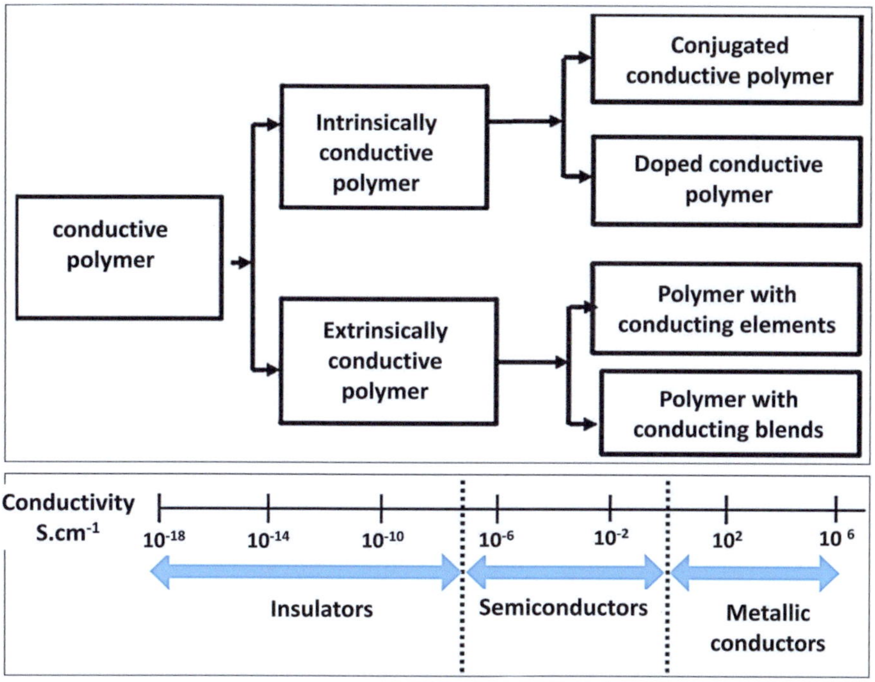

Fig. 7.1 Classification of conductive polymers [13, 14]

Polymer composite preparation involves several techniques, such as melt mixing, in situ polymerization, and solution casting. One reason the result of this research is preferable to metal screens is that the conducting polymers and the nanocomposites from them have high SE. The conductivity of insulative polymers can be altered by creating composites that include conductive fillers, semiconducting fillers like metallic nanoparticles, and carbon-based fillers such as graphene, multi-walled carbon nanotubes (MWCNT), quantum dots, and MXenes [15].

Polymers exhibit a broad spectrum of electrical properties, including insulation, dielectric behavior, semiconductor properties, photoconductivity, metallic conductivity, and superconductivity. However, the experimental techniques and theoretical models designed for conventional inorganic insulators, semiconductors, and conductors usually do not apply to polymers [16]. Like other organic compounds, macromolecules retain their structure and identity in solid form due to weak intermolecular interactions, such as Van der Waals forces or electrostatic interactions [17].

Most polymers are disordered, amorphous, partially crystalline solids, unlike crystalline materials with long-range order and well-defined electronic band structures. Polymer charge carriers have a non-periodic potential that changes over time [18, 19]. Polymers have different physical processes driven by an external electric field in contrast to covalent semiconductors and dielectrics [20]. Because polymers have different electrical properties and complex structures, no uniform methods exist for

Table 7.1 Examples of some conducting and non-conducting polymers

S. No.	Conducting polymer ($S\ cm^{-1}$)	Non-conducting polymer ($S\ cm^{-1}$)	Polymer composites ($S\ cm^{-1}$)
1	Polyacetylene (10^4)	Polyethylene (1×10^{-17})	PET/Gra 3 wt% (0.1)
2	Polypyrrole (10^3)	Teflon (5×10^{-13})	PU/Gra 0.015 wt% (0.001)
3	Polythiophene (10–10^3)	PMMA; PVC (1×10^{-14})	Pvdf/Gra 15 wt% (0.01)
4	Polyaniline (30–200)	PC (5×10^{-15})	HDPE/ Gra 3 wt% (0.00000000025)

measuring conductivity or understanding charge carrier formation and movement, which is non-existent [21]. Each material class requires specialized experimental techniques and theoretical models to assess and interpret observations.

In Table 7.1, examples of some conducting and non-conducting polymers are given. Polymers' electrical conductivity values can be 10^{-14} and 10^{-17} S/cm. Noticeably, the electrical conductivity of undoped conjugated polymers, including polythiophenes and polyacetylenes, is low, within 10^{-10}–10^{-8} S/cm. The electrical conductivity of doped polymers has been found to increase to several orders in magnitude and reach closer to 0.1 S/cm. For instance, the electrical conductivity of Polypyrrole, a conductive polymer, can be enhanced up to 10^3 S/cm when the dopant is biologically active. Others are Polyaniline (PAn) and Polythiophene (PTh). Polymers are generally insulators because most polymers have no free electrons, which facilitates conductivity. However, when electricity is applied, the free-moving electrons cannot move around in the covalent bonds.

7.3 Polymer Composites Using Carbon Fillers

Polymer composites incorporating conductive fillers have become increasingly popular due to their distinctive characteristics. These composite materials' advantages include reduced weight, accessible design and manufacturing processes, cost-effectiveness, excellent corrosion resistance, and remarkable electrical and magnetic properties. The present chapter will explore polymer nanocomposites that contain metal—and carbonaceous nano additives [22].

There are two polymer composites: (1) Conducting polymer mixed with nanofillers and (2) non-conducting polymer with nanofillers.

The widely used filler consists of carbon, conductive polymers, metals, composites, aerogels, and polymer-based foams. Figure 7.2 demonstrates that the nanolaminate configuration exhibits enhanced shielding performance compared to discontinuous composites. For instance, PMMA reinforced with a graphene-related material at similar content levels shows improved results. The shielding effectiveness (SE) of

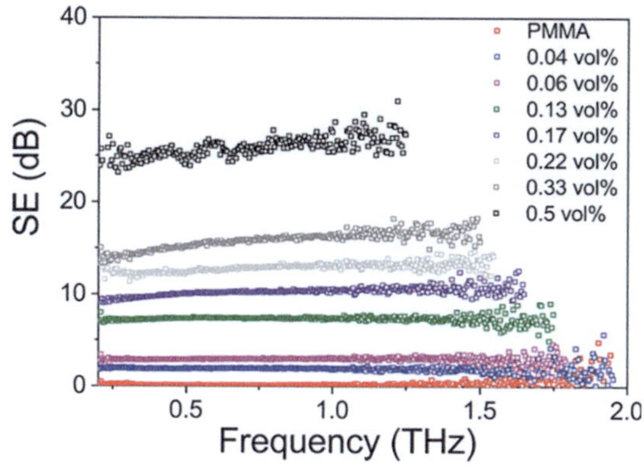

Fig. 7.2 PMMA polymer and graphene made as laminates to block electromagnetic waves [23]

Gr/PMMA (polymethyl methacrylate) nanolaminates increases with the addition of graphene content, reaching values as high as 25 dB at 1 THz for 0.5 vol% [23].

7.3.1 Conductive Polymers with Nanofillers

Conductive polymers play a crucial role in energy storage, optoelectronic applications, and the management of electromagnetic radiation. Furthermore, these polymers can be used as alternate electromagnetic interference (EMI) shielding materials. Conductive polyacetylene polyaniline (PANI) and Polypyrrole are frequently utilized for shielding [9]. These polymers show different chemical and electrical properties, with superior electrical conductivity and chemical stability under certain conditions.

The PANI-PU composite film yields high shielding effectiveness mainly due to Polypyrrole (PPY) elements coated on polyethylene terephthalate fabric. However, composites from conductive polymers often need more mechanical properties and processability [24]. In Fig. 7.3, photographic images of Pani [25] and (b) conductivity measurement of Pani and PMMA composite [26] are shown. The plane is a conductive polymer and can be used as a copolymer in an insulative polymer composite.

7.3.2 Non-conducting Polymer with Nanofillers

Non-conductive polymers are materials that have electrical insulation properties and consist of polymeric substances. These materials are of two types: thermoplastic and thermosetting polymer bonding products [27]. Such materials find their applications in temperature control within semiconductors and other electric appliances. A polymeric system comprises several organic aryl molecules dissolved, mixed, or

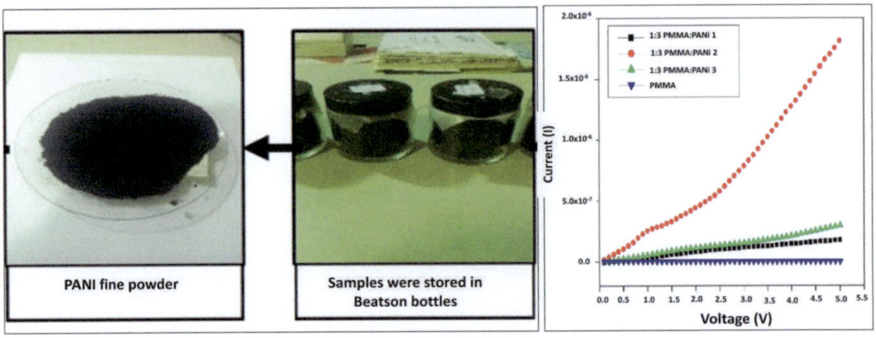

Fig. 7.3 **a** Photographs of Pani [25]; **b** testing the conductivity of Pani and PMMA mixed [26]

incorporated into a polymeric matrix. The polymeric system is thus characterized by its ability to attain electrical conductivity after exposure to irradiation [28].

Non-conducting polymers have some advantages, especially in mechanical stability and flexibility over conducting polymers. Several metallic fillers, such as fibers and nanoparticles, were added to the polymer matrix to enhance interaction with incoming electromagnetic radiation. For example, it has been discovered that the EMI shielding in the system is enhanced by incorporating stainless-steel fibers blended into the polycarbonate (PC), acrylonitrile butadiene styrene (ABS), nylon 6,6, and polyphenylene oxide (PPO) matrices. Aluminum, brass, and copper fibers are some EMI shielding materials that find applications among many other metallic fillers [29]. The addition of flake and powder fillers also made this system relatively inexpensive by adding the same to the EMI shielding composite [30].

Due to the uneven metal fiber dispersion and heavy weight of the polymer/metallic fiber composite, researchers have looked for other materials. Other metal oxide fillers, such as Zinc Oxide (ZnO), Titanium Dioxide (TiO_2), Iron Oxide (Fe_3O_4), and Zirconium Dioxide (ZrO_2), have been studied in Polyvinyl Acetate (PVA) composites for potential use in electromagnetic interference applications [21, 31].

7.4 Polymer Composite Using Different Carbon Fillers

Polymer-based composites with carbon-based fillers are highly effective for electromagnetic interference (EMI) shielding due to the high aspect ratio of the carbon fillers, which boosts the electrical conductivity of the polymer matrix. High specific energy fillers include single-walled carbon nanotubes (SWCNT), multi-walled carbon nanotubes (MWCNT), carbon fibers (CF), graphite fibers (GF), and carbon black (CB) [32].

Studies show that carbon nanotubes (CNTs) can be coated with polymers like polyaniline (PANI) and polyurethane (PU), among others, to improve electrical charge storage and conductivity. CNTs are favored over other carbon fillers due to their lower cost, smaller diameter, easy percolation, excellent conductivity, and high mechanical strength [33].

Carbon black and MWCNT composites have been evaluated with various matrices such as PANI, ABS, HDPE, and PP (Polypropylene) in EMI shielding. The findings indicated that increasing the carbon black content led to a rise in overall SE. Only with conductive carbon blacks can manufacturers find the required combination of high electrical conductivity at low loading, effective electromagnetic signal attenuation, and easy processing.

This means that conductive carbon blacks are ideal for EMI shielding. Carbon black mixed with other conductive additives also aids in EMI shielding. The applications of carbon black include electronics, antennas, sensors, batteries, lighting, electronics, and vehicle interiors, where the applications of their high graphitic ties can be provided to afford thermal management-conductive carbon blacks for demanding conductive applications involving all types of rubber, including fluoro-elastomers and silicones; examples include profiles, seals, fuel hoses, drive belts, wire, cable components, and accessories.

Continuous carbon fiber composites demonstrated better shear strength than their discontinuous counterparts. Flexibly produced graphite sheets exhibited good surface area and shielding properties. It was found that graphite combined with polyphenylene sulfide is a promising choice for EMI shielding, especially at higher temperatures (more than room temperature 25 °C). Graphene-reinforced polymer nanocomposites exhibit exceptional mechanical, chemical, and physical properties, making them effective electromagnetic interference (EMI) shielding materials even at low concentrations [8]. Chen et al. developed graphene/PANI, graphene@Ag/PANI, and graphene@Ni/PANI nanocomposites, with graphene@Ag/PANI showing the highest shielding performance. Polymer matrices commonly used with graphene as the filler include thermoplastic polyurethane (TPU), polyvinylidene fluoride (PVDF), and polydimethylsiloxane (PDMS). Polymer nanocomposites incorporating reduced graphene oxygen (rGO) improve the polymer's electrical and thermal properties while offering structural reinforcement [34]. Figure 7.4 shows cross-section images by light microscopy and Raman map of the multi-layered PVDF (rGO–Fe_3O_4)/M/PC and PVDF (MoS_2–Fe_3O_4)/M/PC film [35].

They are shown to study the interfaces of two different polymers and fillers at the microscopic level to study rheology and miscibility. In Fig. 7.5, total shielding effectiveness (SE_T) versus frequency for polymer composite with polycarbonate, PVDF, CNT, reduce graphene oxide (rGO), Fe_3O_4, and molybdenum disulfide (MoS_2) is illustrated. A series of samples were created to achieve the best shielding polymer nanocomposites by combining various conducting fillers, such as MWCNT RGO, with inorganic fillers like Fe_3O_4 and Mo_2S_2, like magnetic fillers in three combinations. Among them, the polymer composite with PVDF:(rGO–Fe_3O_4):M: PC: CNT gave maximum SE_T values of more than 20 dB for sample types across different frequency bands: X, Ku, and K-bands [35].

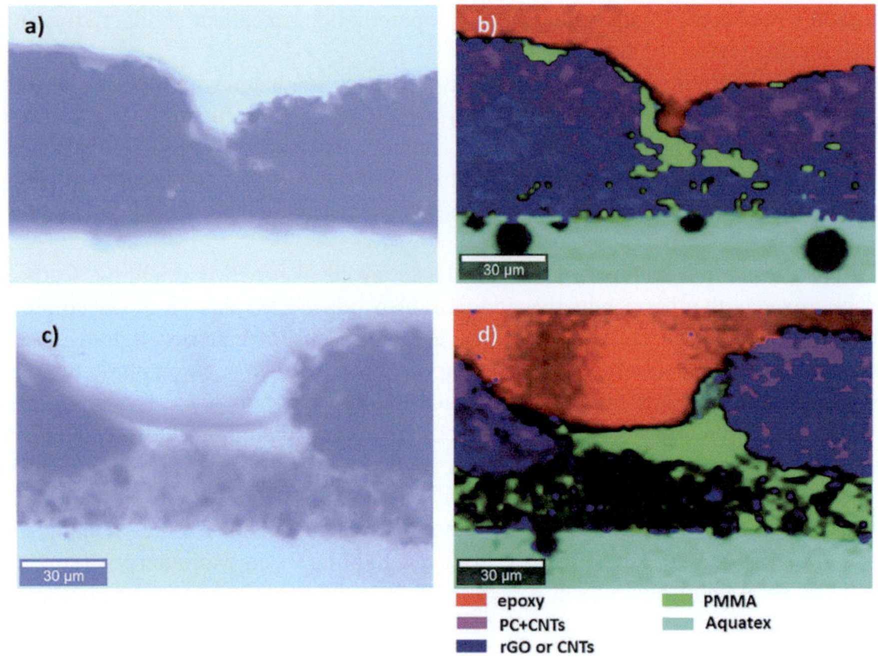

Fig. 7.4 Study showing micro and Raman images of film cross-section PVDF [35]

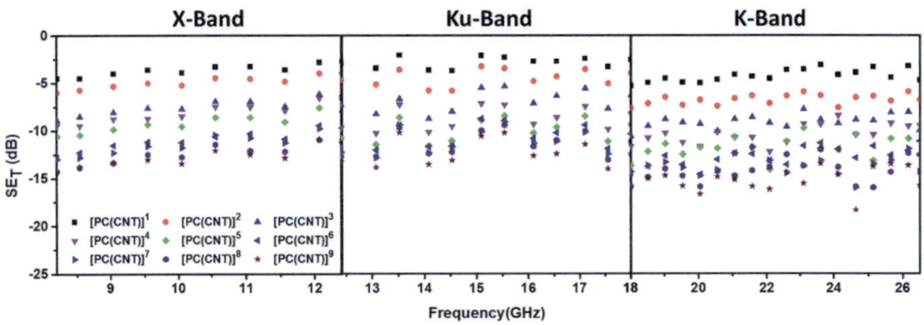

Fig. 7.5 SE_T versus frequency for different samples [35]

7.5 Metallic Alloys and Composites

The synthesis of materials is a highly active field, with many developed materials being used for electromagnetic interference shielding. Steel, Copper and Aluminum's superior electrical conductivity make them the preferred materials for shielding applications. Carbon, metallic filaments, and a range of other fibers are incorporated into matrices to improve the absorption efficiency of electromagnetic waves.

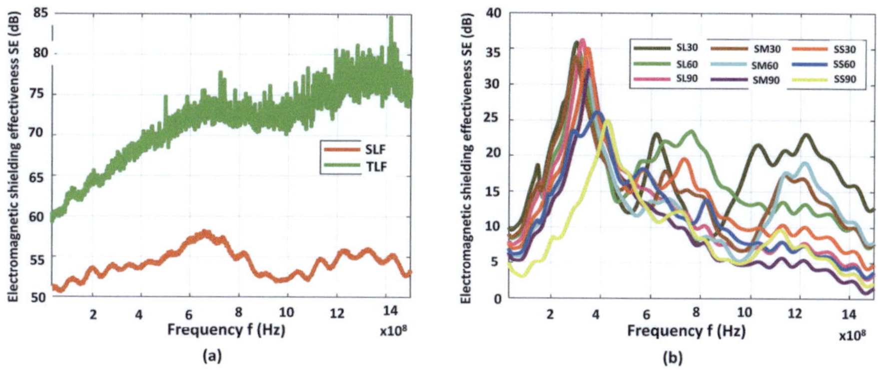

Fig. 7.6 EM shielding effectiveness (EM SE) versus frequency of Cu/Ni NW sample: single layer (SLF) and two layer (TLF) [38]

Utilizing conductive paints and coatings on materials represents a viable method for enhancing their respective shielding capabilities. This can be easily adjusted in thickness, offering notable benefits in weight and cost [36]. Metal foams are also used in shielding applications due to their lightweight. Many of these composites, coatings, or foams exhibit inferior performance in specific energy (SE) relative to metals and must be improved for structural purposes. Consequently, developing novel materials that integrate superior mechanical strength, minimal weight, and exceptional shielding efficacy remains a formidable challenge. Various alloys and composites of Iron, Nickel, Cobalt, Zinc, Aluminum, and Magnesium have been developed to resolve the abovementioned issue. Due to their widespread use and mechanical strength, aluminum alloys and composites have been utilized for electromagnetic interference (EMI) shielding applications [37].

As an example of metallic alloys and composites, Fig. 7.6 illustrates the electromagnetic shielding effectiveness (EM SE) of a Cu/Ni nanowire (NW) sample for both single-layer fabric (SLF) and two-layer fabric (TLF) configurations [38]. It demonstrates the frequency dependence of SE for SLF and TLF samples based on Cu/Ni nanowires. The SLF fabric serves as a precursor for strips designed for subsequent samples. The two-layer sample exhibits the highest SE of 74 dB at 1.5 GHz, with a logarithmic increase in SE with frequency. In contrast, the SLF sample maintains a strong SE of approximately 52 dB at both 30 MHz and 1.5 GHz, demonstrating consistent performance across all frequencies.

7.6 Polymer-Based Composite Foams

Composite foams based on conduction polymer are porous and lightweight, so they can be used in EMI shielding, actuators, and ESD applications [39]. The insulating polymer matrix formed from fillers based on carbon nanofibers, graphene sheets, and CNTs leads to improved EMI shielding effectiveness compared to polystyrene (PS)/

CNFs composite foams [40]. Furthermore, developed synthetic foams include hollow particles covering conductive elements and a second filler into the matrix to improve shielding efficacy. One lightweight material used in EMI applications is graphene/PVDF foam. Furthermore, graphene/PDMS and graphene/PU foams have demonstrated excellent EMI shielding effectiveness [41]. A case study was conducted on incorporating LDPE polymer with MWCNT and graphene with $CoFe_2O_4$ magnetic ferrite nanoparticles, focusing on their SE performance in the X-band [42]. Another case study with α-Fe_2O_3 is presented in Sect. 7.8 [43].

7.7 Case Study 1

7.7.1 Context

Carbon black, MWCNT, graphene, and other conducting carbon families have outstanding aspect ratios, conductivity, thermal stability, and electrical conductivity, making them promising EMI shielding materials. Increased polymer composite electrical conductivity causes dielectric loss. As the medium holds all the fillers, a helpful polymer is chosen. Low-density polyethylene (LDPE) is infrequently employed in commercial polymer applications. LDPE is classified as an engineering plastic due to its lightweight nature, cost-effectiveness, environmental stability, straightforward synthesis, and electrical insulation properties. The main parameters for microwave absorption are dielectric permittivity (ε), magnetic permeability (μ), and conductivity (σ). For EM absorption, researchers prefer dielectric and magnetic lossy materials. Because of its increased magnetic anisotropy and saturation magnetism, $CoFe_2O_4$ absorbs microwaves, particularly in the X-band. The elevated Curie temperature and molecular stability suggest that the absorber nanocomposite has minimal ion interaction with surrounding chemical processes. Ferrites exploit their natural resonance frequency to generate magnetic loss within the GHz spectrum [44, 45].

7.7.2 $CoFe_2O_4$ Preparation

This work prepares $CoFe_2O_4$ nanoparticles via the solution combustion technique. $CoFe_2O_4$ was generated using solution combustion synthesis (SCS) running urea as fuel. In a 100-ml Petri plate, 25 ml of distilled water (DI) was used to dissolve Ferric, Cobalt, and Urea in their stoichiometric ratio. The solution was prepared in a muffle furnace preheated to 500 °C. After several minutes, the water evaporated, resulting in a thick, concentrated gel that ignited and burned, releasing various gases and forming a fluffy, porous mass. This mass was then removed from the furnace, ground, and calcined at 900 °C for three hours to produce phase-pure, well-defined $CoFe_2O_4$ nanoparticles utilized as magnetic fillers in this study.

In the preparation of GNP, 6.67 g APS/g graphite H_2SO_4 medium was used. Previously, researchers created expanded graphite via chemical exfoliation. Graphite powder reacted with APS concentration. The reaction setup was cooled to 0 °C in an ice bath. The solution was magnetically stirred for three days. The production of graphite-peroxy-disulfate $C_{22.5}$ HS_2O_8 $2.5H_2SO_4$ produced a blue intercalated molecule. The result decomposed gradually as the reaction mixture was heated to 40–45 °C.

As various gaseous oxygens emerged between the graphene layers, crystals began to form. The solution, which was initially yellow–brown, was rinsed with distilled water. For 30 min, ultrasonication and microwave irradiation at 80 °C were used to disperse the paste in water and alcohol. After washing and drying, graphene nanoplatelets were stored in an airtight container.

7.7.3 Nanocomposites Fabrication

The melt blended polymer product was compressed in a 2.286×1.016 cm^2 mold using compression molding for one hour at 110 bar pressure and 130 °C. Tetrahydrofuran (THF) was employed to evenly distribute the $CoFe_2O_4$, MWCNT, graphene, and LDPE mixture; it was swirled for sixty minutes at 100 rpm and 60 °C, and the blended mass was then separated through vacuum evaporation in a hot air oven at 60 °C. The resulting dry product was subjected to a melting mixing process at 130 °C and 60 rpm for 20 min for each sample using Twin-screw extruders (Plastic order, Keltron CMEI, MODEL-16 CME SPL East Germany). A constant flow of nitrogen gas was maintained throughout the process to prevent oxidative degradation.

7.7.4 XRD Studies

The X-ray diffraction (XRD) analysis of calcined $CoFe_2O_4$ at 900 °C, as shown in Fig. 7.7, reveals the presence of cubic, single-phase nanoparticles. The peaks observed in the XRD pattern match with the JCPDS (Joint Committee on Powder Diffraction Standards) reference number 22-1086, confirming the crystallographic structure and phase of the material. The XRD data provides insight into the crystal structure and phase purity of the $CoFe_2O_4$ nanoparticles after calcination at the specified temperature.

7.7.5 SEM/EDS

Figure 7.8a–c displays SEM images of $CoFe_2O_4$ nanoparticles prepared using the SCS method at 30, 50, and 100 nm magnifications. The photos reveal polygonal

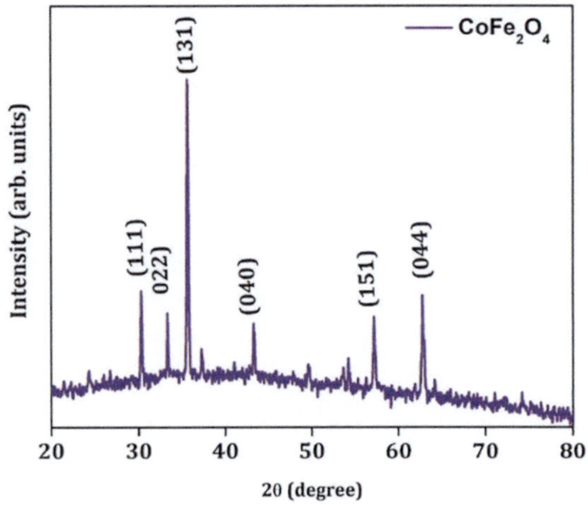

Fig. 7.7 The powder XRD pattern of $CoFe_2O_4$ was prepared through the solution combustion synthesis process, as depicted [42]. Reprinted with permission from Elsevier

crystalline nanoparticles with minimal accumulation. Additionally, the elemental color mapping of $CoFe_2O_4$ shown in Fig. 7.9 demonstrates uniform distribution and presence of Cobalt, Iron, and Oxygen across the areas, supported by the EDS spectra.

The specimen with 5-weight percent $CoFe_2O_4$ was cryogenically cracked. Figure 7.10 illustrates LDPE composites' shape and filler dispersion. SEM- and color-mapped cross-sectional area scanning electron microscopic images indicate that GNP dominates the study sample because of its higher concentration (40%). All around the matrix are also $CoFe_2O_4$ porous agglomerated clusters. Verifying composite manufacture, the color mapping images reveal that magnetic $CoFe_2O_4$ and carbon-based fillers are equally dispersed at all places. Figure 7.11 presents an elemental energy-dispersive X-ray spectroscopy (EDS) color mapping analysis.

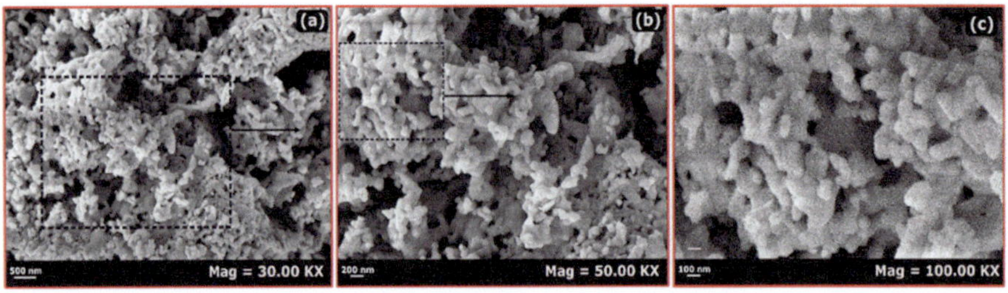

Fig. 7.8 a–c FESEM images of $CoFe_2O_4$ captured at various magnifications [42]. Reprinted with permission from Elsevier

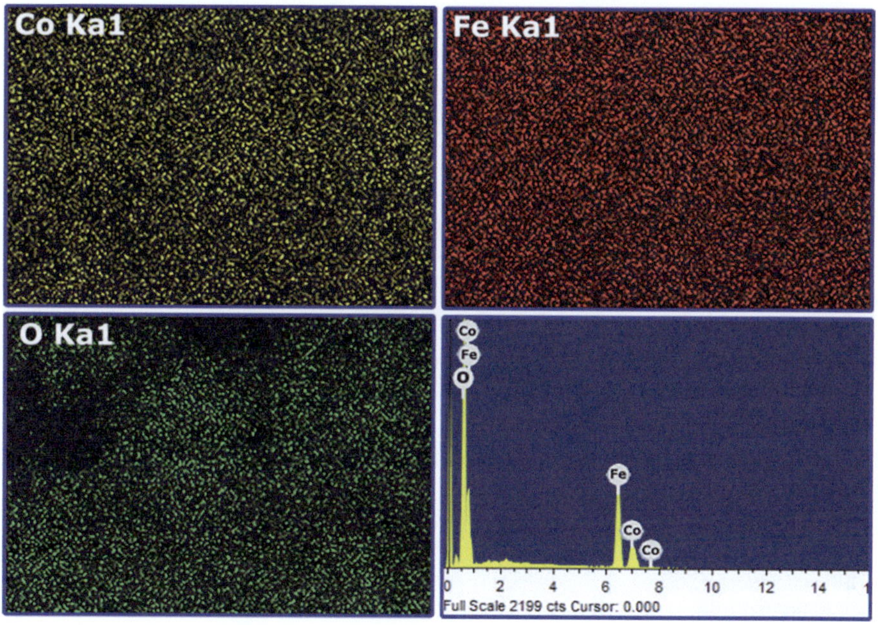

Fig. 7.9 Elemental color mapping images of $CoFe_2O_4$ along with EDS spectra [42]. Reprinted with permission from Elsevier

Fig. 7.10 SEM image of cryogenically fractured LDPE composite [42]. Reprinted with permission from Elsevier

7.7.6 TGA Analysis

Mechanical properties and TGA of CFO/GNP/MWCNT/LDPE composite and pure LDPE are experimentally verified. Figure 7.12 shows the pure LDPEs and the designed polymer composite TGA curves. Temperatures from 25 to 900 °C were used to assess the composite's thermal stability. After 380 °C, pure LDPE decomposed; before that, 3% water vapor was eliminated; nanocomposite decomposed at

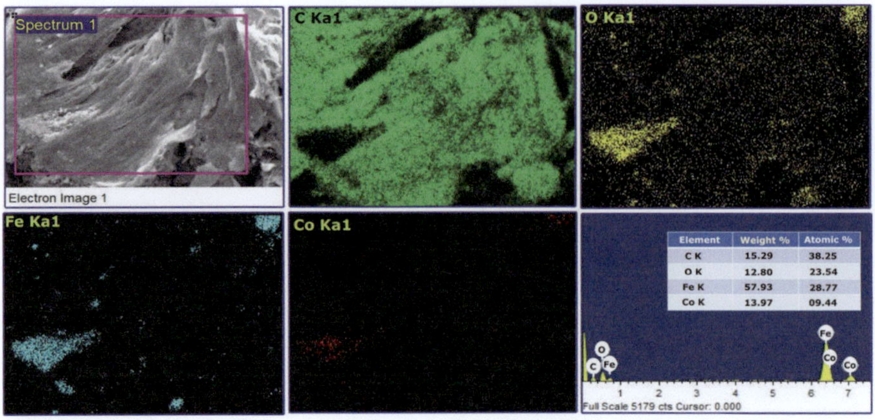

Fig. 7.11 LDPE-based composite elemental color maps [42]. Reprinted with permission from Elsevier

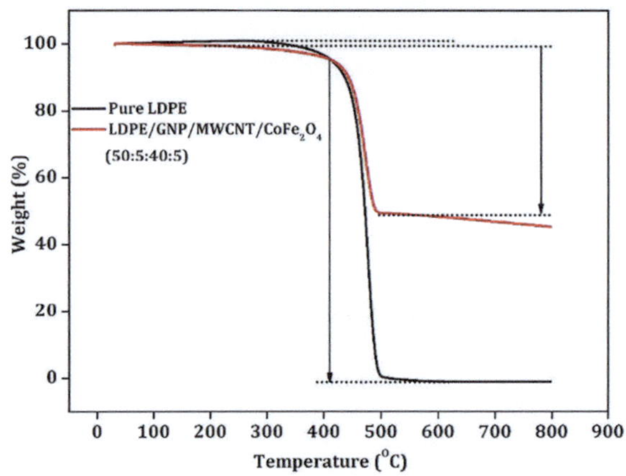

Fig. 7.12 TGA results for LDPE and nanocomposite [42]. Reprinted with permission from Elsevier

430 °C and finished at 492 °C. However, at 490 °C, the nanocomposite material disintegrated to 50% of its initial weight. Residual fillers like ferrite may be stable at this temperature. This indicates that the nanocomposite's thermal retention improved significantly. Combining MWCNT-graphene with $CoFe_2O_4$-filled LDPE polymer improved heat dissipation and temperature storage (Fig. 7.13).

7.7.7 VSM Magnetic Studies

Figure 7.14 illustrates the magnetic hysteresis graph from VSM instrument of $CoFe_2O_4$ nanoparticles at room temperature. Cobalt ferrite ($CoFe_2O_4$) was found

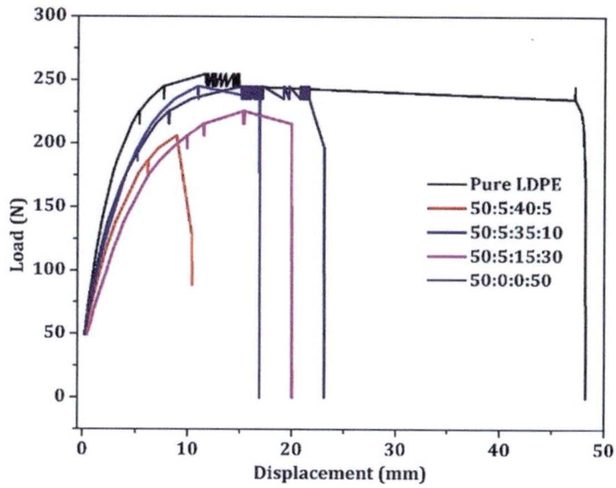

Fig. 7.13 LDPE and LDPE composite mechanical properties based on filler concentration [42]. Reprinted with permission from Elsevier

to be hard ferromagnetic and highly coercive. The MH loop reveals a ferromagnetic sample with a notable hysteresis (H_c) of 1629 Oe, indicating that the $CoFe_2O_4$ produced through SCS is categorized as a semi-hard ferrite. The M_r/M_s ratio for $CoFe_2O_4$ is 0.47, suggesting multi-domain particles exist. The saturation magnetization of $CoFe_2O_4$ is recorded at 83 emu/g, while the remnant magnetization is noted at 39 emu/g. According to Eq. (7.1), the Stoner–Wohlfarth model provides a magnetic anisotropy constant (K) of 0.6 $HAKg^{-1}$, which calculates magnetic anisotropy. H_c is intrinsic coercivity, μ_0 is the permeability of free space, and M_s is the saturation magnetic field.

$$K = \frac{H_c \mu_0 M_s}{2} \tag{7.1}$$

7.7.8 EMI Effectiveness Study

The effectiveness of total shielding includes absorption, multiple reflections, and reflection itself. The shielding effectiveness due to reflection (SE_R), shielding effectiveness due to reflection (SE_{MR}), and shielding effectiveness due to reflection (SE_A) values will be combined to determine the total shielding effectiveness (SE_T):

$$\mathrm{SE(dB)} = -10\log\left(\frac{P_t}{P_i}\right) = -20\log\left(\frac{H_t}{H_i}\right) = -20\log\left(\frac{E_t}{E_i}\right) \tag{7.2}$$

In Eq. (7.2), the symbols (P), (H), and (E) represent the power (W), magnetic field (A m^{-1}), and electric field (V m^{-1}) of the incident electromagnetic wave on the surface of the sample, with subscripts (t) and (i) indicating transmitted and incident

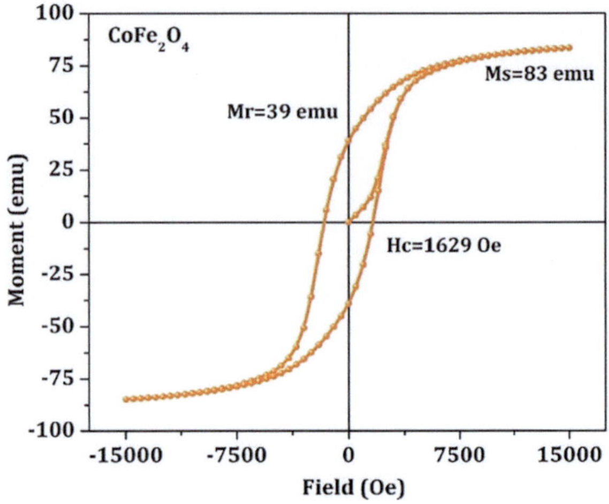

Fig. 7.14 Magnetic hysteresis loop of $CoFe_2O_4$ nanoparticles at ambient temperature (25 °C) [42]. Reprinted with permission from Elsevier

waves, respectively. The shielding effectiveness (SE_T) is defined by Eqs. (7.3) and (7.4). Multiple reflections (SE_{MR}) occur due to numerous reflections at the interfaces of the shielding medium. The significance of SE_{MR} diminishes when the shielding effectiveness of absorption (SE_A) exceeds 10 dB, while SE_{MR} is approximately 1.0 dB.

Shielding effectiveness (SE_T) refers to a material's ability to reduce electromagnetic radiation. SE_T values of − 10 dB and − 20 dB correspond to 90% and 99% effectiveness, respectively. The negative sign indicates a power reduction [46, 47]:

$$SE_T = SE_R + SE_{MR} + SE_A \tag{7.3}$$

$$SE_T = SE_R + SE_A \tag{7.4}$$

Equations (7.5) and (7.6) explain how the scattering S-parameters were used to calculate the overall shielding effectiveness of SE_T and SE_A:

$$SE_T(\text{dB}) = 10\log\frac{1}{|s_{12}|^2} = 10\log\frac{1}{|s_{21}|^2} \tag{7.5}$$

$$SE_A = 10\log\frac{(1-|s_{11}|^2)}{s_{12}^2} \tag{7.6}$$

The S-parameters (S_{11}), (S_{12}), and (S_{21}) denote forward reflection, reverse transmission, and forward transmission, respectively, as measured by the calibrated two-port X-band vector network analyzer (VNA). These parameters are represented as complex numbers at specified frequencies. The eddy current for the samples was calculated using (7.7):

$$\mu''(\mu')^{-2}f^{-1} = \frac{2\pi\mu_0 D^2\sigma}{3} \tag{7.7}$$

In this context, (μ') denotes the real permeability, while (μ'') represents the imaginary permeability component. According to Debye's theory, Equations (7.8) and (7.9) define the complex permittivity that influences AC conductivity and varies with frequency:

$$\varepsilon' = \varepsilon_\infty + \frac{\varepsilon_S - \varepsilon_\infty}{1 + \omega^2\tau^2} \tag{7.8}$$

$$\varepsilon'' = \frac{\varepsilon_s - \varepsilon_\infty}{1 + \omega^2\tau^2}\omega\tau + \frac{\sigma_{AC}}{\omega\varepsilon_0} \tag{7.9}$$

In this context, the dielectric constant at infinite frequency ε_∞, the static dielectric constant ε_S, the permittivity of free space ε_0, the relaxation time τ, and the alternating current conductivity σ_{AC} are the respective parameters in the above equations. The relationship between the real and imaginary permittivity components is expressed through (7.10):

$$\left(\varepsilon' - \frac{\varepsilon_s - \varepsilon_\infty}{2}\right)^2 + \left(\varepsilon''\right)^2 = \left(\frac{\varepsilon_s - \varepsilon_\infty}{2}\right)^2 \tag{7.10}$$

The values of the attenuation constant of the developed new composite are obtained from (7.11):

$$\alpha = \frac{\sqrt{2}\pi f}{c}\sqrt{(\mu''\varepsilon'' - \mu'\varepsilon') + \sqrt{((\mu''\varepsilon'' - \mu'\varepsilon')^2 + (\mu'\varepsilon'' + \varepsilon'\mu'')^2)}} \tag{7.11}$$

A high dielectric coefficient allows good electrical conductivity. Brabender plastic requires melt-mixed graphene in the LDPE composite, maintaining a consistent MWCNT content of 5 wt% across all combinations. At the specified ratio, the graphene and ferrite nanoparticles in LDPE will not blend with MWCNT at concentrations exceeding 5 wt%. An orthogonal electromagnetic wave consists of one electric component and one magnetic component. Electromagnetic waves' magnetic and electric components hit the dipoles of the composite. Therefore, dielectrically lossy and strong magnetic fillers enhance electromagnetic wave blocking, particularly in the 8–12 GHz range. This study measured the SE of obtained polymer composite with varying concentrations of "x" nanoparticles ($x = 0, 5, 10, 20, 30, 40$, and 50 wt%) at a lateral thickness of 3.5 mm in the X-band spectrum.

7.7.9 Shielding Effectiveness Studies

LDPE/MWCNT/GNP/$CoFe_2O_4$ composite EMI SE_T plots are shown in Fig. 7.15. A composite with 5 wt% $CoFe_2O_4$ nanoparticles has a maximum SE of 50 dB. A 0% sample delivered 30.65 dB at 10.3 GHz. SE_T is 30.40 dB for the 10% $CoFe_2O_4$ sample, somewhat less than the 0% sample—the SE max of the 20 wt% sample was 26.72 dB, the 30% sample 26.01 dB, and the 40% sample 25.62 dB. Graphene 50% had 22.49 dB, while $CoFe_2O_4$ 50% had 11.17 dB. All the samples adhered to a specific frequency trend line. The optimal shielding concentration in the LDPE:MWCNT:GNP:$CoFe_2O_4$ composite was 5 wt%. The sample containing the maximum concentration of $CoFe_2O_4$ (50 wt%) composite and graphene possesses negligible value. $CoFe_2O_4$ and graphene cannot achieve maximum synergistic effects; nevertheless, an optimal combination will enhance performance. SE_T performance is forthcoming for 10% and 0% $CoFe_2O_4$.

Figure 7.16 shows SE_A frequency correlation observed using LDPE sample, GNP:MWCNT, and $CoFe_2O_4$ [42]. The sample with 5 wt% showed the highest at 49.55 dB for 10.3 GHz, whereas the sample with 10% $CoFe_2O_4$ recorded 28.9 dB. The SE_A max of 0% $CoFe_2O_4$ sample was 27.7 dB, 20% was 24.24 dB, 30% had 23.6 dB, 40% had 23 dB, 50% graphene had 19.28 dB, and 50% had 8.58 dB. Like SE_T graphs in Fig. 7.15, shielding was a major absorption in the composite sample.

Figure 7.17 shows nanocomposite SE_R graphs—the 50% graphene sample shields 10.3 GHz for 3 dB. Due to favorable fillers and unbound electrons, they scatter EM waves easily. 20% of the $CoFe_2O_4$ samples have 2.86 dB, 0% have 2.83 dB, 30% have 2.6 dB, 40% have 2.54 dB, and 50% have 2.5 dB. At 10.3 GHz, 10% Ferrite has 1.26 dB SE_R maximum, while 5% has 0.36 dB. Our analysis found decreased reflection mechanism contribution for the composite.

It was observed that the sample containing 5% ferrite nanoparticles exhibited a shielding effectiveness (SET) of 50 dB. However, its shielding effectiveness due

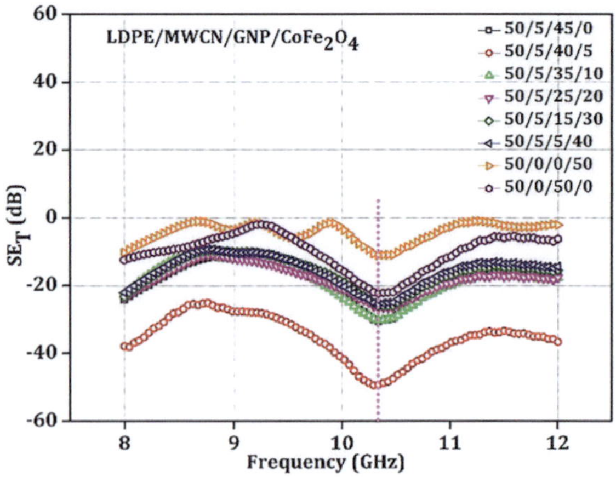

Fig. 7.15 SE_T versus frequency function with a 3.5 mm LDPE sample [42]. Reprinted with permission from Elsevier

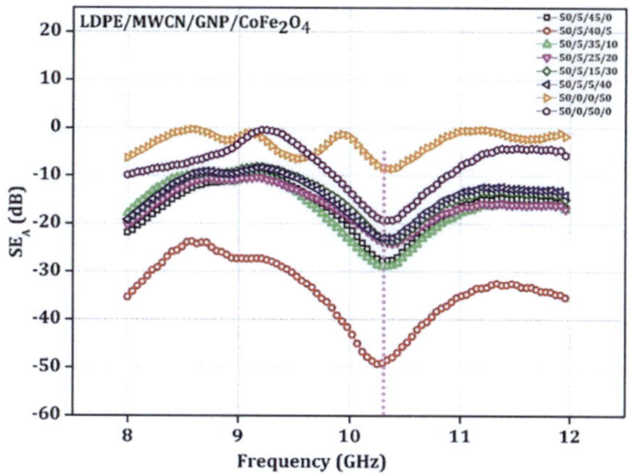

Fig. 7.16 SE_A frequency correlation observed using LDPE sample, GNP:MWCNT, and $CoFe_2O_4$ [42]. Reprinted with permission from Elsevier

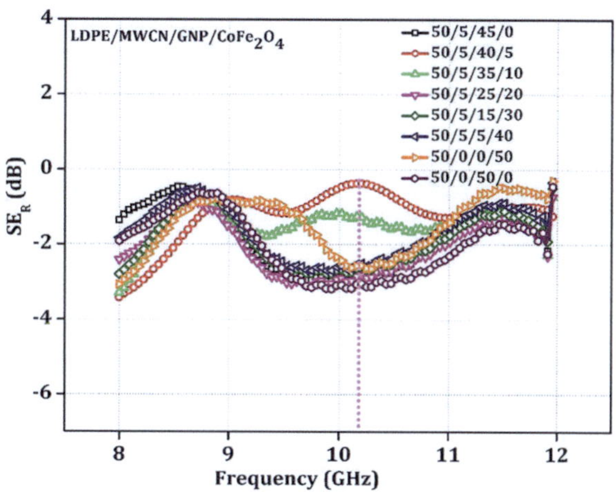

Fig. 7.17 SE_R changes with frequency, LDPE, GNP, and MWCNT percentages in the matrix [42]. Reprinted with permission from Elsevier

to reflection (SER) was much lower, indicating that the composite's filler material acts as an excellent electromagnetic (EM) absorber rather than a reflector. 50:5:40:5 excels in this study. However, the sample with the highest graphene concentration had the highest SE_R value, indicating maximal electrical conductivity. A higher concentration of graphene and an SE_R of 3.7 dB led to an increase in free electrons on the surface, which improved the reflection of incoming waves. However, the maximum SE_T of 23 dB is not reached related to the sample with 5%. $CoFe_2O_4$, at a 50% concentration alone, fails to achieve SE_T (11 dB) or SE_A despite $CoF_{e2}O_4$'s capability to absorb electromagnetic radiation. The ideal concentrations of ferromagnetic and conductive fillers must work together synergistically to create networks of electrical charge conduction within the novel material. SE_A and SE_R together make up the total shielding (SE_T). The graphs show how SE_A and SE_R contribute to various samples. This graph illustrates the contributions of SE_A and SE_R to SE_T in dB and percentage for each sample [48].

It is well known that ε' pertains to the ability for the storage of material charge. The material's real component (ε') disperses electromagnetic waves in all directions, leading to absorption. In creating an electromagnetic shielding material, combining conductive and dielectric properties and transparent features to electromagnetic waves while absorbing or scattering them would be more effective. The higher ε' value can be attributed to the numerous interfaces, imperfections, and boundaries introduced by nanoscale $CoFe_2O_4$ and graphene layers [49].

The increase in ε'' for the $CoFe_2O_4$ of 5% compound suggests dielectric loss resulting from various Debye relaxation mechanisms. In composite materials, dielectric relaxation can lead to a phase lag in real and complex permittivity. The polarization rotations within the integration of graphene and MWCNT improve EM wave absorption, which induces relaxation. These dielectric relaxation phenomena influence the electron's conduction (σ_{AC})-prepared hybrid composite employing dielectric loss ($\sigma_{AC} = \omega\varepsilon_0\varepsilon''$). Consequently, conduction loss contributes to ohmic loss. Both types of losses—ohmic and dielectric relaxation—enhance microwave absorption; as indicated by the increased ε'' values in the 5 wt%, sample reflects improved AC conductivity because of dielectric relaxation and ohmic loss process. The composite's dielectric relaxation ε'' values boost resultant conductivity at 5 wt%. The dielectric loss tangent (tan δε) in nanocomposites reveals the connection between dielectric losses. The Cole–Cole plots provide insights into relaxation, which will be discussed further [50].

Flaws in the graphene and MWCNT graphitic layers of the LDPE composite led to ε'' and AC conductivity issues, which enhanced microwave absorption. Various interfaces and imperfections among $CoFe_2O_4$, MWCNT, Graphene, and LDPE in the composite lead to the scattering and absorption of incoming waves, attributed to the complex permittivity of the sample intern. The findings show that composites filled with 5% $CoFe_2O_4$, which possess elevated real and imaginary permittivity (ε), demonstrate the most effective absorption of EM waves [51].

The complex permeability of the composite is represented as μ' and μ''. The 5% $CoFe_2O_4$ nanocomposite shows the highest magnetic loss. The effective concentration of $CoFe_2O_4$ results in an increase in the real permeability μ'. Fillers such as Graphene/MWCNT/LDPE do not have any magnetic components, so they don't contribute to the μ'. Here, μ'' graphs demonstrate a combined magnetic loss. The composite with 5% $CoFe_2O_4$ is notable for having the highest values—the sample with 5 wt% has more SE_A, marked by a significant magnetic loss in both μ'' and μ'. The resonance of $CoFe_2O_4$ domain walls typically occurs around 210 MHz within the frequency range. As a result, the main contributors to magnetic loss in LDPE composites are ferromagnetic resonance and eddy current losses. The inclusion of $CoFe_2O_4$ magnetic nanoparticles has improved the absorption of EM waves. $CoFe_2O_4$ promotes heat loss when absorbing EM waves due to crystal vibrations in ferromagnetic $CoFe_2O_4$. While electromagnetic absorption and magnetic loss may occur, MWCNT and Graphene contribute to dielectric losses. The composite samples showed a notable loss of magnetic material. Materials that exhibit significant magnetic loss have higher tanδ_μ values in 5 wt% $CoFe_2O_4$, resulting in a maximum SE_A of 49 dB. The conduction network aids in distributing EM wave power from $CoFe_2O_4$ to different regions of the composite at that frequency [52].

The LDPE composite with 5% $CoFe_2O_4$ filling creates a skin effect due to the opposing eddy currents that the changing magnetic fields produce from alternating current (σ_{AC}). Various phenomena, including natural ferromagnetic resonance, are associated with the altered graphs. Since it has high magnetic anisotropy, nano-sized $CoFe_2O_4$ is a hard ferritic material with a natural resonance frequency in the MHz range. As a result, the ferromagnetic resonance frequency was adjusted to the GHz range using MWCNT, Graphene, and LDPE. A peak in resonance frequency was observed at 10.3 GHz for the LDPE composite filled with 5% $CoFe_2O_4$ [53]. Recent electromagnetic absorption investigations using cobalt ferrite and its composite are compared to the present work. To understand this study, see Table 7.2.

Figure 7.18 illustrates a proposed mechanism schematic. The LDPE:MWCNT:GNP:$CoFe_2O_4$ composite protects against electromagnetic radiation through absorption and conduction losses. MWCNT and Graphene contribute to dielectric components and conductivity losses, while $CoFe_2O_4$ causes magnetic loss, which leads to EM wave reflection, absorption, multiple internal reflections, and heat transmission effects. The developed material captures electromagnetic fields, converting them to thermal energy to dissipate energy. $CoFe_2O_4$ demonstrates considerable magnetic loss due to the uniform distribution of graphene and multi-walled carbon nanotubes (MWCNT) [78].

This study focuses on a polymer composite made with LDPE, examining prime composition with MWCNT, Graphene, and $CoFe_2O_4$ nanoparticles. The $CoFe_2O_4$, produced through a solution combustion synthesis (SCS) method, shows notable ferromagnetic coercivity, contributing to magnetic losses. Graphene, obtained via chemical exfoliation and multi-walled carbon nanotubes created through chemical vapor deposition, leads to dielectric and conduction losses. Material characterization was used to assess the quality and composition at the nanoscale. The research included an analysis of the nanofiller phase and its morphology. We explored how permittivity and permeability affect electromagnetic shielding. The losses in our composition, attributed to eddy currents, magnetism, and dielectric properties, enhance shielding effectiveness. The chosen fillers in this study display various structural deficiencies that promote internal scattering and absorptions. We examined multiple ratios of materials to identify the optimal concentration of each type. A specific ratio of $CoFe_2O_4$ and LDPE: MWCNT: GNP: $CoFe_2O_4$:: 50:5:40:5 yielded synergistic effects, resulting in exceptional X-band magnetic, dielectric, and ohmic losses. This composite sample attenuated 50 dB at 10.3 GHz, making it suitable for electromagnetic interference shielding and stealth technology applications [79].

Table 7.2 Absorption values of various reported materials [42]

S. No.	Material	SE value	Frequency (GHz)	Reference
1	PVDF:PANI:CFO	63 dB	8.2–12.4	[54]
2	$Ti_3C_2T_x$:CoFe	36.29 dB	8.56	[55]
3	$CoFe_2O_4$:CNTs	34.6 dB	10.0–17.1	[56]
4	CoFe@C:CNT:rGO	40 dB	5.17	[57]
5	CFO@CNTs:PVDF	26 dB	11.26	[58]
6	$CoFe_2O_4$	40 dB	2–18	[59]
7	$CoFe_2O_4$:rGO:SiO_2	24.8	2–8	[60]
8	$CoFe_2O_4$:polypyrrole	19 dB	3.8	[61]
9	$NiCoFe_2O_4$:RGO	36.3 dB	12.4–18	[62]
10	$CoFe_2O_4$	31.2 dB	8.1	[63]
11	$CoFe_2O_4$:CNTs	37.39 dB	5.2	[64]
12	GNS@$CoFe_2O_4$:CNT	29.1 dB	10.34	[65]
13	$CoFe_2O_4$:polyaniline	21.5 dB	12.4–18	[66]
14	$CoFe_2O_4$:MWCNTs	50.80 dB	3.36	[67]
15	$CoFe_2O_4$:NG:PMMA	36.7 dB	2–18	[68]
16	$CoFe_2O_4$:PPy	43.85 dB	5.7	[69]
17	RGO:$CoFe_2O_4$:MWCNTs	46.8 dB	11.6	[70]
18	$CoFe_2O_4$:PPy:rGO	50.1 dB	8–18	[71]
19	$Co_{1-x}Ni_xFe_2O_4$	37.66 dB	16.24	[72]
20	CCNTs:CFO	14 dB	2–18	[73]
21	$CoFe_2O_4$@rGO	67.58 dB	8–18	[74]
22	LPA:SWCNT:$CoFe_2O_4$	30.7 dB	12.9	[75]
23	GN@$CoFe_2O_4$:PANI	47.7 dB	14.9	[76]
24	$CoFe_2O_4$:C NFs	14.0 dB	4.4–18.0	[77]
25	$CoFe_2O_4$:MWCNT:GNP:LDPE	50 dB	8.2–12.4	**Present work**

Reprinted with permission from Elsevier

7.8 Case Study 2

This study focuses on the effective attenuation of electromagnetic waves achieved through the synergistic effects of α-Fe_2O_3 and multi-walled carbon nanotubes (MWCNT) and graphene in low-density polyethylene (LDPE)-based composites for electromagnetic interference (EMI) applications. Combining these materials enhances the composites' shielding effectiveness and overall performance, making them suitable for various EMI shielding applications.

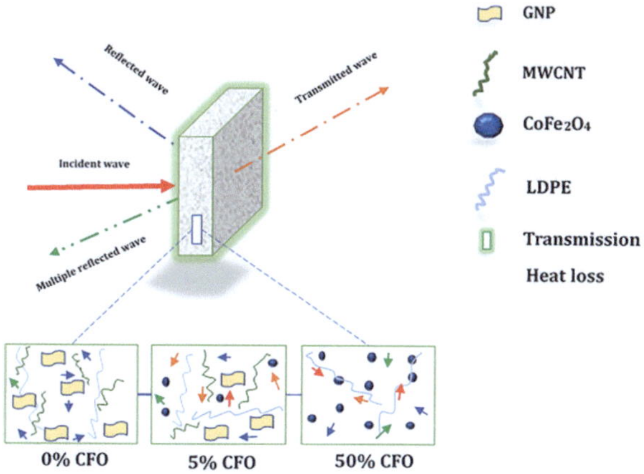

Fig. 7.18 Schematic diagram of a composite featuring various fillers [42]. Reprinted with permission from Elsevier

7.8.1 Context

This research focuses on a polymer nanocomposite incorporating magnetic and conductive fillers to protect against electromagnetic interference. The fillers used in low-density polyethylene (LDPE) polymer matrices include Alfa-ferrite (α-Fe_2O_3) nanoparticles, with multi-walled carbon nanotube (MWCNT) concentrations starting at 5%, along with various amounts of graphene nanoplatelets (GNP). X-ray diffraction, scanning electron microscopy, color mapping, energy-dispersive X-ray spectroscopy, and thermogravimetric analysis characterize the nanofillers and polymer matrices. A vector network analyzer is used to measure EMI shielding effectiveness. The polymer composite with LDPE polymer: Graphene: MWCNTs: α-Fe_2O_3 nanoparticles was analyzed in the X-band frequency range with increased magnetic, dielectric, and ohmic losses. An impressive attenuation of 99.99% was achieved with just 5% α-Fe_2O_3. The shielding mechanism is discussed in terms of permittivity, conductivity, and attenuation [80].

An effective composite will combine magnetic loss permeability (μ), and electrical conductivity (σ_{AC}). Metallic atoms have many free-charge carriers, allowing them to reflect electromagnetic waves and provide electromagnetic solid shielding efficiently. However, chemical resistance, processability, weight, and cost can limit their application. Ferrites act as intrinsic electromagnetic absorbers, but Snoek's limit constrains their performance in multi-gigahertz excitations. A good shielding material should have a high dielectric constant, dielectric loss tangent, magnetic loss tangent, permeability, and conductivity. To achieve optimal synergy, fillers must be evenly distributed throughout the composite [81].

To enhance electromagnetic interference (EMI) shielding, materials such as Iron, Cobalt, Nickel, Ferrites, metallic nanoparticles, and conductive carbon fillers are incorporated. The ideal nanofiller improves the absorption of magnetic components while minimizing dielectric loss for effective shielding. Carbon nanodiamonds and carbon nanofillers, including carbon black, graphene, and multi-walled carbon nanotubes, are known for enhancing the migration of electrical charges, making their use necessary. Conductive fillers can reflect electromagnetic waves, thanks to the presence of free electrons. Magnetic and dielectric lossy fillers require a polymeric medium for application. A semiconductor polymer can effectively shield against electromagnetic interference. Conducting polymers like Polyaniline, Polypyrrole, and Polyacetylene can reflect electromagnetic waves without absorbing them. Furthermore, materials such as epoxy, polymethyl methacrylate (PMMA), polyethylene (PE), and polycarbonate (PC) can achieve electrical conductivity by adding a small percentage of conductive fillers, making them suitable for EMI shielding [82].

7.8.2 Polymer Nanocomposite Preparation

Polymer nanocomposites were created through melt blending using a melt mixing process in a twin-screw extruder. This process involved specific percentages of polymer and nanofiller by weight, conducted at a temperature of 130 °C and a rotational speed of 100 rpm for 20 min. After the blending, a square sample measuring $2.286 \times 1.016 \times 0.35\ cm^3$ was compressed at a pressure of 110 bars in a hydraulic press at a temperature of 110 °C. Table 7.3 shows a polymer composite that includes various fillers.

An X-band waveguide is set up in the ZNB-40 model VNA. A WR-90 waveguide, measuring 2.286 cm × 1.016 cm for the band, was used to examine the prepared composite's electromagnetic interference properties. Two-port short-open-load and pre-calibration techniques were employed to mitigate reflection and absorption losses in the lines.

Table 7.3 Polymer nanocomposite preparation table [43]

S. No.	Polymer (wt%)	Conductive filler (wt%)		Magnetic filler (wt%)
	LDPE	MWCNT	GNP	α-Fe_2O_3
1	50	5	45	0
2	50	5	40	5
3	50	5	35	10
4	50	5	25	20
5	50	5	15	30
6	50	5	5	40
7	50	0	0	50
8	50	0	50	0

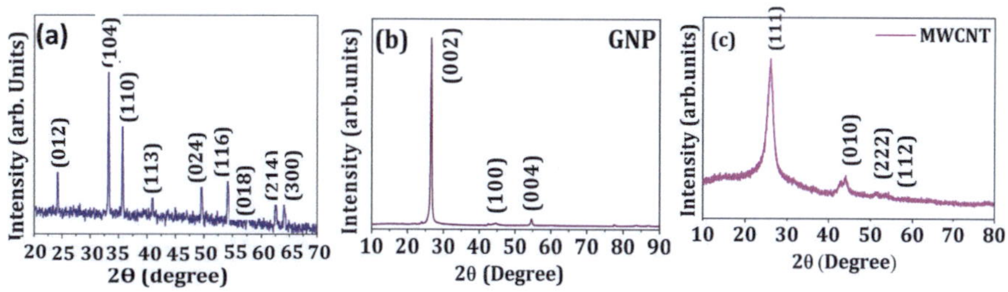

Fig. 7.19 Powder XRD pattern of α-Fe_2O_3 GNP, MWCNT [43]

7.8.3 XRD Studies

The XRD results obtained are depicted in Fig. 7.19.

7.8.4 Raman Spectroscopy

This study examined the carbon derivatives using Raman spectroscopy. The synthesized graphene nanoplatelets (GNPs) produced through the chemical exfoliation process show distinct peaks at the D (1350 cm^{-1}), G (1582 cm^{-1}), and 2D (2700 cm^{-1}) bands, as illustrated in Fig. 7.20a. The G-band is linked to the E^2g phonon near the center of the Brillouin zone; its sharpness suggests that carbon atoms are interconnected by employing the sp^2 hybridization process, leading to a hexagonal structural appearance. The appearance of the D-band in the Raman spectra indicates the vibrational mode of sp^2 carbon atom rings. It is reduced due to defect-induced strong sp^2 covalent bonds from newly formed edges and sp^3 hybridized atoms. An I_G/I_D ratio of 2.60 was calculated to assess the average sp^2 domain size, characterized by notable structural defects and disorders in GNP. Meanwhile, carboxylic, epoxy, and oxygen-functionalized side groups were synthesized. The position, shape, and FWHM indicate a second order of the 2D band in the GNP, composed of graphene layers. The Raman spectra of the multi-walled carbon nanotube (MWCNT) skeleton is presented in Fig. 7.20b. The G-and D-bands are found at 1340 cm^{-1} and 1581 cm^{-1}, respectively. The difference of the D-band indicates the sp^3 hybridization occurred in carbon bonds, while the G-band reflects sp^2 hybridization due to MWCNT defects. No single-walled carbon nanotubes (SWCNT) were demonstrated. The I_G/I_D ratio of 0.97 indicated structural imperfections, suggesting that the sample was disordered due to the displacement of graphene layers.

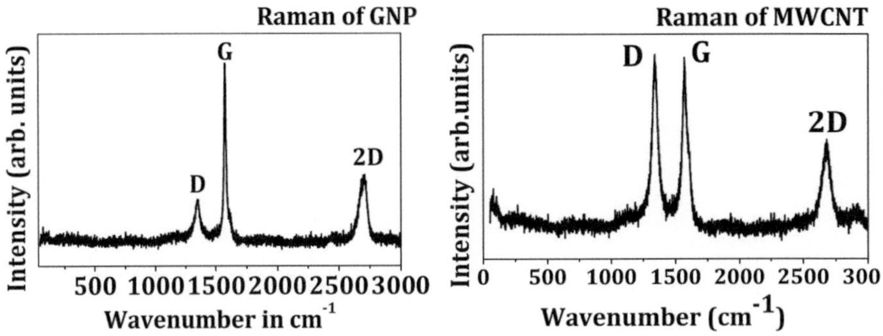

Fig. 7.20 Raman spectra of **a** GNP and **b** MWCNT [43]

7.8.5 SEM and EDS

SEM was used at various magnifications to examine the prepared fillers' structure and morphology. Figure 7.21a–d displays SEM images of α-Fe_2O_3 nanoparticles at various magnifications. Nanoparticles of α-Fe_2O_3 exhibit crystalline, agglomerated formations, polygonal forms, and distinct particle boundaries. Solution combustion-derived nanoparticles accumulate particles of all sizes and shapes. The release of gaseous byproducts during product production causes surface gaps and holes.

The EDS provides a clear view of the distribution of elements. In Fig. 7.22, the elemental mapping from the EDS confirms this distribution. Iron and oxygen are uniformly spread throughout the α-Fe_2O_3 crystals, and no impurities were detected, which suggests the product is pure [51, 52].

7.8.6 TGA (Thermogravimetric Analysis)

Samples of pure LDPE and α-Fe_2O_3: CNT: GNP: LDPE composites were examined for thermal stability using TGA, as shown in Fig. 7.23. Pure LDPE shows rapid decomposition starting at 380 °C. Before reaching around 432 °C, there was a weight loss of about 3% due to the evaporation of absorbed and coordinated water molecules. Following this, the degradation of the composite sample began. The thermal stability of the fillers resulted in MWCNT decomposing in the range 400–500 °C, compared to 380 °C for the pure form. The weight loss was minimal at 50% sample loss because of the residual α-Fe_2O_3 during thermal degradation.

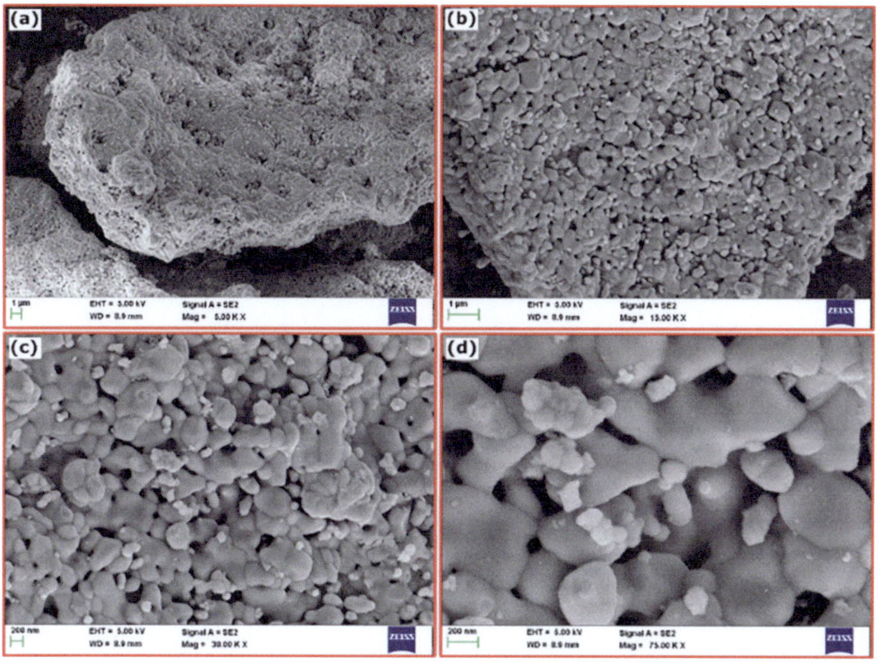

Fig. 7.21 **a–d** FESEM of α-Fe_2O_3 at different magnifications [43]

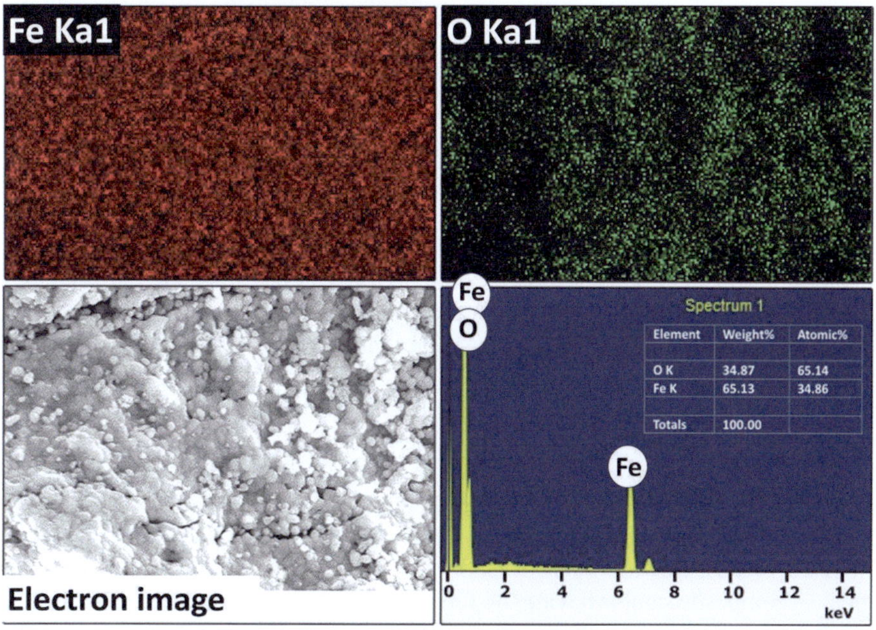

Fig. 7.22 Elemental color mapping of α-Fe_2O_3 and EDS [43]

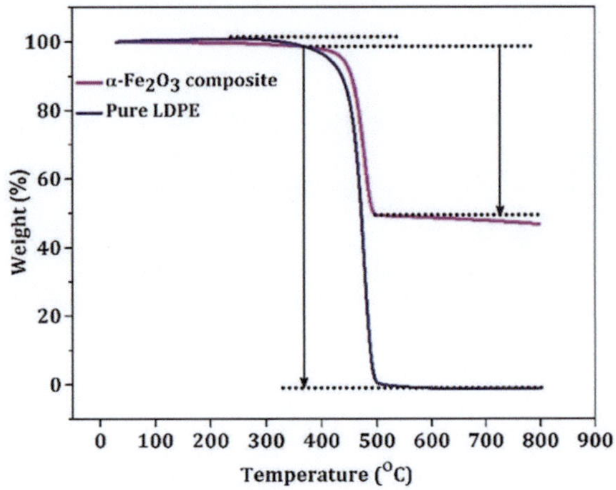

Fig. 7.23 TGA results for LDPE and nanocomposite are shown [43]

7.8.7 EMI Shielding Study

This study addresses the shielding mechanism. Integrating suitable filler and ferrite enhances the shielding efficacy of LDPE composites for X-band frequencies. Microwave absorption necessitates the electrical conductivity of nanocomposite samples, which is associated with dielectric loss. Material designers deliberately generate electrical and magnetic fields using electromagnetic waves. Electric and magnetic dipoles must be uniformly distributed within the matrix. This study incorporates magnetically lossy (α-Fe_2O_3) and dielectrically lossy (MWCNT/graphene) materials into LDPE. The pure phase of α-Fe_2O_3 exhibited considerable coercivity, as demonstrated by XRD and EDS analysis. This study established the SET for several combinations to comprehend the variable interactions necessary for attaining synergism with critical filler concentration. Pre-prepared samples of LDPE, MWCNT, GNP, and α-Fe_2O_3 nanocomposites ($x = 0$, 5, 10, 20, 30, 50, and 50% weight %) were evaluated at a thickness of 3.5 mm [83].

Figure 7.24 shows that a sample containing 5% α-Fe_2O_3 embedded in a polymer matrix achieves a maximum shielding efficacy of 40 dB at a frequency of 10.3 GHz. In comparison, a sample with 0% ferrite provides a shielding efficacy of 34.7 dB. The samples with 10 wt% and 20% weight percentages show 24 dB and 23.8 dB of SE, respectively, whereas the sample with a 50% weight percentage resulted in a sound level of 11.98 dB. On the other hand, pure LDPE without any fillers shows minimal attenuation, recorded at just 1 dB. This suggests that the presence of medium-fillers contributed to the losses observed. Each composite sample has a dependency on EM absorption at a given frequency. The critical concentration, which includes 5% α-Fe_2O_3, 5% MWCNT, 40% GNP, and 50% LDPE, exhibited a synergistic effect. A composite sample having 0% ferrite with only conducting materials showed subsequent consecutive shielding effectiveness due to the lack of EM absorbing filler enhancing SE, unlike the 5% sample. The samples with 10 and 20% α-Fe_2O_3 weight percentages demonstrated sub-optimal shielding performance, measuring 23.8 dB at 10.3 GHz and slightly reduced effectiveness at specific frequencies. The 10% sample showed better synergy for improved electromagnetic shielding than the 20% sample. The sample with 50% ferrite by weight recorded a minimum SE of 12 dB at a frequency of 10.3 GHz [84].

Figure 7.25 illustrates the shielding efficiency due to absorption (SE_A). The sample with 5% α-Fe_2O_3 reaches a maximum Specific Absorption Rate (SAR) of 39.19 dB at a 10.3 GHz frequency. In contrast, a 0% ferrite sample results in a signal attenuation of 34.3 dB. In comparison, the samples containing 10 weight % and 20 weight % of α-Fe_2O_3 recorded SE of 23.29 dB and 20 dB, correspondingly. A similar pattern was noted in later specimens. The specimen having 50% α-Fe_2O_3, lacking conductive incorporation, exhibited a low shielding value of about 11.38 dB. The significance of using conductive materials in improving dielectric and magnetic loss parameters at optimal concentrations is highlighted. Achieving the right balance of conductive and magnetic fillers is crucial for maximizing shielding efficiency through absorption [85] (Fig. 7.26).

In a composition with 45% ferrite, adding 5% MWCNT improves the results by 14.25 dB, indicating that at least 5% MWCNT can provide a 3 dB enhancement in

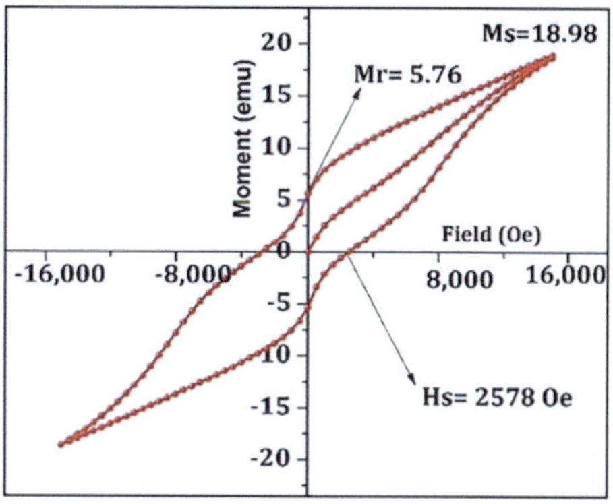

Fig. 7.24 MH loop of α-Fe_2O_3 nanoparticles at 25 °C [43]

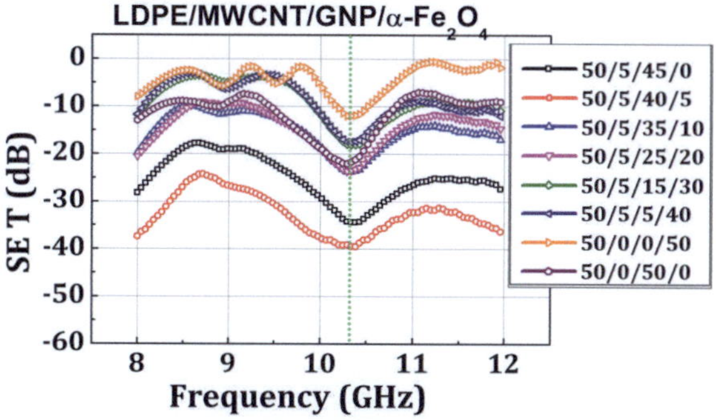

Fig. 7.25 Comparison of SE_T to frequency for nanocomposite specimen with LDPE [43]

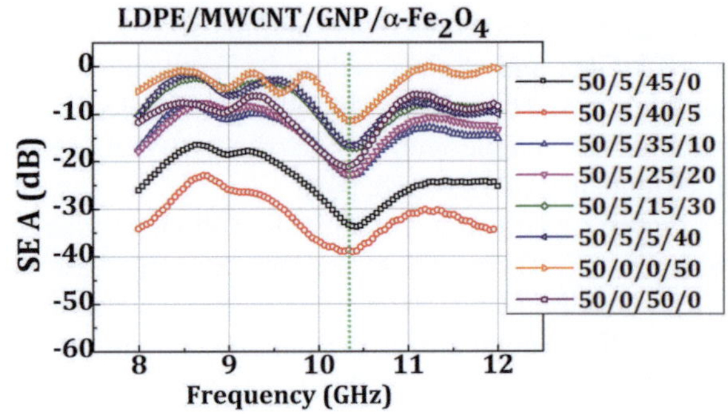

Fig. 7.26 SE_A versus frequency for 3.5 mm thick LDPE: MWCNT: GNP: α-Fe_2O_3 nanocomposite [43]

specimens. The specimen that lacks incorporation of α-Fe_2O_3 and MCNT, which contains 50% of GNP, achieved a SE_A of 21.1 dB, suggesting that the conductive filler has notable SE_A properties.

The SE_R graphs are illustrated in Fig. 7.27. A specimen containing GNP of 50 reaches an SE_R of about 1.13 dB, while a specimen without ferrite has 1 dB of SE_R at the same frequency. The 5% ferrite specimen has a SE_R of 0.58 dB, and the specimen without any filler, which is pure LDPE, displayed a minimal SE_R of 0.6 dB, showcasing the transparent characteristics of EM waves. The efficiency of shielding through reflection is negligible. The nanomaterial composite substantiates this assertion. Consequently, the proposed material exhibits a higher SE_A relative to reflection. This indicates that the electromagnetic pulse is transformed into thermal energy within the proposed nanocomposite material, validating the material's efficacy as an absorber [86].

A polymer nanocomposite containing ferrite nanoparticle filler exhibits enhanced electromagnetic wave absorption, optimizing its performance within the composite. This is due to the generation of polarons and bi-polarons, which help absorb electromagnetic waves. With varying compositions, the interaction between the conductive

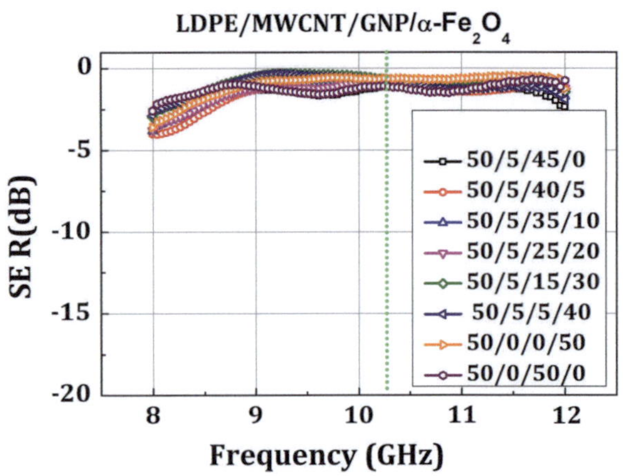

Fig. 7.27 Comparison of SE_R versus frequency with variable weight percentages [43]

Table 7.4 Literature survey table [43]

S. No.	Material	SE Max (dB)	Thickness (mm)	Frequency (GHz)	Reference
1	α-Fe_2O_3:PCL	16.37	7	1–4	[88]
2	α-Fe_2O_3:Fe_3O_4	20	4	3.76–8.15	[89]
3	Fe:PVDF	40.21	2	8–12	[90]
4	γ-Fe_2O_3	11.2	6	8.2–12.4	[91]
5	Fe_3O_4:GNP	9.6	5	8–12	[92]
6	α-Fe_2O_3:PPy:CDCA	39.4	2	8–13	[93]
7	rGO:Fe_2O_3:PU	37.4	10	8.2–12.4	[94]
8	γ-Fe_2O_3:Fe_2O_3:PPy	22.5	1.5	12.4–18	[95]
9	HDPE:CNT	21.8	5	8.2–12.4	[96]
10	γ-Fe_2O_3:C nanorod:carbon sphere	8.11	5	2–18	[97]
11	LDPE:MWCNT:GNP:α-F_2O_3	40	3.5	8–12	This work

fillers, MWCNT and GNP, improves absorption, while GNP and MWCNTs significantly boost the conduction of electrical charges. Consequently, the insulative LDPE shows non-conductive behavior; as a result, electrical charges on the composite reflected EM waves from surface impedance mismatching as depicted in Fig. 7.27. The best combination of filler ratio was determined to be a ferrite of 5% combined with conductive materials, which created a practical synergistic effect and several percolation thresholds [32].

According to the high-frequency principle, the complex permittivity and permeability of a composite within a specific electromagnetic spectrum determine the efficiency of its shielding. The degree of polarization within the substance, representing its ability to store electrical charges densely, is closely correlated with the value of ε'. On the other hand, the dissipation of electrical energy in the sample is represented by ε''. As is common in industrial materials, this directly affects the idea of the dielectric loss tangent. The rise in ε' is due to the increased availability of polarons and bi-polarons present inside the MWCNT and GNP specimens [87] (Table 7.4).

7.8.8 Shielding Mechanism

Figure 7.28 shows the shielding mechanism of typical absorbers. The LDPE/MWCNT/GNP matrix attenuated mostly due to conduction loss. The schematic diagram shows that the absorption process and reflection characteristics of α-Fe_2O_3 and MWCNT-GNP will be impacted by shielding. α-Fe_2O_3 enhances absorption, scattering, and transmission features, while MWCNT-GNP enhances reflection,

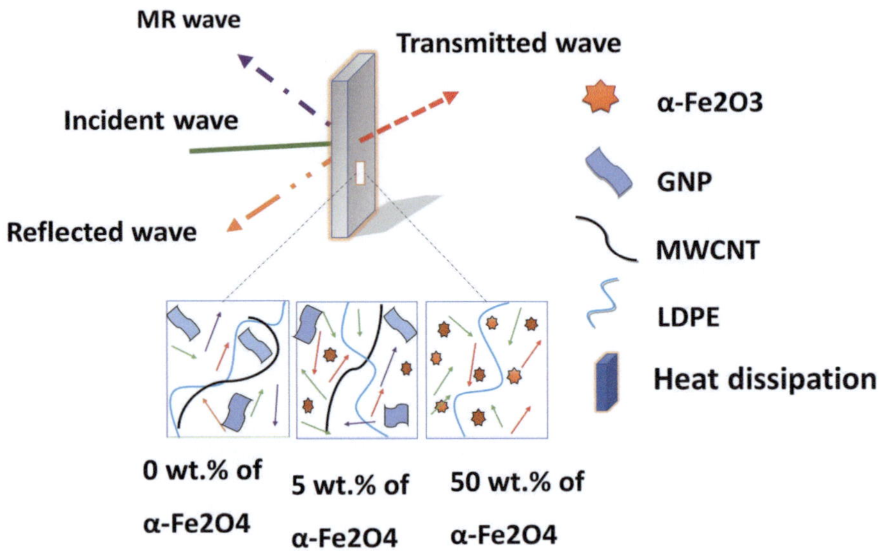

Fig. 7.28 EMI shielding mechanism [43]

absorption, and scattering. On the other hand, LDPE serves solely as a matrix for enclosing nanomaterials, giving the necessary shape, and is electromagnetically transparent.

7.9 Conclusions

Polymers play a crucial role in electromagnetic interference (EMI) shielding by providing lightweight, flexible, and cost-effective solutions. However, most polymers are inherently non-conductive and electromagnetically transparent. Conductive fillers like carbon-based materials (MWCNT, graphene), metal particles, or magnetic oxides (e.g., α-Fe_2O_3) are often incorporated into the polymer matrix to overcome this limitation. These fillers enhance the shielding effectiveness through mechanisms like reflection, absorption, and scattering of electromagnetic waves. The choice of polymer and filler materials, along with their distribution and interaction, determines the shielding performance, making polymer-based composites versatile for various EMI shielding applications in electronics and communication systems.

As demonstrated in Shielding results, the shielding mechanism of typical absorbers highlights the critical contributions of different materials in the LDPE/MWCNT/GNP matrix. The primary attenuation occurs through conduction loss, driven by the matrix's composition. $CoFe_2O_4$ and α-Fe_2O_3 significantly enhance absorption, scattering, and transmission properties, while MWCNT-GNP further improves reflection, absorption, and scattering behavior. LDPE, though electromagnetically transparent, serves an essential role by acting as the matrix that supports and shapes the nanomaterials. These combined properties provide an effective shielding solution with a balanced performance across reflection, absorption, and scattering.

References

1. Kumari, S., Dalal, J., Kumar, A., Pal, R., Chahal, R., Ohlan, A.: Enhanced microwave absorption properties of conducting polymer@graphene composite to counteract electromagnetic radiation pollution: green EMI shielding††Electronic supplementary information (ESI) available. RSC Adv. **14**(1), 662–676 (2024). https://doi.org/10.1039/d3ra07245b
2. Chakraborty, T., Saha, S., Gupta, K., Dutta, S., Mahapatra, A.S., Mondal, D., Pradhan, A.K., Chakraborty, C., Das, S., Sutradhar, S.: An effective microwave absorber using multi-layer design of carbon allotrope based Co2Z hexaferrite-polymer nanocomposite film for EMI shielding applications. Chem. Eng. J. **487**, 150323 (2024)
3. Guo, Z., Ren, P., Yang, F., Wu, T., Zhang, L., Chen, Z., Ren, F.: Robust multifunctional composite films with alternating multilayered architecture for highly efficient electromagnetic interference shielding, Joule heating and infrared stealth. Compos. B Eng. **263**, 110863 (2023)
4. Kumari, N., Meena, S., Singhal, R., Choubey, R.K., Dwivedi, U.K.: Chapter 8—advanced material for EMI shielding. In: Verma, S., Srivastava, A.K. (eds.) Advanced Radiation Shielding Materials, pp. 195–226. Elsevier (2024)
5. Wang, Y., Liang, L., Li, Y., Liu, C., Wu, Z., Li, G., Tian, Y., Zhu, B.: Facile construction of low-cost and high-efficiency microwave absorbent of Co/C/CG composite with dual-enhancement performance. Diam. Relat. Mater. **126**, 109038 (2022)
6. Raju, P., Thirupathi, A., Neelima, P., Kanakadurga, M.: Ni0.48Cu0.12Zn0.4Fe2O4+ paraformaldehyde nanocomposites as microwave absorbent dominant materials. Solid State Commun. **351**, 114790 (2022)
7. Calheiros Souto, L.F., Soares, B.G.: Electromagnetic wave absorption, EMI shielding effectiveness and electrical properties of ethylene—vinyl Acetate (EVA)/Polyaniline (PAni) blends prepared by in situ polymerization. Synth. Met. **298**, 117441 (2023)
8. Chen, J., Liu, Y.-L., Sun, D.-X., Qi, X.-D., Yang, J.-H., Wang, Y.: Recent progress in structural design of graphene/polymer porous composites toward electromagnetic interference shielding application. Chem. Eng. J. **495**, 153586 (2024)
9. Iqbal, S., Ahmad, S.: Chapter 16—conducting polymer composites: an efficient EMI shielding material. In: Joseph, K., Wilson, R., George, G. (eds.) Materials for Potential EMI Shielding Applications, pp. 257–266. Elsevier (2020)
10. Madhusudhan, C.K., Faisal, M., Maruthi, N., Raghavendra, N., Mahendra, K., Abdul Kadar, C.H.: 24—functionalized conducting polymer nanocomposites for EMI shielding applications. In: Patel, G., Deshmukh, K., Hussain, C.M. (eds.) Advances in Functionalized Polymer Nanocomposites, pp. 959–1000. Woodhead Publishing (2024)
11. Hamidinejad, M., Salari, M., Ma, L., Moghimian, N., Zhao, B., Taylor, H.K., Filleter, T., Park, C.B.: Electrically and thermally graded microcellular polymer/graphene nanoplatelet composite foams and their EMI shielding properties. Carbon **187**, 153–164 (2022)
12. Ma, L., Hamidinejad, M., Wei, L., Zhao, B., Park, C.B.: Absorption-dominant EMI shielding polymer composite foams: microstructure and geometry optimization. Mater. Today Phys. **30**, 100940 (2023)
13. Kaur, G., Adhikari, R., Cass, P., Bown, M., Gunatillake, P.: Electrically conductive polymers and composites for biomedical applications. RSC Adv. **5**(47), 37553–37567 (2015)
14. Tadesse, M.G., Ahmmed, A.S., Lübben, J.F.: Review on conductive polymer composites for supercapacitor applications. J. Compos. Sci. **8**(2), 53 (2024)

15. Kruželák, J., Kvasničáková, A., Hložeková, K., Hudec, I.: Progress in polymers and polymer composites used as efficient materials for EMI shielding. Nanoscale Adv. **3**(1), 123–172 (2021)
16. Kozak, N., Matzui, L., Vovchenko, L., Kosyanchuk, L., Oliynyk, V., Antonenko, O., Nesin, S., Gagolkina, Z.: Influence of coordination complexes of transition metals on EMI-shielding properties and permeability of polymer blend/carbon nanotube/nickel composites. Compos. Sci. Technol. **200**, 108420 (2020)
17. Ahmed, I., Khan, A.N., Jan, R., Gul, I.H.: Structure—properties relationships of graphene and spinel nickel ferrites based poly(vinylidene fluoride) hybrid polymer nanocomposites for improved dielectric and EMI shielding characteristics. Mater. Res. Bull. **148**, 111687 (2022)
18. Li, R., Ding, L., Gao, Q., Zhang, H., Zeng, D., Zhao, B., Fan, B., Zhang, R.: Tuning of anisotropic electrical conductivity and enhancement of EMI shielding of polymer composite foam via CO_2-assisted delamination and orientation of MXene. Chem. Eng. J. **415**, 128930 (2021)
19. Glowacki, I., Jung, J., Rybak, A., Ulanski, J.: Conductivity measurements. In: Reference Module in Materials Science and Materials Engineering. Elsevier (2016)
20. Shelly, D., Jin, F.-L., Lee, S.-Y., Park, S.-J.: Emerging trends in MXene-polymer composites for electromagnetic shielding applications. Compos. Commun. **51**, 102056 (2024)
21. Hazarika, D., Karak, N.: Chapter 1—fundamentals of polymeric nanostructured materials. In: Bajpai, A.K., Saini, R.K. (eds.) Advances in Polymeric Nanomaterials for Biomedical Applications, pp. 1–40. Elsevier (2021)
22. Al-Saleh, M.H.: Influence of conductive network structure on the EMI shielding and electrical percolation of carbon nanotube/polymer nanocomposites. Synth. Met. **205**, 78–84 (2015)
23. Pavlou, C., Pastore Carbone, M.G., Manikas, A.C., Trakakis, G., Koral, C., Papari, G., Andreone, A., Galiotis, C.: Effective EMI shielding behaviour of thin graphene/PMMA nanolaminates in the THz range. Nat. Commun. **12**(1), 4655 (2021)
24. Ma, X., Shen, B., Zhang, L., Liu, Y., Zhai, W., Zheng, W.: Porous superhydrophobic polymer/carbon composites for lightweight and self-cleaning EMI shielding application. Compos. Sci. Technol. **158**, 86–93 (2018)
25. Banjar, M.F., Joynal Abedin, F.N., Fizal, A.N.S., Muhamad Sarih, N., Hossain, M.S., Osman, H., Khalil, N.A., Ahmad Yahaya, A.N., Zulkifli, M.: Synthesis and characterization of a novel nanosized polyaniline. Polymers **15**(23), 4565 (2023)
26. Abu Hassan Shaari, H., Ramli, M.M., Abdullah, M.M.A.B., Mohtar, M.N., Abdul Rahman, N., Ahmad, A., Osman, N.H., Rusydi, F.: Effects of polymerization time towards conductivity and properties of poly(methyl methacrylate)/polyaniline (PMMA/PANi) copolymer. Sustainability **14**(14), 8940 (2022)
27. Orasugh, J.T., Ray, S.S.: Functional and structural facts of effective electromagnetic interference shielding materials: a review. ACS Omega **8**(9), 8134–8158 (2023)
28. Zachariah, S.M., Grohens, Y., Kalarikkal, N., Thomas, S.: Hybrid materials for electromagnetic shielding: a review. Polym. Compos. **43**(5), 2507–2544 (2022)
29. Vijay, S.J., Tugirumubano, A., Go, S.H., Kwac, L.K., Kim, H.G.: Chapter 15—carbon fiber-reinforced polymer-metal wire mesh hybrid composite for EMI shielding. In: Joseph, K., Wilson, R., George, G. (eds.) Materials for Potential EMI Shielding Applications, pp. 237–256. Elsevier (2020)
30. Thomassin, J.-M., Jérôme, C., Pardoen, T., Bailly, C., Huynen, I., Detrembleur, C.: Polymer/carbon based composites as electromagnetic interference (EMI) shielding materials. Mater. Sci. Eng. R. Rep. **74**(7), 211–232 (2013)
31. Nath, K., Bhattacharyya, S.K., Das, N.C.: Chapter 10—biodegradable polymeric materials for EMI shielding. In: Joseph, K., Wilson, R., George, G. (eds.) Materials for Potential EMI Shielding Applications, pp. 165–178. Elsevier (2020)
32. Gao, Q., Wang, X., Schubert, D.W., Liu, X.: Review on polymer/MXene composites for electromagnetic interference shielding applications. Adv. Nanocomposites **1**(1), 52–76 (2024)
33. Zhang, X., Zhang, X., Yang, M., Yang, S., Wu, H., Guo, S., Wang, Y.: Ordered multilayer film of (graphene oxide/polymer and boron nitride/polymer) nanocomposites: an ideal EMI shielding material with excellent electrical insulation and high thermal conductivity. Compos. Sci. Technol. **136**, 104–110 (2016)

34. Bhattacharjee, Y., Biswas, S., Bose, S.: Chapter 5—Thermoplastic polymer composites for EMI shielding applications. In: Joseph, K., Wilson, R., George, G. (eds.) Materials for Potential EMI Shielding Applications, pp. 73–99. Elsevier (2020)
35. Sushmita, K., Formanek, P., Fischer, D., Pötschke, P., Madras, G., Bose, S.: Ultrathin structures derived from interfacially modified polymeric nanocomposites to curb electromagnetic pollution. Nanoscale Adv. **3**(9), 2632–2648 (2021)
36. Gao, Y., Huang, L., Zheng, Z.J., Li, H., Zhu, M.: The influence of cobalt on the corrosion resistance and electromagnetic shielding of electroless Ni–Co–P deposits on Al substrate. Appl. Surf. Sci. **253**(24), 9470–9475 (2007)
37. Los, P., Lukomska, A., Jeziorska, R.: Metal-polymer composites for electromagnetic interference shielding applications. Polimery **61**(10), 663–669 (2021)
38. Palanisamy, S., Tunakova, V., Hu, S., Yang, T., Kremenakova, D., Venkataraman, M., Petru, M., Militky, J.: Electromagnetic interference shielding of metal coated ultrathin nonwoven fabrics and their factorial design. Polymers **13**(4), 484 (2021)
39. Sankaran, S., Deshmukh, K., Ahamed, M.B., Khadheer Pasha, S.K.: Recent advances in electromagnetic interference shielding properties of metal and carbon filler reinforced flexible polymer composites: a review. Compos. Part A: Appl. Sci. Manufact. **114**, 49–71 (2018)
40. Yang, Y., Li, J., Wang, Z., Ju, X., Duan, H., Tang, Y.: Structural design and performance regulation of green electromagnetic interference shielding conductive polymer composites: a review. Adv. Nanocomposites **1**(1), 290–303 (2024)
41. Liu, J.-T., Zheng, Y.-C., Hou, X., Feng, X.-R., Jiang, K., Wang, M.: Structured carbon for electromagnetic shielding and microwave absorption from carbonization of waste polymer: a review. Chem. Eng. J. **496**, 154013 (2024)
42. Praveen, M., Karthikeya, G.S., Krishna, R.H., Mamatha, G.M., Manjunatha, C., Khosla, A., Nagabhushana, B.M.: The role of magnetic nano $CoFe_2O_4$ and conductive MWCNT/graphene in LDPE-based composites for electromagnetic interference shielding in X-band. Diam. Relat. Mater. **130**, 109501 (2022)
43. Manjappa, P., Rajan, H.K., Mahesh, M.G., Sadananda, K.G., Channegowda, M., Shivashankar, G.K., Mutt, N.B.: Effective attenuation of electromagnetic waves by synergetic effect of α-Fe_2O_3 and MWCNT/graphene in LDPE-based composites for EMI applications. Materials **15**(24), 9006 (2022)
44. Yang, C.P., Smith, P.A., Krupka, J., Button, T.W.: The losses of microwave ferrites at communication frequencies. J. Eur. Ceram. Soc. **27**(8), 2765–2770 (2007)
45. Song, N.-N., Yang, H.-T., Liu, H.-L., Ren, X., Ding, H.-F., Zhang, X.-Q., Cheng, Z.-H.: Exceeding natural resonance frequency limit of monodisperse Fe_3O_4 nanoparticles via superparamagnetic relaxation. Sci. Rep. **3**(1), 3161 (2013)
46. Nath, K., Ghosh, S.K., Das, N.C.: 13—electrical and electronic applications of polymer-graphene composites. In: Rahaman, M., Nayak, L., Hussein, I.A., Das, N.C. (eds.) Polymer Nanocomposites Containing Graphene, pp. 343–377. Woodhead Publishing (2022)
47. Nasreen Taj, M., Daruka Prasad, B., Narapareddy, R., Nagabhushana, H., Ramakrishna, G., Mahesh, B., Dadami, S.T.: PANI-molybdate nanocomposites: structural, morphological and dielectric properties for the effective electromagnetic interference (EMI) shielding applications in X-band. Appl. Surf. Sci. Adv. **7**, 100203 (2022)
48. Hsiao, S.-T., Ma, C.-C.M., Tien, H.-W., Liao, W.-H., Wang, Y.-S., Li, S.-M., Huang, Y.-C.: Using a non-covalent modification to prepare a high electromagnetic interference shielding performance graphene nanosheet/water-borne polyurethane composite. Carbon **60**, 57–66 (2013)
49. Zhang, C.-S., Ni, Q.-Q., Fu, S.-Y., Kurashiki, K.: Electromagnetic interference shielding effect of nanocomposites with carbon nanotube and shape memory polymer. Compos. Sci. Technol. **67**(14), 2973–2980 (2007)
50. Shen, L., Wang, F.Q., Yang, H., Meng, Q.R.: The combined effects of carbon black and carbon fiber on the electrical properties of composites based on polyethylene or polyethylene/polypropylene blend. Polym. Testing **30**(4), 442–448 (2011)

51. Chung, D.D.L.: Carbon materials for structural self-sensing, electromagnetic shielding and thermal interfacing. Carbon **50**(9), 3342–3353 (2012)
52. Al-Ghamdi, A.A., El-Tantawy, F.: New electromagnetic wave shielding effectiveness at microwave frequency of polyvinyl chloride reinforced graphite/copper nanoparticles. Compos. A Appl. Sci. Manuf. **41**(11), 1693–1701 (2010)
53. Wang, M., Tang, X.-H., Cai, J.-H., Wu, H., Shen, J.-B., Guo, S.-Y.: Construction, mechanism and prospective of conductive polymer composites with multiple interfaces for electromagnetic interference shielding: a review. Carbon **177**, 377–402 (2021)
54. Dabas, S., Chahar, M., Thakur, O.P.: Electromagnetic interference shielding properties of $CoFe_2O_4$/polyaniline/poly(vinylidene fluoride) nanocomposites. Mater. Chem. Phys. **278**, 125579 (2022)
55. Zhou, C., Wang, X., Luo, H., Deng, L., Wang, S., Wei, S., Zheng, Y., Jia, Q., Liu, J.: Interfacial design of sandwich-like CoFe@$Ti_3C_2T_x$ composites as high efficient microwave absorption materials. Appl. Surf. Sci. **494**, 540–550 (2019)
56. Liu, G., Wang, L., Zhang, H., Du, Z., Zhou, X., Wang, K., Sun, Y., Gao, S.: Cage-structured $CoFe_2O_4$@CNTs from Fe–Co-MOF confined growth in CNTs for high electromagnetic wave absorption performances. Compos. Commun. **27**, 100910 (2021)
57. Wang, Y., Wang, H., Ye, J., Shi, L., Feng, X.: Magnetic CoFe alloy@C nanocomposites derived from ZnCo-MOF for electromagnetic wave absorption. Chem. Eng. J. **383**, 123096 (2020)
58. Kazmi, S.J., Nadeem, M., Warsi, M.A., Manzoor, S., Shabbir, B., Hussain, S.: PVDF/CFO-anchored CNTs ternary composite system with enhanced EMI shielding and EMW absorption properties. J. Alloy. Compd. **903**, 163938 (2022)
59. Jiang, P., Xu, Q., Tran, N., El-Shafay, A.S., Mohanavel, V., Abdelrahman, A., Ravichandran, M.: Boosted microwave absorption properties of $CoFe_2O_4$ with extraordinary 3D morphologies. Ceram. Int. **48**(10), 13541–13550 (2022)
60. Jaiswal, R., Agarwal, K., Pratap, V., Soni, A., Kumar, S., Mukhopadhyay, K., Eswara Prasad, N.: Microwave-assisted preparation of magnetic ternary core-shell nanofiller ($CoFe_2O_4$/rGO/SiO_2) and their epoxy nanocomposite for microwave absorption properties. Mater. Sci. Eng.: B **262**, 114711 (2020)
61. Zhang, L., Stalin, N., Tran, N., Mehrez, S., Badran, M.F., Mohanavel, V., Xu, Q.: Development of high-efficient double-layer microwave absorbers based on 3D cabbage-like $CoFe_2O_4$ and cauliflower-like polypyrrole. Ceram. Int. **48**(11), 16374–16385 (2022)
62. Dalal, J., Malik, S., Dahiya, S., Punia, R., Singh, K., Maan, A.S., Dhawan, S.K., Ohlan, A.: One pot synthesis and electromagnetic interference shielding behavior of reduced graphene oxide nanocomposites decorated with $Ni0.5Co0.5Fe_2O_4$ nanoparticles. J. Alloy. Compd. **887**, 161472 (2021)
63. Jafarian, M., Kashani-Bozorg, S.F., Amadeh, A.A., Atassi, Y.: Hierarchical brain-coral-like structure (3D) vs rod-like structure (1D): effect on electromagnetic wave loss features of $SrFe_{12}O_{19}$ and $CoFe_2O_4$. Ceram. Int. **47**(21), 30448–30458 (2021)
64. Yuan, Y., Wei, S., Liang, Y., Wang, B., Wang, Y., Xin, W., Wang, X., Zhang, Y.: Solvothermal assisted synthesis of $CoFe_2O_4$/CNTs nanocomposite and their enhanced microwave absorbing properties. J. Alloy. Compd. **867**, 159040 (2021)
65. Ren, F., Guo, Z., Shi, Y., Jia, L., Qing, Y., Ren, P., Yan, D.: Lightweight and highly efficient electromagnetic wave-absorbing of 3D CNTs/GNS@$CoFe_2O_4$ ternary composite aerogels. J. Alloy. Compd. **768**, 6–14 (2018)
66. Gandhi, N., Singh, K., Ohlan, A., Singh, D.P., Dhawan, S.K.: Thermal, dielectric and microwave absorption properties of polyaniline–$CoFe_2O_4$ nanocomposites. Compos. Sci. Technol. **71**(15), 1754–1760 (2011)
67. Ashfaq, M.Z., Ashfaq, A., Majeed, M.K., Saleem, A., Wang, S., Ahmad, M., Hussain, M.M., Zhang, Y., Gong, H.: Confined tailoring of $CoFe_2O_4$/MWCNTs hybrid-architectures to tune electromagnetic parameters and microwave absorption with broadened bandwidth. Ceram. Int. **48**(7), 9569–9578 (2022)
68. Yadav, P., Rattan, S., Tripathi, A., Kumar, S.: Tailoring of complex permittivity, permeability, and microwave-absorbing properties of $CoFe_2O_4$/NG/PMMA nanocomposites through swift heavy ions irradiation. Ceram. Int. **46**(1), 317–324 (2020)

69. Ren, X., Wang, J., Yin, H., Tang, Y., Fan, H., Yuan, H., Cui, S., Huang, L.: Hierarchical $CoFe_2O_4$@PPy hollow nanocubes with enhanced microwave absorption. Appl. Surf. Sci. **575**, 151752 (2022)
70. Zhang, K., Gao, X., Zhang, Q., Li, T., Chen, H., Chen, X.: Preparation and microwave absorption properties of asphalt carbon coated reduced graphene oxide/magnetic $CoFe_2O_4$ hollow particles modified multi-wall carbon nanotube composites. J. Alloy. Compd. **723**, 912–921 (2017)
71. Ding, L., Zhao, X., Huang, Y., Yan, J., Li, T., Liu, P.: Ultra-broadband and covalently linked core–shell $CoFe_2O_4$@PPy nanoparticles with reduced graphene oxide for microwave absorption. J. Colloid Interface Sci. **595**, 168–177 (2021)
72. Tong, Z., Yao, Q., Deng, J., Cheng, L., Chuang, T., Wang, J., Rao, G., Zhou, H., Wang, Z.: Effects of Ni-doping on microstructure, magnetic and microwave absorption properties of $CoFe_2O_4$. Mater. Sci. Eng. B **268**, 115092 (2021)
73. Feng, J., Wang, Y., Hou, Y., Li, J., Li, L.: Synthesis and microwave absorption properties of coiled carbon nanotubes/$CoFe_2O_4$ composites. Ceram. Int. **42**(15), 17814–17821 (2016)
74. Zhang, M., Zhang, J., Lin, H., Wang, T., Ding, S., Li, Z., Wang, J., Meng, A., Li, Q., Lin, Y.: Designable synthesis of reduced graphene oxide modified using $CoFe_2O_4$ nanospheres with tunable enhanced microwave absorption performances between the whole X and Ku bands. Compos. B Eng. **190**, 107902 (2020)
75. Li, G., Sheng, L., Yu, L., An, K., Ren, W., Zhao, X.: Electromagnetic and microwave absorption properties of single-walled carbon nanotubes and $CoFe_2O_4$ nanocomposites. Mater. Sci. Eng. B **193**, 153–159 (2015)
76. Liu, P., Huang, Y., Zhang, X.: Synthesis, characterization and excellent electromagnetic wave absorption properties of graphene@$CoFe_2O_4$@polyaniline nanocomposites. Synth. Met. **201**, 76–81 (2015)
77. Li, Y., Yuan, M., Liu, H., Sun, G.: In situ synthesis of $CoFe_2O_4$ nanocrystals decorated in mesoporous carbon nanofibers with enhanced electromagnetic performance. J. Alloy. Compd. **826**, 154147 (2020)
78. Liang, C., Gu, Z., Zhang, Y., Ma, Z., Qiu, H., Gu, J.: Structural design strategies of polymer matrix composites for electromagnetic interference shielding: a review. Nano-Micro Lett. **13**(1), 181 (2021)
79. Antunes, M.: Recent trends in polymeric foams and porous structures for electromagnetic interference shielding applications. Polymers **16**(2), 195 (2024)
80. Sathish Kumar, K., Rengaraj, R., Venkatakrishnan, G.R., Chandramohan, A.: Polymeric materials for electromagnetic shielding—a review. Mater. Today: Proc. **47**, 4925–4928 (2021)
81. Ryan, E.A., Seibers, Z.D., Reynolds, J.R., Shofner, M.L.: Electrically conducting polymers and composites for applications in space exploration. J. Appl. Polym. Sci. **141**(15), e55225 (2024)
82. Wang, Y., Zhao, W., Tan, L., Li, Y., Qin, L., Li, S.: Review of polymer-based composites for electromagnetic shielding application. Molecules **28**(15) (2023)
83. Magnetic characterization. In: Wang, Z.L., Liu, Y., Zhang, Z. (eds.) Handbook of Nanophase and Nanostructured Materials, pp. 595–625. Springer US, Boston, MA (2002)
84. Varghese, J., Joseph, N., Jantunen, H., Behera, S.K., Kim, H.T., Sebastian, M.T.: Microwave materials for defense and aerospace applications. In: Mahajan, Y., Roy, J. (eds.) Handbook of advanced ceramics and composites: defense, security, aerospace and energy applications, pp. 1–48. Springer International Publishing, Cham (2019)
85. Abbasi, H., Antunes, M., Velasco, J.I.: Effects of carbon nanotubes/graphene nanoplatelets hybrid systems on the structure and properties of polyetherimide-based foams. Polymers **10**(4), 348 (2018)
86. Elmas, O.: Effects of electromagnetic field exposure on the heart: a systematic review. Toxicol. Ind. Health **32**(1), 76–82 (2013)
87. Wang, L., Ma, Z., Zhang, Y., Chen, L., Cao, D., Gu, J.: Polymer-based EMI shielding composites with 3D conductive networks: a mini-review. SusMat **1**(3), 413–431 (2021)
88. Mensah, E.E., Abbas, Z., Azis, R.S., Ibrahim, N.A., Khamis, A.M., Abdalhadi, D.M.: Complex permittivity and power loss characteristics of α-Fe_2O_3/polycaprolactone (PCL) nanocomposites: effect of recycled α-Fe_2O_3 nanofiller. Heliyon **6**(12), e05595 (2020)

89. Tong, G., Wu, W., Guan, J., Qian, H., Yuan, J., Li, W.: Synthesis and characterization of nanosized urchin-like α-Fe_2O_3 and Fe_3O_4: microwave electromagnetic and absorbing properties. J. Alloy. Compd. **509**(11), 4320–4326 (2011)
90. Gargama, H., Thakur, A.K., Chaturvedi, S.K.: Polyvinylidene fluoride/nanocrystalline iron composite materials for EMI shielding and absorption applications. J. Alloy. Compd. **654**, 209–215 (2016)
91. Singh, K., Ohlan, A., Kotnala, R.K., Bakhshi, A.K., Dhawan, S.K.: Dielectric and magnetic properties of conducting ferromagnetic composite of polyaniline with γ-Fe_2O_3 nanoparticles. Mater. Chem. Phys. **112**(2), 651–658 (2008)
92. Bhaskaran, K., Bheema, R.K., Etika, K.C.: The influence of Fe_3O_4@GNP hybrids on enhancing the EMI shielding effectiveness of epoxy composites in the X-band. Synth. Met. **265**, 116374 (2020)
93. Wan, C., Li, J.: Synthesis and electromagnetic interference shielding of cellulose-derived carbon aerogels functionalized with α-Fe_2O_3 and polypyrrole. Carbohyd. Polym. **161**, 158–165 (2017)
94. Chen, K.-Y., Gupta, S., Tai, N.-H.: Reduced graphene oxide/Fe_2O_3 hollow microspheres coated sponges for flexible electromagnetic interference shielding composites. Compos. Commun. **23**, 100572 (2021)
95. Varshney, S., Singh, K., Ohlan, A., Jain, V.K., Dutta, V.P., Dhawan, S.K.: Synthesis, characterization and surface properties of Fe_2O_3 decorated ferromagnetic polypyrrole nanocomposites. J. Alloy. Compd. **538**, 107–114 (2012)
96. Moazen, S., Sahebian, S., Haddad-Sabzevar, M.: Low percolation behavior of HDPE/CNT nanocomposites for EMI shielding application: random distribution to segregated structure. Synth. Met. **281**, 116900 (2021)
97. Wu, G., Cheng, Y., Ren, Y., Wang, Y., Wang, Z., Wu, H.: Synthesis and characterization of γ-Fe_2O_3@C nanorod-carbon sphere composite and its application as microwave absorbing material. J. Alloy. Compd. **652**, 346–350 (2015)

Chapter 8
PCB-Based Absorbers

Abstract PCB-based absorbers are an alternative to material absorbers. Here, periodic unit cells would be designed on a specific electrically small dielectric substrate. This chapter delves into PCB-based absorbers' design, construction, and characterization. A classification of PCB-based absorbers is also explored in the chapter. A detailed analysis of each of the ten classes of absorbers is also illustrated. Each absorber class is introduced with generic schematics, which demonstrate the specific features of a particular absorber class. Numerous design examples are introduced for each class of absorber. The contrast between the features of PCB-based absorbers and their material components is also explained in this chapter.

8.1 Introduction

In the previous chapters, material-oriented design approaches for high-shielding effectiveness were explored. Numerous material synthesis and characterization methodologies were also examined. The concoction of different types of materials would yield high shielding effectiveness. The ingredients that make up the composite material would exhibit properties and characteristics that are not readily available in nature. For instance, low-density polyethylene (LDPE) alone would dB offer an insertion loss close to 1–2 in the X- and Ku-bands. On the other hand, when this polymer is blended with other composite mixtures with specific proportions, the engineered material delivers high absorption. In other words, the transmission and reflection characteristics would be drastically changed.

Chapters 5, 6, and 7 illustrated numerous composites that deliver close to 100% absorption. In some of the applications, the chemical method of material engineering might not be suitable for large-scale production. Most shielding effectiveness demonstrations were restricted to laboratory prototypes; industrial production and analytical design for a specific frequency spectrum are poorly established. In other words, the material scientists must run over many iterations to land up with the appropriate set of chemicals within the polymer base. The optimized set of chemicals may not be operational in other bands. Hence, the lack of standardization of the design process leads

S. K. Koul et al., *Microwave Absorbers*, Engineering Materials,
https://doi.org/10.1007/978-981-96-9596-6_8

to alternate shielding methods. The engineered material might not offer wide-angle incidence or other similar features. It must also be noted that material engineering primarily relies on inducing or enhancing dielectric and magnetic loss within the material, which, in turn, alters or enhances the shielding metrics. The printed circuit board (PCB)-based absorbers discussed in this chapter contrast this approach.

One of the popular PCB-based absorbers is the frequency selective surface (FSS)-based design. This chapter delves into the various aspects of the FSS-based absorbers operating in different spectra and engineered for multiple applications. Any FSS is built upon a primary PCB on a specific dielectric material; typically, these boards have two-sided designs. One of the sides of these boards would be a particular geometry designed on metal. Industrial boards have copper-based designs. This geometry would exhibit certain properties favorable for shielding.

The other layer of the board would be typically a complete copper pour. This is the generic layout for most of the FSS-based designs. It must be noted that these FSS-based absorbers operate in one of the directions as absorbers. To elaborate, when a plane wave is incident on the FSS structure, customized metallic pattern side, the board behaves as an absorber. In other words, when Electromagnetic (EM) energy is incident on the pattern layer, the transmission and reflection would be close to zero or $-\infty$ dB. On the other hand, when energy is incident on the other side of the absorber, it behaves as a perfect reflector, as the design is a complete copper pour. Material designs investigated in the previous chapters do not encounter this problem. The absorption phenomenon in the periodic structure is due to the resonance effect. The metal geometry designed on one of the layers of the substrates would act as a resonator. This resonator prohibits transmission and reflection of the incident EM energy. Numerous classes of examples illustrate this concept throughout this chapter.

FSS-based absorbers rely heavily on software tools to quantify performance metrics. In other words, FSS designs are analytically compared to experimental works in material engineering. Typically, an FSS is initially characterized by its unit cell, a specific metallic geometry that eventually would be plugged into one of the sides of the dielectric substrate. The unit cell is designed using an industry-standard full-wave solver such as Ansys® HFSS® (High-Frequency Structure Simulator) or CST® microwave Studio®.

In the full-wave solver, the patterned geometry on the dielectric is maintained. A waveguide setup with periodic boundary conditions on either side of the unit cell is established. A plane wave of a certain frequency is incident upon the structure. The setup's S-parameters $|S_{11}|$ and $|S_{21}|$ are analyzed for desired behavior. The geometry on the top layer of the dielectric would be tuned and optimized to relocate the absorption frequency. The absorption bandwidth could also be altered by varying this geometry. In addition to incorporating thicker dielectric layers, it must be noted that the thickness of the dielectric substrate would typically be a fraction of the lowest operating wavelength. The commercially available, inexpensive substrates generally design absorbers in the microwave bands. Once these structures are created, they will be fabricated using industry-standard photolithography.

The measurement technique discussed in Sect. 4.2 would not characterize the FSS-based absorbers. The primary reason is that these are PCB-based. Absorbers are sensitive to a wide range of parameters, including angle of incidence, operation frequency, and incident energy polarization. To accommodate these parameters, the FSS absorbers would be kept in front of two wideband horn antennas, which are kept next to each other. One of the horns acts as a transmitter, and the other horn antenna acts as a receiver. With this arrangement, the absorber could be characterized. The error due to path loss could be quantized using an electrically large metallic sheet. Each frequency's values must be subtracted from the receiver horn values.

8.2 Classifications of FSS Absorbers

FSS-based absorbers could be classified into numerous types. There is no strict classification of FSS absorbers. The present section discusses one way to categorize PCB-based absorbers. The parameters discussed in the previous section come in handy to categorize absorbers based on numerous performance metrics and other relevant parameters. Absorbers, in general, could be classified based on the following metrics.

- **Bandwidth of operation**: The absorption bandwidth is an important metric when designing and deploying PCB-based absorbers [1–15].
- **Narrowband** designs typically exhibit absorption bandwidth in the 2–10% range.
- **Wideband designs** are electrically thick and deliver absorption bandwidth in the 3–100% range. Multi-octave super wideband absorbers are challenging to design with PCB-based methodology. The absolute bandwidth is usually dictated by the application in hand; multiband absorbers exhibit narrow bandwidths at different resonant frequencies.
- **Frequency of operation**: absorbers could be classified based on their operating frequency [16–24]. The limitation of the photolithography process governs the lower limit on the feasible designs. On the other hand, the higher frequency limit is judged by the process limitation of the integrated circuits. Ultra-High-Frequency (UHF) bands are a feasible choice for designing absorbers in the lower frequency. Any frequency band below this would lead to electrically and physically large structures. These PCB-based structures would be difficult to realize with commercial technologies.

In other words, multiple PCBs would be essential to delivering the desired absorption metrics. This would be an expensive process. Terahertz-based absorbers have very low electrical dimensions. Hence, they need to be manufactured with alternate precision technologies. FSS-based absorbers would be functional in the UHF to D-band. PCB-based designs engineered for absorption would be challenging to realize beyond these bands.

- **Application of active circuitry**: One of the striking advantages of realizing PCB-based designs is the application of active circuitry. The absorption characteristics of PCB-based design could be altered by applying active circuitry, as illustrated in [25–29]. Here, the metallic geometry on one of the sides of the PCB would be carefully integrated with active devices. These devices could be p–i–n diodes, transistor chips, etc., and depending on the control signal or activation signal to these devices, the metrics of the absorber could be altered. The primary absorption frequency could be changed by adjusting the electrical lengths within the structure of the FSS; other operating metrics of the absorbers could be altered or reconfigured by the application of active circuits. This dynamic variation of key absorption metrics is possible only with PCB-based board designs. This feature starkly contrasts the bulk material design, which offers static characteristics at a specific frequency.

- **Angle of incidence**: PCB-based absorbers could be classified based on their response to the wide angle of incidence. Most real-world deployment scenarios demand a wide incidence angle, meaning that the absorber realized at a certain resonant frequency must deliver similar characteristics when illuminated with incoming energy from various angles. Numerous research articles are available in the literature on this subject [30–36]. This class of absorbers exhibits desirable absorption characteristics within $\pm$ 60°; a detailed analysis will be presented in Sect. 8.3.7.
- **Polarization insensitivity**. Another important class of PCB absorbers is the class of absorbers that are agnostic to the polarization of the incident energy. Most of the real-world scenarios demand polarization insensitivity. The fundamental reason for this requirement is that the incident energy in the real world is never linearly polarized. The incident EM radiation would be a composite signal with wide bandwidth and electrical polarization. The PCB-based absorber must cater to these types of incoming signals. Hence, researchers design and develop absorbers with identical absorption characteristics for TE and TM waves. Several designs have been demonstrated in [37–42].
- **Structure**: PCB-based absorbers can be classified based on their mechanical structure. Commercial PCB-based designs on a substrate with a thickness greater than or equal to 300 μm could be considered rigid [1–42]. If the thickness is less than this value, the absorber offers flexibility. Flexible PCB-based absorbers are essential for designing and realizing conformal structures. Many real-world applications dictate the need for conformal absorbers, as a product on which the absorbers need to be integrated would be non-standard geometry. Absorbers incorporating alternate processes to photolithography could be used to realize conformal absorbers [43–46].
- **Transparency**: PCB-based absorbers can also be classified based on optical properties. Certain applications might warrant optical transparency without significant compromise in the absorption characteristics in the microwave bands. Transparent conductive sheets could be used to realize these designs, as demonstrated in [47–51].

8.3 Glimpse of FSS Absorbers

Each type of absorber will be discussed in detail in the subsequent subsections.

8.3.1 *Narrowband Absorbers*

Narrowband absorbers are essential in stealth applications to block and absorb a specific band of frequencies. Most military applications operate with narrow transmission bandwidths; the flying target would be typically attached to the stealthy absorber. These absorbers have a capacity in the range of 1–5%. The absorptivity in the operating band would be higher than 80%. The design of narrowband absorbers would involve a single dielectric substrate with metal patterns on either side of the substrate; a popular unit cell that serves as a narrowband absorber is illustrated as an absorber in Fig. 8.1.

The unit cell is typically a fraction of the operating wavelength at a certain band within the absorption spectrum, typically 0.1–0.3 λ. Manufacturing the PCB will be challenging if the sub-wavelength size exceeds this value. On the other hand, if the unit cell is close to half wavelength size, then the FSS phenomenon would cease to exist.

Hence, the size of the unit cell is an important dimension to consider before designing a narrowband, PCB-based absorber. The copper geometry would be one of the layers of a low-loss dielectric. A simple square loop is shown in the schematics. The metal geometry would be discontinuous in practical designs to mimic the inductance and capacitance effects. The top pattern metal layer may or may not involve lumped resistors. The application of chip resistors would slightly enhance absorption bandwidth. The low-loss dielectric's thickness would also be a small fraction of the operating wavelength. The bottom layer of the illustrated absorber would be a complete copper structure. The exploded view of the typical narrowband absorber gives a clearer insight into the construction, as shown in Fig. 8.2.

The industry-standard chemical etching process and the standard pick-and-place process for placing chip resistors would be sufficient to fabricate the narrowband absorbers. The unit cell would be initially analyzed for its S-parameters in a full-wave solver. The behavior of the unit cell and that of the scaled version would be

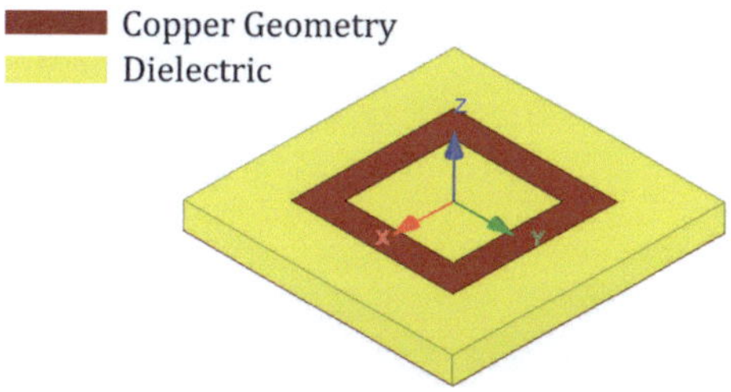

Fig. 8.1 Schematic of a unit cell for narrowband absorption

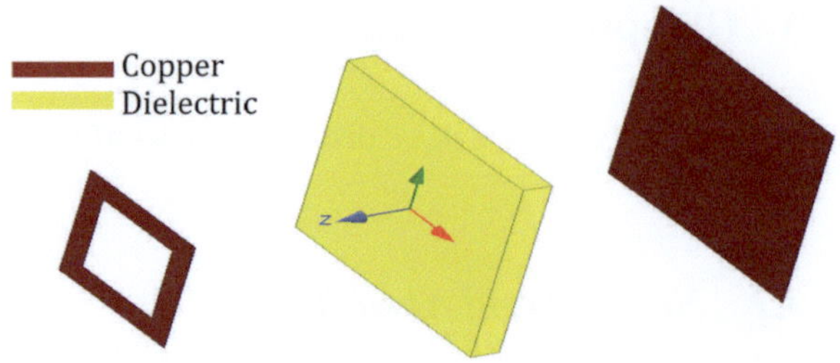

Fig. 8.2 Schematics of a unit cell for narrowband absorption (exploded view)

identical, assuming that the unit cells would be repeated infinitely on both axes. In practice, the overall length and width of the absorber would be several wavelengths long. The wavelength computed at the lowest operating frequency and absorber with 5 λ × 5 λ would deliver the desired absorption characteristics. The dimensional requirement for the overall absorber is decided by the application at hand. A software model of the overall absorber is illustrated in Fig. 8.3. The example is, for a 10 λ × 10 λ size, assuming the unit cell was designed for 0.1 λ. Narrowband absorbers would deliver insertion or transmission loss lower than − 10 dB with a reflection coefficient less than − 10 dB. In other words, the absorber would neither transmit nor reflect the incident energy in the narrow operation bandwidth. Several designs have existed in the literature, as evident from the sources [1–4]; only a sample among myriad strategies of narrowband operation is illustrated here.

Dipole-like structures integrated with metallic slots are proposed in [1]. Here, the PCB-based structure is transparent to microwaves at certain frequencies and acts as an absorber for several other frequency bands. Lumped chip resistors dissipate energy in the lossy layer of the proposed design. The resistor acts like the lossy dielectric in material engineering, aiding energy dissipation. The design uses an F4B substrate with a thickness of 20 mils, and the dielectric loss tangent is close to zero. The fabricated prototype is shown in Fig. 8.4, and its corresponding experimental absorption is demonstrated in Fig. 8.5.

The minimal dielectric loss tangent proves that the absorption is primarily due to the geometry of the periodic unit cells. A narrowband resistorless design is proposed

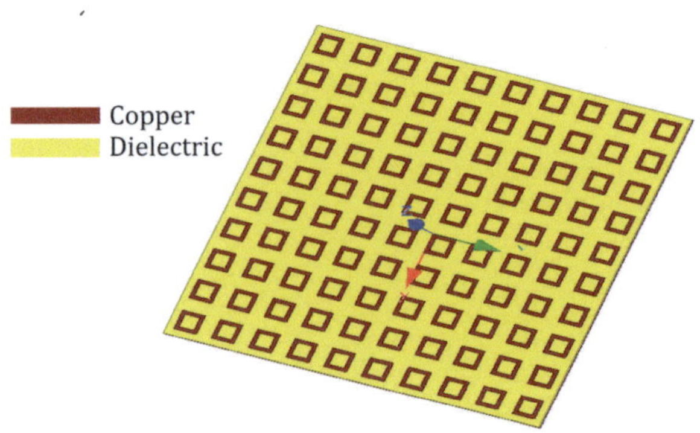

Fig. 8.3 Schematics of a narrowband absorber (simulation model)

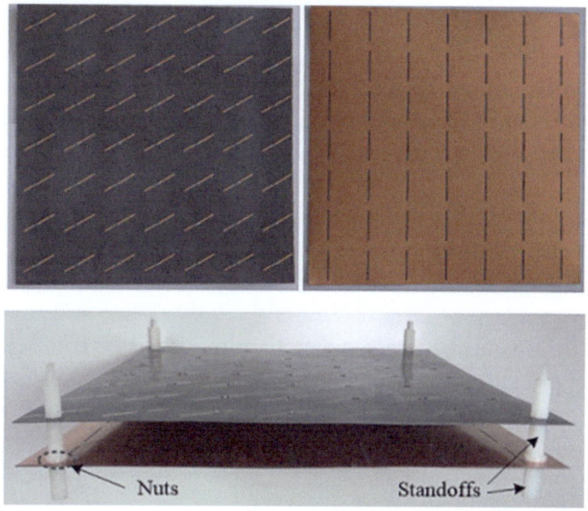

Fig. 8.4 Photograph of a narrowband absorber [1]

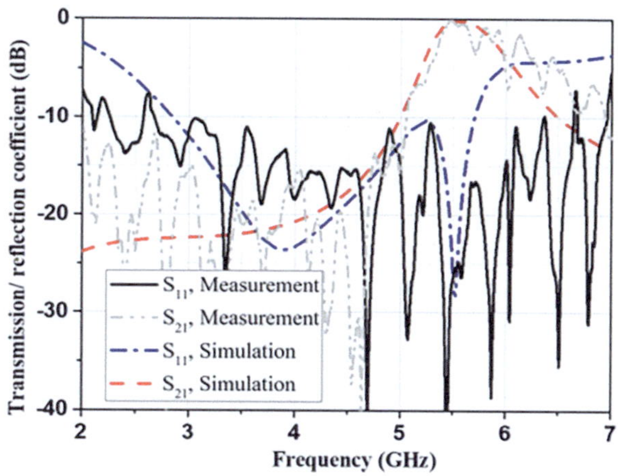

Fig. 8.5 Absorption characteristics of the absorber [1]

in [2]; the design is realized on the inexpensive FR-4 substrate with a thickness of 0.8 mm, and the top layer consists of two concentric circles, which act as a spatial resonator. The bottom layer of the substrate has a complimentary split ring resonator kind of geometry. The slots in the ground plane of the absorber allow for transmission windows at certain bands; the transmission and absorption spectra are completely controlled by the design of the unit cells. The authors have experimentally verified the FSS behavior in the desired bands.

Another variation of FSS-based absorbers is illustrated in [3]. On the other hand, inexpensive broadband resistors are used in the lossy layer. A slot is created in the lossless layer to engineer the transmission bands. A sub-wavelength spacing is made between the lossy and lossless layers, which aids in the bandwidth enhancement. The distance between the layers is proportional to the absorption bandwidth. The spacing could be realized using dielectric spacers of precise dimensions. An ultra-narrowband FSS structure operating in the Ka-band is illustrated in [4]. The design works on strong resonance, as dictated by the choice of the dimensions within the unit cell. Even though the design demonstrated is for a transmission window, the same design logic could be applied to realize a narrow ultra-narrowband absorber.

8.3.2 Multiband Absorbers

Multiband absorbers are another important PCB-based absorber with a wide range of applications in the defense and aerospace sectors. Most PCB-based absorber designs find applications in stealth technology. The military radars operate at very specific bands. As high-power amplifiers are involved in the transmitter, harmonic transmission and out-of-band leakage are strongly possible. If these signals are transmitted to the target, the fundamental frequency might be absorbed, but the radar signature on the adjacent or harmonic bands might be compromised.

To mitigate this effect, multiband absorbers could be designed. These designs extend the design principles used in narrowband absorbers. The generic structure of a multiband PCB-based absorber looks like that of a narrowband design. The only change would be in the construction of the unit cell. The unit cell would be a combination of different pieces of geometry that deliver or promote absorption at certain specific bands. A few design examples are presented in this subsection to illustrate the process.

A dual-band absorber is presented in [5], which exhibits absorption bandwidth bands in 4–10 and 13–19 GHz bands. It must be noted that the spacing between multiple layers controls the absorption bandwidth within each band. Spacing plays a critical role in deciding the band rejection frequencies. With multiple layers, bandwidth enhancement within each absorption band could be realized. The design uses an F4B substrate with a low dielectric constant and ultra-low loss. The experimental verification is performed on a 270 mm × 270 mm sample, which translates to 3.6 $\lambda \times 3.6\ \lambda$ where λ is calculated at 4 GHz, the least operating frequency. The same sample would be electrically large, 17.1 $\lambda \times 17.1\ \lambda$. λ is computed at 19 GHz. A triple-band absorber is demonstrated in [6]; here, a lossy resistor is integrated with one of the layers, which, in turn, is co-designed with a slotted lossless layer. The overall thickness of the absorber is 7 mm; the absorption bands are 7–7.5, 8.9–9.4, and 12.6–14.2 GHz. The absorption bands and the bandwidths within each band could be altered with the geometry.

The concept of T-shaped coupled slots with chip resistors is illustrated in [7]. The design exhibits dual-band absorption. Due to wideband chip resistors, the bandwidth in each band is high. An equivalent circuit analysis is also demonstrated, which draws an analogy between the physical structure and an electrical circuit equivalent. Another popular approach to designing multiband absorbers is incorporating lumped inductors and capacitors to realize the design, as illustrated in [8]. These lumped components reduce the absorber's physical footprint. In principle, most designs discussed in this subsection could be redesigned with lumped components. Spatial bandpass filter designs are illustrated in [9, 10].

8.3.3 *Wideband Absorbers*

Wideband absorbers are another important class of absorbers for numerous reasons. If the absorption bandwidth exceeds 50%, it could be classified as a wideband absorber. Transmission and reflection must be less than − 10 dB throughout the operating spectrum without transmission windows. In the middle of the band, fabrication technologies limit the absorption bandwidth. Most wideband absorber designs have a wide bandwidth ranging from 30 to 300% [11–15].

Multi-octave bandwidth would be extremely challenging to obtain for various reasons. For instance, if an absorber must be operational at 3.5 and 60 GHz, the corresponding sub-wavelength structure would be 8.5 mm and 500 μm, respectively. These unit cells must be interlaced with periodicity, which would be impossible to realize. Hence, the achievable bandwidth range is governed by the structural feasibility and the topology of the placement of the optimized unit cells. An illustrative schematic for a wide band unit cell is displayed in Fig. 8.6.

The construction is almost identical to that of a narrowband design. The thickness of the substrate used determines the absorption bandwidth; thicker substrates tend to deliver a relatively wider absorption bandwidth. The dielectric may be interlaced with additional unit cell structures to achieve a wide bandwidth. This subsection discusses a few design examples.

A three-layer design with aperture-coupled resonators is presented in [11]. Here, the absorption bandwidth is greater than 100%. The multi-layer design operates at closely spaced bands, resulting in a wide absorption bandwidth. The last layer of the design has 400 resistors soldered onto the dielectric board. The design has been experimentally verified. In conjunction with chip resistors, the lumped capacitors are engineered for active impedance matching in the 1.6–2.3 GHz band [12]. The lumped circuit elements offer an electrically compact solution compared to the pure microstrip transmission lines. The design illustrates a hybrid use of lumped capacitors with the transmission line geometry.

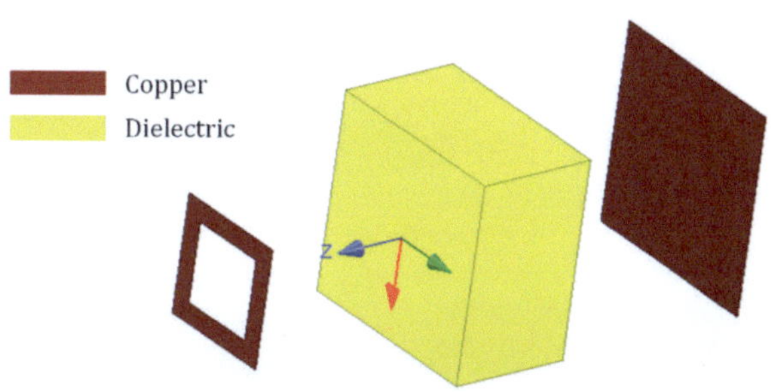

Fig. 8.6 Schematics of a unit cell for wideband characteristics (exploded view)

An ultra-wideband absorber with low insertion and low Infrared (IR) emissivity is showcased in [13]. The absorption spectrum runs from 2.16 to 8 GHz, and the Infrared (IR) emissivity is less than 0.23 in the entire IR band. Combined with the strategic placement of chip resistors, sinusoidal microstrip lines lead to an extremely wide operation bandwidth. IR coating is realized at the plane of incidence to reduce the thermal signature of the absorber. The fabricated prototype and key experimental results are recreated in Figs. 8.7 and 8.8, respectively.

A multi-resonant design is proposed in [14], which delivers an absorption bandwidth greater than 150%; this phenomenon could be attributed to the closely spaced resonances within the absorber. Optimization techniques could be employed to realize the design of absorbers; one such example is illustrated in [15]. Here, the

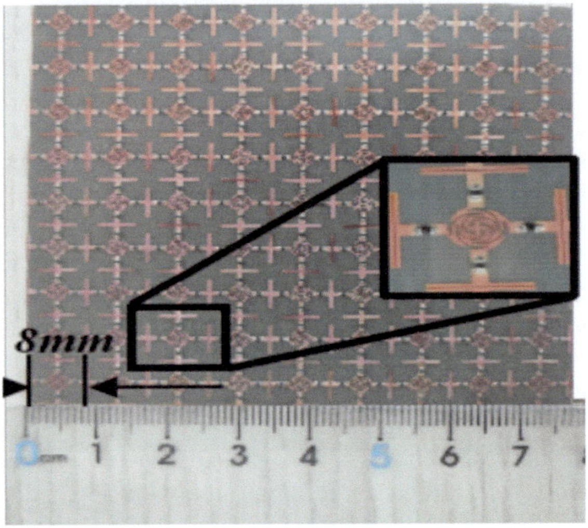

Fig. 8.7 Ultra-wideband absorber with low IR emissivity [13]

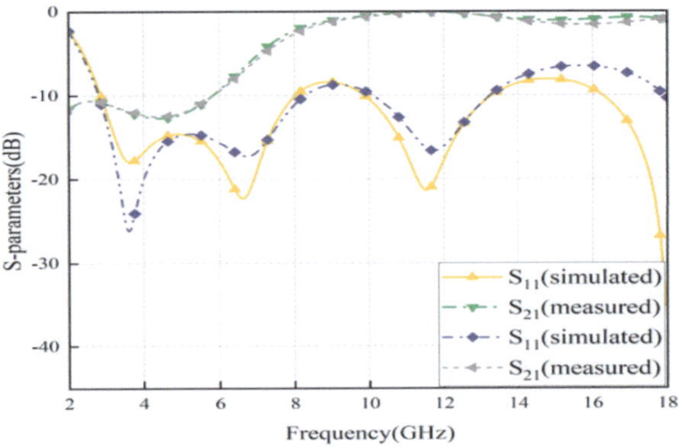

Fig. 8.8 Absorption characteristics of the ultra-wideband absorber with low IR emissivity [13]

constraints of low transmission and low reflection are fed to the algorithm in conjunction with the full-wave simulation results. The design demonstrates ultra-wideband behavior with more than 80% absorption in the operating band.

8.3.4 Low-Frequency Absorbers

Another class of microwave absorbers is the low-frequency absorbers, typically absorbers designed under 1 GHz bands, which could be categorized as low-frequency absorbers. The PCB-based metamaterial or FSS-based absorbers are not quite popular in the low frequencies. The primary reason behind this observation is that the unit cell that operates in these bands would be in the range of 0.1 λ, and the size of the unit cell itself would be around 1.5 m. To design a complete absorber, which is usually 10 λ × 10 λ, the absorber size would be 150 m × 150 m. Commercially available photolithography techniques have inherent limitations on the maximum achievable size of 1 m x 1 m. Multiple boards must be electrically connected if PCBs are needed beyond these values. Multiple board designs for realizing absorbers would lead to discontinuity at the interface of these boards, which in turn leads to leakage of incident energy.

The other alternate method to realize PCB-based designs with such large dimensions is to learn the copper geometry using laser machining and then paste the patterned geometry onto a suitable substrate. This fabrication method leads to errors when mounting the copper unit cells on the dielectric substrate. Deviations in the topology of the unit cells would lead to deterioration in the performance of the low-frequency absorbers. Hence, most of the PCB-based FSS absorbers avoid the low-frequency bands, 1–500 MHz, purely due to the challenge involved in fabrication.

It must also be noted that the free space measurement technique would require large anechoic chambers to characterize these low-frequency absorbers. Nevertheless, there are a few design examples on this subject in the literature. For instance, a fractal structure-based Ultra-High-Frequency (UHF) absorber is designed and experimentally characterized in [16]. The proposed absorber has a very low bandwidth with an operating frequency of 878 MHz. The authors have fabricated a 1.26 m × 1.26 m absorber. Another fractal design variant is illustrated in [17]. A two-layer UHF absorber is proposed in [18].

8.3.5 High-Frequency Absorbers

As established earlier, designing high-frequency absorbers is equally challenging as its low-frequency counterpart. The unit cell is the fundamental building block of any absorber. Similar to the other classes of absorbers, these unit cells would be a fraction of the least operating wavelength, which is of the order of a few microns. The analytical simulations would be operational up to a few hundred THz, beyond which the high-frequency analysis principles must be re-engineered, as the electromagnetic wave would exhibit a thermal phenomenon.

The design of absorbers for thermal phenomena is quite different from the RF and microwave approaches. Thus, the concept of metamaterial or periodic structures works up to a few hundred THz. As the sub-wavelength values reach a micron or lower, commercial photolithography technologies will fail to deliver these structures. The primary limitation of PCB-based fabrication processes is that the minimum achievable dimension is 100 μm in a high-resolution, expensive PCB manufacturing unit. The distance between copper patterns must be maintained beyond this value.

The requirement for the design and realization of the THz absorber is the microfabrication technology, which is common and popular for realizing integrated chips. The only difference is that, instead of active circuits or various layers within the chip, passive periodic, conductive patterns would deliver the absorption effect. Several design examples have been investigated in the literature. For instance, a switchable Terahertz absorber in the range of 8.71–9.75 THz is proposed in [19]. The overall size of the chip is $6\ \mu m \times 6\ \mu m$. The design consists of a stacked layer of VO_2 and graphene on the dielectric substrate. The bias voltage controls the conductivity of the VO_2 layer, which alters absorption characteristics. Other variants of this process of THz absorbers are demonstrated in [20–24].

8.3.6 Active Absorbers

Microwaves and other high-frequency absorbers could be classified based on the utility of active circuitry within their respective geometries. Applying active circuits would enable agility regarding the absorption spectrum and other associated performance metrics. Active components include diodes and transistors. These active circuits must be biased at a suitable voltage to be functional.

The design and realization of active absorbers are very complex. The reason is that the active component must be mounted onto the unit cell. The electrically large absorber would have numerous unit cells in both orthogonal directions. Hence, the application of active circuits is an engineering choice between the necessity for controlling the absorption spectrum and the challenges involved in design and deployment.

Figure 8.9 illustrates a typical schematic of an active absorber, like the generic structure presented in Sect. 8.3.1. The only difference between a passive and active absorber is the strategic placement of the active components, such as diodes and transistors. In the current schematic, two active circuits are integrated within the primary metal geometry of the absorber. The bias lines for controlling the parameter of the active circuit are assumed in the figure and not explicitly shown.

By applying the control signal, the capacitance or inductance of the active circuit could be varied, in principle. With the capacitance variation, the primary geometry's overall impedance could be varied, resulting in absorption resonance and bandwidth

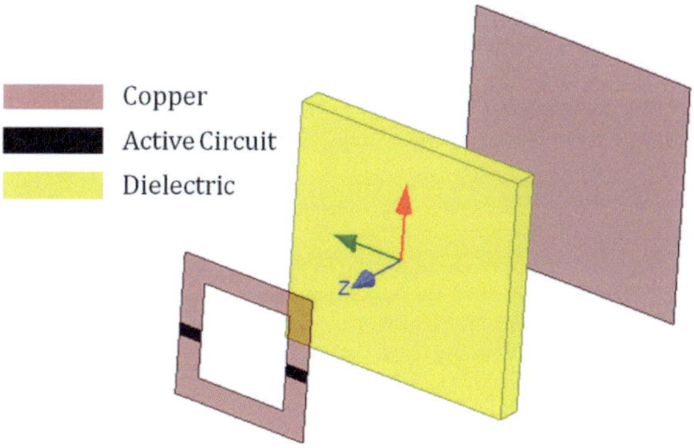

Fig. 8.9 Generic schematic of an active absorber (exploded view)

variation. Numerous examples have been established in the literature on this subject [25–29]. A wide tunable pass band window within an operating absorption spectrum is presented in [25]; the active absorber operates in the 6.4–14.1 GHz range. The agility of the passband transmission window is achieved by applying varactor diodes with the bias voltage variation. The capacitance of the varactor component varies, leading to variation in the transmission window. The bias line design is challenging as these metal geometries hinder the absorber's performance.

The absorption spectrum's reconfigurability is achieved by applying p–i–n diodes and varactor diodes, as demonstrated in [26]. Control of the bandwidth with a known central absolute frequency is difficult to achieve. A THz-based micro-fabrication technology-oriented design is reported in [27]; here, the conductivity of VO_2 could be varied based on the thermal excitation given to the VO_2 geometry in the stack-up. The proposed design is recreated in Fig. 8.10, and the corresponding performance metrics are shown in Fig. 8.11.

An ultra-wideband absorber using p–i–n diodes is demonstrated in [28]. A continuously tunable absorber is presented in [29]; the absorption peak could vary with the bias voltage's magnitude. The authors demonstrated the work through experimental verification and characterization.

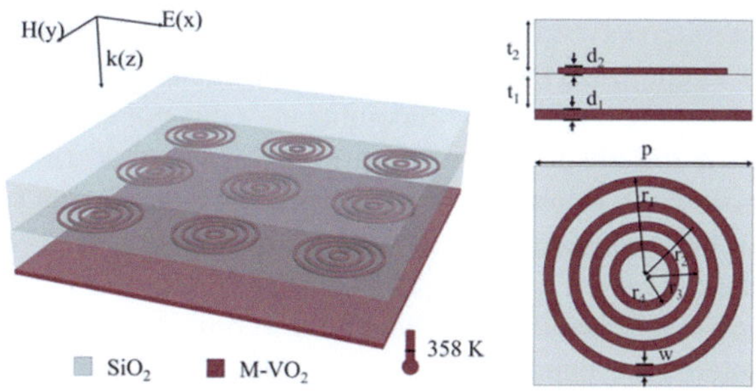

Fig. 8.10 Example of an active absorber [27]

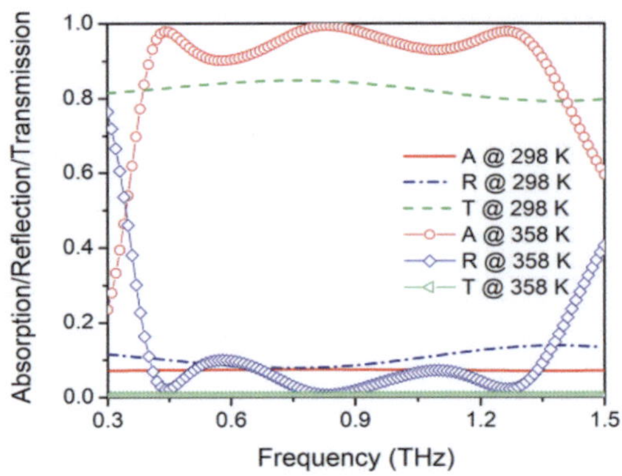

Fig. 8.11 Variation in the absorption spectrum [27]

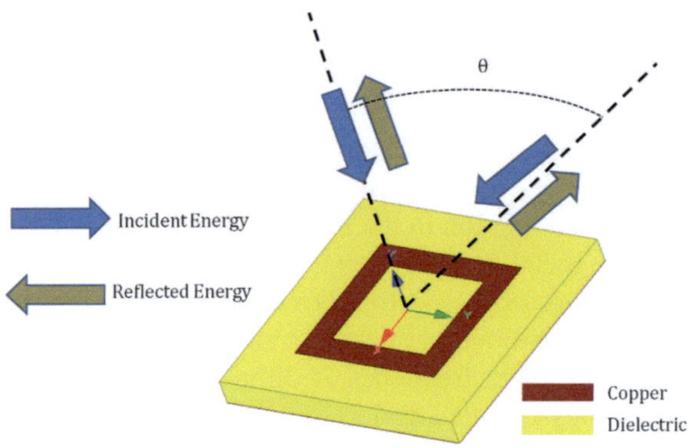

8.3.7 *Wide-Angle Incidence Absorbers*

Another important and popular choice of absorbers is the wide-angle incidence absorbers. These are absorbers with absorption characteristics like various angles of incidence. This requirement is essential for most of the real-world scenarios. Absorbers, in general, are used for stealth and other military applications, especially in microwave and millimetre wave bands. The fundamental premise of an absorber is to minimize reflections scattered from the target, but the problem in the real world is that the angle of the incoming signal is unknown.

If the absorber performs well for a limited range of angles, there is a strong possibility of reflection at undesired angles, leaving room for detection. To mitigate this effect, wide-angle absorbers are essential. A generalized schematic to illustrate the concept of wide-angle absorption is shown in Fig. 8.12. The absorber type shown in the schematics is for illustrative purposes only and may not work for any operating microwave band. The geometry of the absorber construction remains like that of the other classes of absorbers; the geometry of the periodic unit cells, in addition to the topology of the absorber, would decide the wide-angle absorption characteristics.

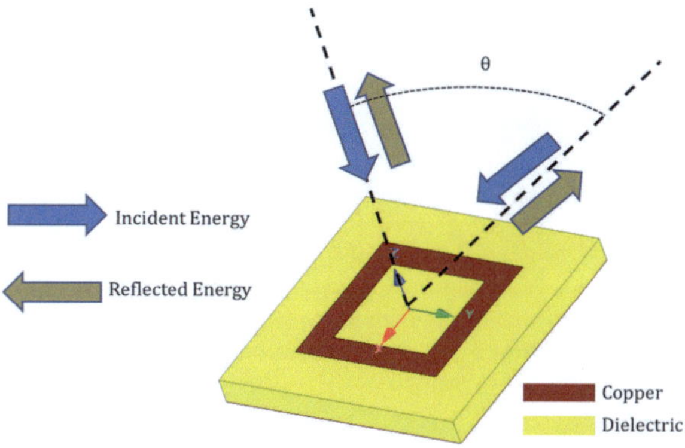

Fig. 8.12 Generic schematic of a wide-angle absorber

As the figure illustrates, a plane wave of a certain frequency and polarization is incident on the absorber surface. The boresight axis of the absorber is marked as 0° incidence. As the angle "θ" is increased, absorption characteristics are noted. If the absorption characteristics remain invariant with angular variation, the absorber could be termed a wide-angle or angularly stable absorber.

Typically, the maximum achievable angle could be in the range of ± 40° to ± 60°. Beyond these values, the stack-up of the absorber would dominate the performance metrics. It must be noted that the absorbers do not deliver absorption characteristics when the EM energy is incident orthogonal to the FSS-based absorbers. Numerous articles have been published in recent years on this class of absorbers; a few of these are discussed in the subsection.

An orthogonally placed metal geometry integrated with Polyurethane (PU) form layer is presented in [30]. The design presented has wide-angle stability with an exceptionally high bandwidth of 9.4 GHz. The high bandwidth and angular agnosticity could be attributed to applying PU foams impregnated with carbon and iron particles. Another variant of wide-angle wideband design is proposed in [31]. A hierarchical impedance matching method is presented in [32]. The design offers wideband starting from 1.46 GHz to 19 GHz. This multi-layer design has almost identical absorption characteristics when illuminated with ± 50° angles.

A wide-angle, stable design is showcased in [33]. An all-metallic sheet with wide-angle absorption characteristics is reported in [34]. The design consists of shaped slots on a 50 μm thick copper sheet. The design operates as a pure absorber at 23 GHz with a very narrow absorption bandwidth. The measurement setup and the fabricated prototype of the metallic wide-angle design are shown in Fig. 8.13. The angle invariant behavior is illustrated in Fig. 8.14.

A stacked design with integrated chip resistors to achieve wide-angle characteristics is reported in [35]. A multi-layer multiband design with ± 40° angular stability is presented in [36].

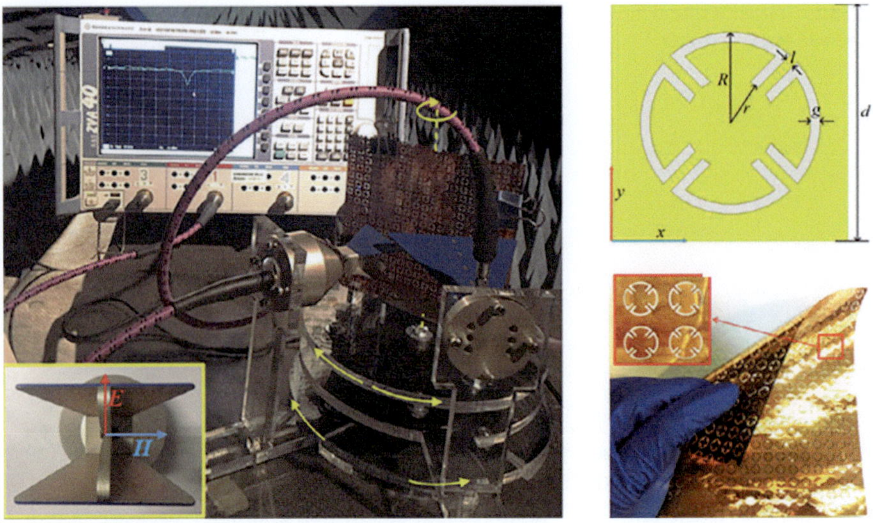

Fig. 8.13 Example of a wide-angle absorber [34]

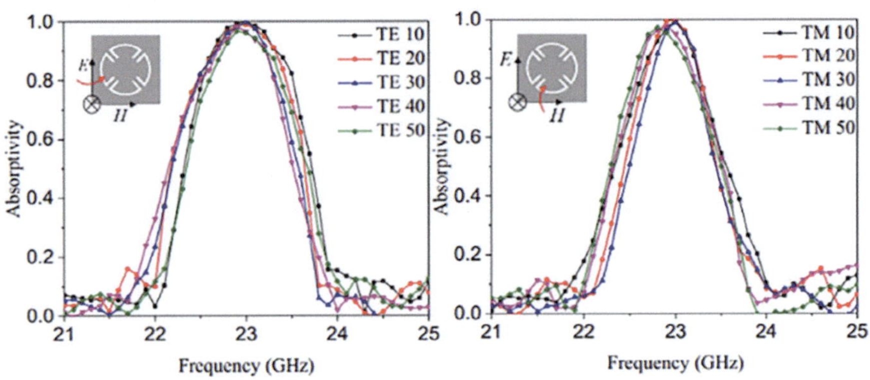

Fig. 8.14 Absorption characteristics of wide-angle design [34]

8.3.8 Polarization-Insensitive Absorbers

Absorbers could be classified based on the variation of the absorption characteristics depending on the incident energies' polarization. This class of absorbers is also widely popular among researchers as the design and experimental verification steps are relatively easier than other absorbers classes. A generalized schematic of the polarization-insensitive absorber is illustrated in Fig. 8.15. The schematic of the polarization-insensitive absorber is like that of the ones presented in the preceding subsections. The absorber is a two-layer PCB, where the top layer of the substrate has a periodic structure, and the bottom layer is a complete copper cladding.

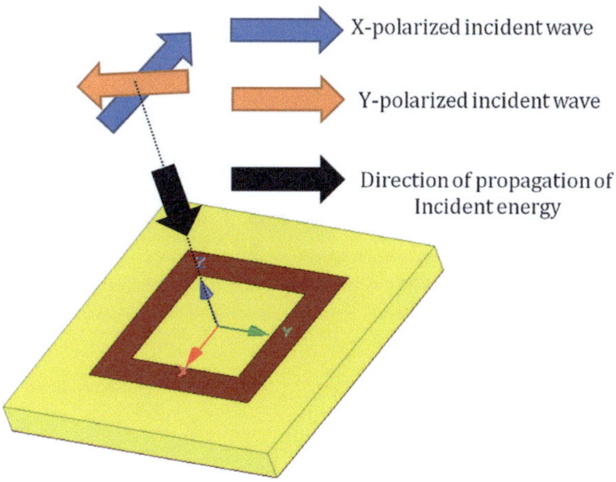

Fig. 8.15 Generic schematic of a polarization-insensitive absorber

Like the wide-angle requirement in the real world, the incident energy's polarization is also unknown. To deliver absorption for any incident polarization, absorbers must be engineered to have identical absorption spectra for any incident polarization. The schematics illustrate the electric field polarization of the incident TEM waves. The direction of energy propagation is the direction of the Poynting vector; the electric field of the incident EM energy could be in any random orientation. If the absorber is operational for two orthogonal polarizations, then the absorber could be termed polarization-insensitive. Designing symmetric structures is one of the simplest and most effective strategies to realize polarization-agnostic behavior. If the unit cell is rotationally symmetric, the chances of polarization invariability are high.

A nanomaterial-based absorber combined with a metamaterial absorber is presented in [37]. The unit cell has a symmetric structure. A closely spaced dual-band absorber that operates in 2.13 and 2.52 GHz bands is proposed in [38]. The design has three concentric ring-like structures, manifesting identical characteristics with various incident polarization. The periodic structure and its measured characteristics are shown in Figs. 8.16 and 8.17.

A concentric connected ring unit cell operating in C-, X-, and Ku-bands is experimentally verified in [39]. A wideband polarization-insensitive absorber is showcased

Fig. 8.16 An example of a polarization-insensitive absorber [38]

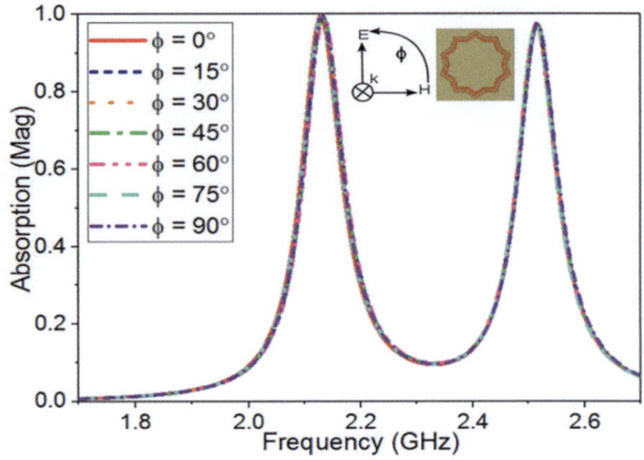

Fig. 8.17 Polarization invariant behavior of the absorber [38]

in [40]. A multiband design with a symmetrical structure demonstrates polarization invariant behavior [41]. A wide band, 3D structure is presented in [42].

8.3.9 Conformal Absorbers

Conformal absorbers are a small subset of flexible absorbers. Figure 8.18 illustrates a generalized schematic of a conformal absorber; here, the layer-based absorber must be able to conform to a given non-uniform surface without significant degradation in the absorption metrics. Electrically thin substrates designed with double-sided periodic unit cells would be able to deliver mechanical elasticity in addition to high absorption within a few operating bands.

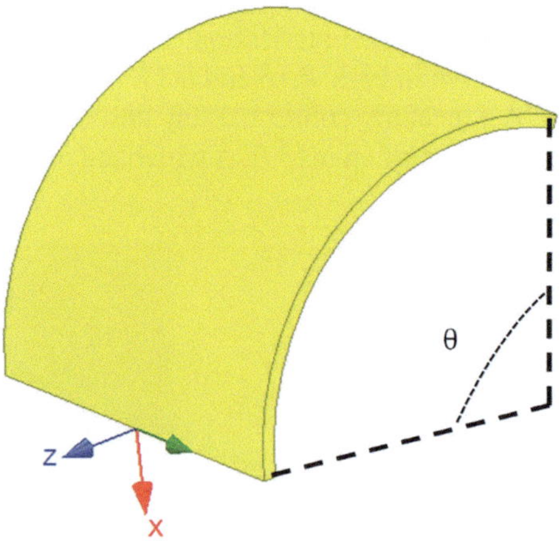

Fig. 8.18 Schematic of a conformal absorber

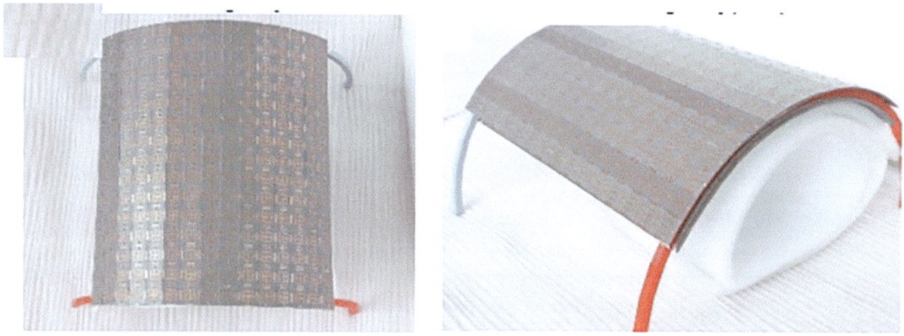

Fig. 8.19 A conformal absorber example [45]

There are few published articles in this class of absorbers due to the complexity involved in experimental realization and characterization. Today, a polymer composite with an integrated metamaterial structure is presented [43]. As the polymer exhibits high elasticity, the absorber is flexible. It must be noted that the metal pattern on the flexible substrate must closely retain its geometry to avoid the tuning of the absorption spectrum. If the top layer of the absorber is inelastic, then the copper geometry might snap. If the copper geometry snaps, the discontinuity will completely deteriorate the absorption spectrum. If the copper geometry stretches, then the inductive and capacitive effects of the copper geometry would be altered; this, in turn, would be manifested in the absorption spectrum.

An electrically thin substrate with a thickness of 70 μm is reported in [44]. This process is commercially viable as the industry-standard photolithography technique is used to realize the same. Chip resistors are integrated into the metal geometry. The solid parts might not be stable for flexible and conformal designs. Another multiband flexible design is reported in [45]; retention of absorption performance at higher frequencies is challenging when mechanical bending is introduced to the structure. The design and experimental results shown in Figs. 8.19 and 8.20 [46] also demonstrate another interesting conductive interlocking structure. This design offers conformity without metrics degradation.

8.3.10 Transparent Absorbers

Transparency is another feature that microwave absorbers can feature. A generalized schematic is displayed in Fig. 8.21. The design would have a unit cell, like other classes of absorbers. The only constraint is that a transparent conductive sheet would replace the metal geometry. The dielectric base onto which the unit cell would be patterned must also be optically transparent. Optically transparent absorbers are relatively new, so the number of research articles published is comparatively lower.

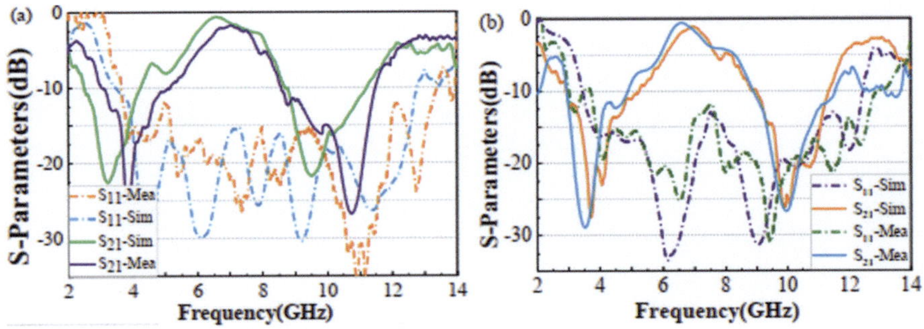

Fig. 8.20 Absorption characteristics of the conformal absorber [45]

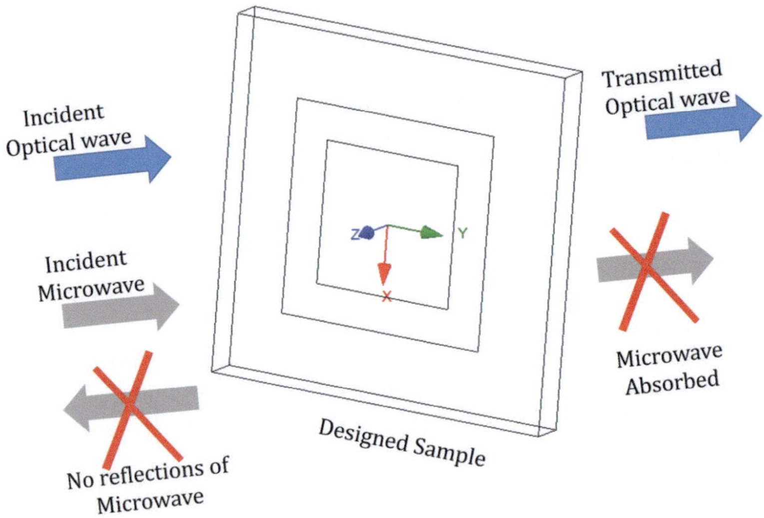

Fig. 8.21 General schematic of an optically transparent microwave absorber

Indium Tin Oxide (ITO) is designed on a Polyethylene Terephthalate (PET) substrate, as illustrated in [47]. The design achieves a high microwave absorption bandwidth with optical transparency greater than 60%. In another design variant, chip resistors are integrated with ITO sheets to achieve microwave absorption [48]. As the physical footprint of the chip resistors is insignificant compared to the whole ITO sheet, the optical transparency is not compromised. Another design is presented in [49]. The application of silver nanowires is reported in [50]. The fabricated sample and its associated experimentally verified characteristics are shown in Fig. 8.22.

Screen printing technology could realize optically transparent microwave absorbers [51]. This layered design, along with the measured optical transparency, is showcased in Fig. 8.23.

(a)

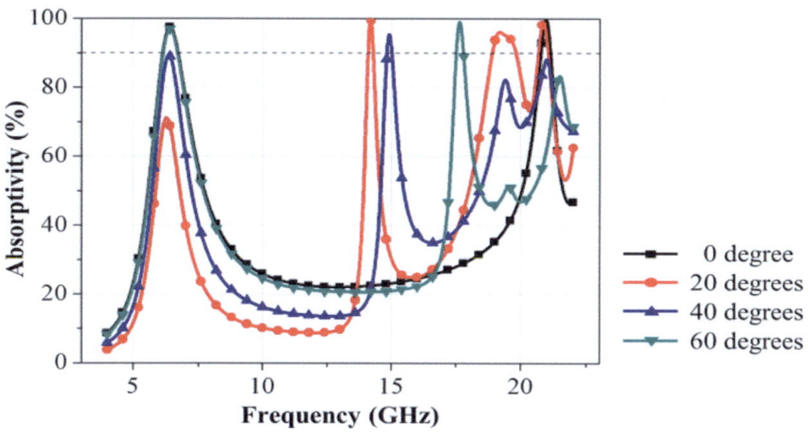

(b)

Fig. 8.22 **a** Fabricated transparent microwave absorber; **b** absorption characteristics of the transparent absorber [50]

8.4 Conclusions

PCB-based absorbers are an important class of absorbers. In this chapter, the definition of a PCB-based absorber with the expected technical requirements was introduced. A classification of periodic structure-based absorbers was introduced. A comprehensive overview of different types of PCB-based absorbers was introduced in this chapter. A technical analysis with detailed nuances for each class of absorbers was introduced to the readers.

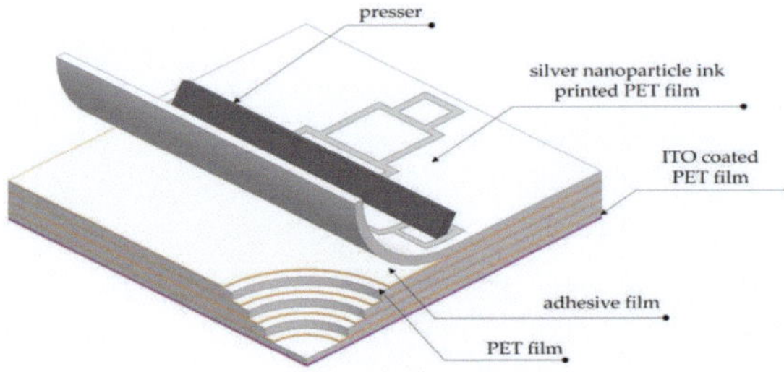

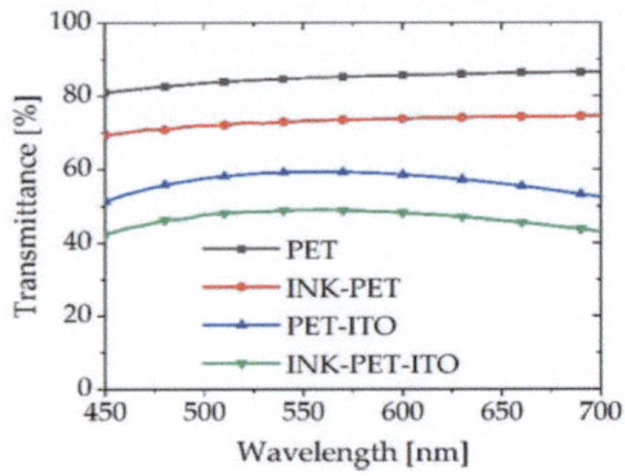

Fig. 8.23 **a** Screen printing process to realize optically transparent microwave absorber; **b** measured optical transparency [51]

References

1. Wang, Z., et al.: A high-transmittance frequency-selective rasorber based on dipole arrays. IEEE Access **6**, 31367–31374 (2018)
2. Zargar, M.M., Rajput, A., Saurav, K., Koul, S.K.: A polarization-insensitive frequency selective surface based rasorber with narrow-band absorption between two transmission bands. In: 2020 50th European Microwave Conference (EuMC), pp. 503–505. (2021)
3. Qu, M., Sun, S., Deng, L., Li, S.: Design of a frequency-selective rasorber based on notch structure. IEEE Access **7**, 3704–3711 (2019)
4. Kim, J., Bang, S., Jeong, J., Wi, S.H., Oh, J.: Enhancing selectivity in 5G/6G ultranarrowband applications: plug-and-play frequency-selective surfaces with field-manipulating Vias. IEEE Trans. Antennas Propag. **72**(8), 6783–6788 (2024)
5. Wu, K., Xu, W., Yang, Y., Li, P.: A dual-polarized C-X-Ku-band frequency-selective rasorber with wide high-transmission passband. IEEE Trans. Electromagn. Compat. **66**(4), 1295–1299 (2024)
6. Zhou, Q., Guo, M., Moghadas, H., Wu, Z., Liu, P., Daneshmand, M.: A frequency selective rasorber with three transmission bands and three absorption bands. IEEE Access **7**, 160973–160981 (2019)
7. Guo, M., Chen, Q., Sun, Z., Sang, D., Fu, Y.: Design of dual-band frequency-selective rasorber. IEEE Antennas Wirel. Propag. Lett. **18**(5), 841–845 (2019)
8. Shang, Y., Shen, Z., Xiao, S.: Frequency-selective rasorber based on square-loop and cross-dipole arrays. IEEE Trans. Antennas Propag. **62**(11), 5581–5589 (2014)
9. Ye, H., Dai, W., Chen, X., Zhang, H., Bie, S., Jiang, J.: High-selectivity frequency-selective rasorber based on low-profile bandpass filter. IEEE Antennas Wirel. Propag. Lett. **20**(2), 150–154 (2021)

10. Li, Y., Zhong, Y., Wang, M., Chen, K., Ren, P., Xiang, Z.: A miniaturized and highly stable frequency-selective rasorber incorporating an embedded transmission window. Micromachines **15**(8). https://doi.org/10.3390/mi15080980
11. Xiu, X., Che, W., Yang, W., Han, Y., Xue, Q.: A highly selective rasorber based on second-order resonance. IEEE Antennas Wirel. Propag. Lett. **19**(2), 223–227 (2020)
12. Li, Y., Ren, P., Chen, R., Xu, B., Xiang, Z., Wang, M.: Design of miniaturized frequency selective rasorber based on complementary bandpass FSS for 5G applications. IEEE Trans. Electromagn. Compat. **66**(2), 636–639 (2024)
13. Song, H., et al.: An ultra-wideband frequency selective rasorber with low infrared emissivity. Materials **17**(14). https://doi.org/10.3390/ma17143414
14. Yao, Z., Xiao, S., Li, Y., Wang, B.Z.: On the design of wideband absorber based on multilayer and multiresonant FSS array. IEEE Antennas Wirel. Propag. Lett. **20**(3), 284–288 (2021)
15. Ding, K., Yang, Y.: Reverse design of a sandwich microwave broadband metamaterial absorber based on resistive film-PMI-copper. J. Electron. Mater. **53**(6), 3211–3222 (2024)
16. Venneri, F., Costanzo, S., Borgia, A., Buonanno, G.: Fractal metamaterial surfaces for UHF sensing applications. IEEE Access **12**, 65810–65824 (2024)
17. Venneri, F., Costanzo, S., Massa, G.D.: Fractal-shaped metamaterial absorbers for multireflections mitigation in the UHF band. IEEE Antennas Wirel. Propag. Lett. **17**(2), 255–258 (2018)
18. Costa, F., Genovesi, S., Monorchio, A., Manara, G.: Low-cost metamaterial absorbers for sub-GHz wireless systems. IEEE Antennas Wirel. Propag. Lett. **13**, 27–30 (2014)
19. Wang, Q., et al.: Switchable quadruple narrowband to broadband terahertz perfect absorber based on graphene and VO2 metamaterials. Diam. Relat. Mater. **142**, 110832 (2024)
20. Bilal, R.M.H., Borgese, M., Genovesi, S., Manara, G., Costa, F.: D-band absorber comprising tantalum nitride-based resistively-loaded high impedance surfaces. In: 18th European Conference on Antennas and Propagation (EuCAP), pp. 1–5. (2024)
21. Ding, L., Wu, Q.Y.S., Teng, J.H.: Polarization independent broadband terahertz antireflection by deep-subwavelength thin metallic mesh. Laser Photonics Rev. **8**(6), 941–945 (2014)
22. Watkins, C.A., Liu, Y., Feng, P.X.L.: Metasurfaces designed on silicon nitride thin films for enhanced mid-infrared light absorption. In: IEEE Research and Applications of Photonics in Defense Conference (RAPID), pp. 1–2. (2023)
23. Korani, N., Danaie, M.: A plasmonic terahertz perfect absorber based on L-shaped graphene patches and gold rods. Appl. Phys. A **129**(11), 806 (2023)
24. Sabaruddin, N.R., et al.: Designing a broadband terahertz metamaterial absorber through bi-layer hybridization of metal and graphene. Plasmonics (2024)
25. Zhang, L.-M., Ding, X., Bozzi, M.: Active frequency-selective rasorber with wide range passband continuous tunability. AEU-Int. J. Electron. C. **179**, 155318 (2024)
26. Fan, Y., Li, D., Xing, J., Ma, H., Ang, L.K., Li, E.P.: A trifunctional active frequency selective surface for 6G shielding applications. IEEE Trans. Electromagn. Compat. 1–10 (2024)
27. Mou, N., Tang, B., Li, J., Dong, H., Zhang, L.: Switchable ultra-broadband terahertz wave absorption with VO2-based metasurface. Sci. Rep. **12**(1), 2501 (2022)
28. Farooq, U.B., Jha, K.R., Sharma, S.K.: Simplified biasing AFSS microwave absorber. In: IEEE Wireless Antenna and Microwave Symposium (WAMS), pp. 1–5. (2024)
29. Xing, J., Li, D., Fan, Y., Zhang, L., Li, E.P.: Tunable coherent perfect absorption based on varactor-embedded metamaterial absorber. IEEE Trans. Antennas Propag. **72**(4), 3461–3471 (2024)
30. He, L., et al.: Wide-angle microwave absorption performance of polyurethane foams combined with cross-shaped metamaterial absorber. Results Phys. **11**, 769–776 (2018)
31. Sood, D., Tripathi, C.C.: A wideband wide-angle ultrathin low profile metamaterial microwave absorber. Microw. Opt. Technol. Lett. **58**(5), 1131–1135 (2016)
32. Qin, J., Sun, M., Wang, M., Guan, M., Chen, A.: A hierarchical impedance matching approach for ultrawideband absorbers with wide angular stability. IEEE Trans. Antennas Propag. **72**(5), 4116–4128 (2024)

33. Malekara, A., Khalilzadegan, A., Ghobadi, C., Nourinia, J.: Wide-angle, dual-polarized frequency selective rasorber based on the electric field coupled resonator using characteristic mode analysis. J. Appl. Phys. **133**(16), 164504 (2023)
34. Wang, S., Zeng, L., He, L., Mao, J., Chen, Y.: All-metal wide-angle and polarization-independent microwave metamaterial absorber. Electron. Lett. **60**(7), e13173 (2024)
35. Begaud, X., Lepage, A.C., Varault, S., Soiron, M., Barka, A.: Ultra-wideband and wide-angle microwave metamaterial absorber. Materials **11**(10). https://doi.org/10.3390/ma11102045
36. Zargar, M.M., Rajput, A., Saurav, K.: Miniaturized design of dual transmission frequency selective rasorber with wide angular stability. IEEE Open J. Antennas Propag. **5**(4), 922–932 (2024)
37. Hossain, I., et al.: Synthesis and characterization of flexible $CaCoAlFe_2O_4$ based microwave substrate for triband polarization-insensitive metamaterial absorber. Opt. Laser Technol. **180**, 111429 (2025)
38. Kaur, K.P., Upadhyaya, T.: Performance evaluation of wide-angle ultrathin microwave metamaterial absorber with polarization independence. Adv. Electromagnet. **7**(4), 71–77 (2018)
39. Routray, P., Ghosh, D.: Design and modeling of a miniaturized multilayer metamaterial absorber with oblique angle and polarization insensitivity. IEEE Lett. Electromagn. Compat. Pract. Appl. **6**(1), 29–34 (2024)
40. Bandyopadhyay, B., Shahi, A.K., Dileep, A., Saikia, M., Srivastava, K.V.: Polarization insensitive broadband frequency selective rasorber with improved selectivity for stealth applications. In: 18th European Conference on Antennas and Propagation (EuCAP), pp. 1–5. (2024)
41. Muthukrishnan, K., Narasimhan, V.: An ultra-thin triple-band polarization-independent wide-angle microwave metamaterial absorber. Plasmonics **14**(6), 1983–1991 (2019)
42. Thi Quynh Hoa, N., Huu Lam, P., Duy Tung, P.: Wide-angle and polarization-independent broadband microwave metamaterial absorber. Microwave Opt. Technol. Lett. **59**(5), 1157–1161 (2017)
43. Anjali, M., Sahoo, R., Stephen, L., Krishnamurthy, C.V., Subramanian, V.: Flexible bandwidth-enhanced metamaterial absorbers with epoxy/graphene nanoplatelets-silver nanowire polymer composites as substrates. Compos. Sci. Technol. **249**, 110492 (2024)
44. Raj, A., Chaudhary, V., Kumar Gangwar, R., Kumar Chaudhary, R.: Experimental and analytical analysis of conformal and polarization insensitive electromagnetic absorber. AEU—Int. J. Electron. Commun. **179**, 155340 (2024)
45. Fu, Y., et al.: Conformal frequency selective rasorber in S, C, X-band with low backward-scattering. Opt. Express **32**(10), 16879–16890 (2024)
46. Fu, C., et al.: Ultrawideband microwave absorber conformable on nondevelopable curved surfaces based on the hauberk structure. IEEE Trans. Antennas Propag. **72**(6), 5057–5065 (2024)
47. Bai, X., Mei, Z., Zhang, J., Xu, W., Lin, W., Niu, T.: An ultrawideband, wide-angle, and transparent microwave absorber using indium tin oxide conductive films. IEEE Antennas Wirel. Propag. Lett. **23**(5), 1543–1547 (2024)
48. Lee, M., Park, J., Kim, K., Kim, J., Park, E., Lim, S.: Broadband and wide-incidence-angle transparent metamaterial absorber using eight-resistive-arm cell. IEEE Antennas Wirel. Propag. Lett. **23**(6), 1754–1758 (2024)
49. Khan, H.A., Majeed, A., Zahra, H., Kakepoto, F.G., Abbas, S.M., Alathbah, M.: Transparent conformal metasurface absorber for ultrawideband radar cross section reduction. J. Phys. D Appl. Phys. **57**(13), 135105 (2024)
50. Bark, S., Kim, J., Lee, M., Lim, S.: Optically transparent dual-band metamaterial absorber using Ag nanowire screen-printed second-order cross-fractal structures. Fractal Fractional **8**(3). https://doi.org/10.3390/fractalfract8030153
51. Choi, J., Lim, D., Lim, S.: Screen-printed metamaterial absorber using fractal metal mesh for optical transparency and flexibility. Fractal Fractional **8**(5). https://doi.org/10.3390/fractalfract8050284

Chapter 9
Future Scopes and Research Gaps

Abstract The advancement of technology in the Fourth Industrial Revolution has led to a growing need for EMI shielding materials to protect gadgets and people from electromagnetic energy. Research is ongoing to find new materials, such as metal, carbon, and MXene, or combinations that can effectively block EMI. With the demand for smaller, multitasking devices rising, EMI protection materials must evolve to provide additional features beyond shielding. Balancing EMI shielding with strain sensing, humidity sensing, and thermal management in flexible materials needs further research for improvements.

9.1 Introduction

The development of 5G and 6G technology has raised concerns about electromagnetic pollution, prompting a demand for materials that resist electromagnetic interference (EMI). Research in EMI protection has focused on properties like thickness, lightness, flexibility, and cost [1]. Scientists are now working on multifunction EMI shielding materials with broader spacecraft and military equipment applications. These materials can also be used in clothing to provide thermal protection and monitor vital signs [2]. "Smart clothing" is being developed to shield users from electromagnetic waves and monitor their health. While progress has been made in EMI shielding materials, more research is needed on materials that filter EMI. This research aims to advance understanding in this area and suggest areas for further study [3].

S. K. Koul et al., *Microwave Absorbers*, Engineering Materials,
https://doi.org/10.1007/978-981-96-9596-6_9

9.2 Current Trends in EMI Shielding

EMI shielding materials include metals, carbon-based substances, and MXene materials, either single or combined. Metal-based materials like Silver, Copper, Aluminum, Nickel, and Steel are preferred for their electrical conductivity and abundant free electrons for wave reflection and absorption. MXene advancements are ongoing [4].

9.2.1 Metal-Based EMI Shielding Materials

Due to their conductivity, metals like Silver, Copper, and Aluminum are used for EMI shielding. Traditional materials include magnesium sheets, alloys, and micro-composites. Thin films can be created using deposition methods, with NiFe–Cu/Ti layers being adequate [5]. Metal foams are developed to improve EMI shielding efficiency. Aluminum foams show increased performance over time. Popular EMI shielding textiles use metal-coated fabrics, like copper, on modal fabric [6].

Coating AgNPs on AgnPWT enhances EMI shielding by forming a conductive Ag network. SSF/PET yarn with 40% SSF achieves an EMI SE of 40 dB. Metals and polymers combine to create conductive composites that absorb EM waves. With good adhesion, nickel-coated carbon fibers in ABS composites achieve EMI SE values of up to 30 dB in the 30–1000 MHz range [7].

Zhang et al. developed a P@Ni–Co hybrid membrane with high electrical conductivity and magnetic properties, achieving an EMI SE of 77.8 dB. Metal nanomaterials play a critical role in polymer nanocomposites, boosting conductivity and EMI shielding. The MSMP method produces nanocomposites with strong electrical conductivity and EMI shielding; 9.57 wt% silver-coated paper achieves 48.6 dB [8].

AgNWs are used in transparent EMI shields. CA/AgNWs/PU film achieved 31.3 dB EMI SE with 81% transmittance [9]. 2D metal materials improve EMI shielding with reflectivity and absorption. Silver nanoplates in PLA showed strong shielding at low silver content. Yet, metal materials can be heavy, inflexible, and vulnerable to damage [10].

In Fig. 9.1, the study demonstrates that incorporating SF composite material into cold-mixed and cold-laid asphalt significantly enhances its microwave absorption properties and accelerates the curing process. Tests found optimal mineral aggregate gradation and emulsified asphalt ratio. SF content's effect on mixing performance and microwave absorption was analyzed [11].

Fig. 9.1 Microwave oven: **a** appearance; **b** internal structure [11]

9.2.2 Carbon-Based EMI Shielding Materials

Researchers are exploring carbon materials like carbon black, graphite, carbon fibers, carbon nanotubes (CNT), and graphene for making EMI shielding materials. CNT and graphene show potential for high SE and flexibility [12].

Various studies have utilized carbon nanotubes (CNTs) and graphene for EMI shielding materials. Wang achieved an EMI SE of 30.89 dB in X-band with a Poly(vinylidene fluoride)/MWCNTs composite. Li used MWCNT and WPU, Jia blended CNT/PE, Kar wrapped MWCNTs in PMMA for high conductivity in PVDF-ABS, and Ren demonstrated a segregated MWCNT structure, showing various strategies for improving EMI shielding [3].

In Fig. 9.2, EM absorbers and breast cancer detection were used. Researchers have added additives like Ni, NdFeB magnetic powders, and Au nanoparticles to improve EMI shielding in CNT-based composites. Park's group enhanced performance by plating Ni on MWCNTs in Ni–MWCNTs/HDPE and Ni–MWCNTs/epoxy composites. Li et al. created a composite with absorbent and reflective layers using MWCNTs and NdFeB powders. Lei et al. coated Au nanoparticles on CNTs in Au@CNT/SA/PDMS composites for better conductivity—Yang et al. densified CNTs for improved EMI SE through thermal-induced evaporation and filtration processes [13].

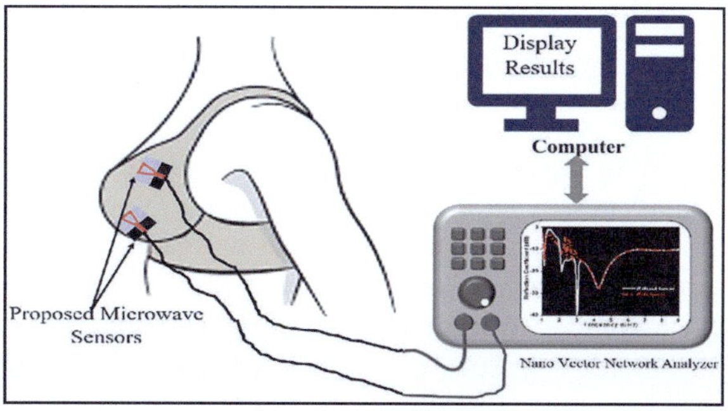

Fig. 9.2 Proposed smart bra for breast cancer monitoring detection [14]

9.2.3 MXene-Based EMI Shielding Materials

MXene is a 2D material with high electron density and exceptional mechanical properties used in EMI shielding [9]. Koo's 2016 study on MXene films, specifically $Ti_3C_2T_x$, $Mo_2TiC_2T_x$, and $Mo_2Ti_2C_3T_x$, demonstrated their effectiveness in EMI shielding, with $Ti_3C_2T_x$ films achieving an EMI SE of 48–92 dB within a thickness range of 1.5–45 μm. The success of $Ti_3C_2T_x$ films can be attributed to their multi-layered structure, increased absorption of EM waves, and polarization losses due to abundant functional groups on the surface. Ongoing research has further improved EMI shielding performance through novel techniques such as treating MXene films with hydrazine solution, enhancing thickness, porous structure, hydrophobic properties, and oxidation stability [15]. Additionally, Ti_3CNT_x nanosheets have shown promise as effective EMI shielding materials, with annealing treatments improving their electrical conductivity [5].

9.3 Beyond the EMI Shielding Function

9.3.1 Strain Sensing Function

Materials crucial for robotics and wearable electronics provide tactile feedback and monitor health. However, they are vulnerable to electromagnetic interference (EMI). Research focuses on multifunctional materials with strain sensing and EMI shielding capabilities to address this. These materials can protect equipment and humans from harmful waves while sensing strain [16]. Resistive-type strain sensors are commonly used, with thin conductive layers or nanomaterials embedded into stretchable substrates. Textiles are ideal for wearable devices due to their flexibility and washability, though they have some limitations. Zhang et al. made an MXene-coated cotton fabric sensor for superior EMI shielding, strain sensing, and detecting motion. This sensor showed a high gauge factor, stability, and durability, making it ideal for wearable applications [8].

9.3.2 Humidity Sensing Function EMI

Shielding materials with flexible, wearable properties now benefit pregnant women, military personnel, and electronics industry workers. These materials include features for monitoring humidity and skin moisture. Humidity sensors track sweat levels across the body, providing insight into metabolism, aging, and wound healing. 2D MXene's hydrophilic surface and conductivity make it a promising option for humidity sensors, potentially surpassing MoS_2 and graphene. CS/$Ti_3C_2T_x$ film excels in EMI shielding and humidity sensing, boasting exceptional performance in electrical conductivity and EMI shielding in X-band frequencies [17].

9.3.3 *Temperature Sensing Function*

There is a rising need for temperature-sensitive EMI shielding due to critical monitoring in control systems. Researchers have developed composite structures that detect temperature changes, allowing EMI shielding to protect electronic devices from harsh conditions. Farid El-Tantawy developed conducting elastomer blends with temperature sensing and EMI shielding qualities in 2005. Other research has also investigated elastomer blends for temperature sensors and EMI shielding, exhibiting NTC and PTC effects at specific temperatures. Southwest University in China innovated a foam-structured composite with superior temperature sensing and EMI shielding abilities using thermo-expandable microspheres in a PDMS and CNT matrix [18].

Astronauts in space face significant radiation risks from galactic cosmic rays (GCRs), which include highly energetic ions (HZE particles) and high-intensity solar particle events (SPEs). A year on the Moon could result in absorbed radiation doses in the range 100–120 mGy, while a Mars mission could reach around 450 mGy. For comparison, six to twelve months on the International Space Station (ISS) would involve 30–120 mGy doses. During intense SPEs, astronauts may be exposed to proton doses as high as 1 Gy, risking acute radiation sickness (ARS). Though spacecraft and spacesuit shielding absorb some SPE radiation, not all events are predictable, posing threats during extravehicular activities.

GCR exposure is particularly concerned due to HZE particles' high energy and penetration, which allows them to pass through shielding and release secondary neutrons, adding further risk. This space radiation causes DNA damage, oxidative stress, and chronic tissue inflammation, linked to various health issues such as cancer, cardiovascular and neurodegenerative diseases, and conditions typically associated with aging. Additionally, some astronauts report visual disturbances likely caused by ionizing radiation effects on sensory perception. Non-targeted effects (NTEs) of radiation further amplify risks, as even low doses of high-linear energy transfer (LET) radiation, such as HZE particles, may lead to long-term genomic instability and altered cell signaling; in Fig. 9.3, radiation protection of spaceflight is given [19].

9.3.4 *Heating Function*

In cold environments, electronics may malfunction due to issues like component breakage. Researchers are investigating combining heating and EMI protection in one material to address this. Multipurpose EMI shielding materials can function effectively in freezing temperatures, preventing distortion and frost buildup for

Fig. 9.3 Radiation protection of spaceflight [20]

smooth operation. Conductive cloth or e-textile EMI shielding is a versatile option. Jia et al. created MXene-decorated wood-pulp cloth, which can block EMI and generate heat when activated electrically. The material's EMI shielding effectiveness increased significantly with MXene concentration. Its Joule-heating property enables it to melt snow or de-ice roads with low power due to MXene nanosheets' unique electron–phonon interaction capability [21].

9.3.5 Other Functions

EMI materials with thermal control properties can be used for multiple purposes, such as blocking active electromagnetic waves. Nanomaterials that block active waves are effective at moving heat, making them suitable for thermal control. Films made with MXene and PVDF exhibited superior EMI shielding effectiveness, with one film reaching 48.47 dB and another at 44.4 dB with 19.5% MXene content, 27 μm width, and in-plane thermal conductivity of 4.57 W/mK. The PVA/MXene film also had high EMI SEand similar characteristics, while the MXene/MMT hybrid film showed fire resistance and enhanced thermal conductivity. Composites with PMCNT and PMGNP had heating properties, solvent resistance, and liquid-repelling abilities [21, 22].

Morphological microwave imaging can reconstruct biological objects in the body to show their condition. A new technique uses UWB microwave applicators to locate signals from a micro tag mimicking an action potential. The applicators focus radiation into a brain region with the micro tag, modulated by a low-frequency electrical signal from a photodiode [23]. In Fig. 9.4, human brain modeling, functional microwave imaging, and numerical reconstructions simulating three circular targets resembling the human head are shown.

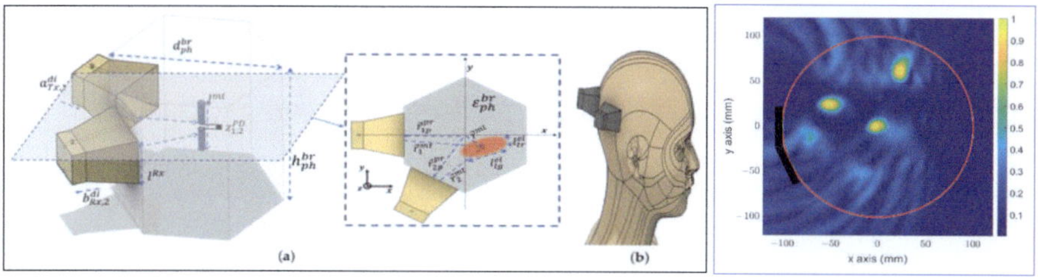

Fig. 9.4 Human brain modeling, functional microwave imaging, and numerical reconstructions simulating three circular targets resembling human head [23]

9.4 Perspectives and Suggestions

Studying materials with strong EMI shielding characteristics is crucial due to the current and future need for emitting electromagnetic waves. Research is focused on improving electromagnetic wave absorption, reducing reflection, and developing multifunctional EMI shielding materials with additional functionalities like strain sensing, humidity sensing, and thermal conductivity, requiring further exploration. Recent advancements in EMI shielding were reviewed, highlighting potential uses like self-cleaning and gas sensing [24].

Future research areas for multifunctional EMI shielding materials in materials science and applications involve utilizing composite materials to combine nanomaterial traits for diverse properties. MXene is crucial in this research due to its superior features and EMI shielding capabilities. Building nanocomposites with MXene, graphene, CNTs, and other nanomaterials can enhance material functionalities while maintaining excellent EMI shielding performance. Improving the oxidation stability of MXene nanocomposites can ensure their longevity in various applications. Cellulose nanofiber composites are natural, strong, and lightweight, with low thermal expansion and high modulus [25].

Multifunctional EMI shielding textiles offer a versatile solution for various applications, including bright clothes, sensors, actuators, and energy production. These textiles provide seamless integration for EMI shielding, improving noninvasive healthcare monitoring and offering potential energy harvesting capabilities for industries such as space exploration, robotics, and remote locations. This innovative technology can enhance performance in military, industrial, healthcare, and other settings [26]

Multifunctional EMI shielding materials are suggested to address various practical needs, including transparent shields with touchscreen capabilities for display applications. Device touchscreens can allow harmful EM waves to pass through, causing malfunctions and health risks. EMI shielding in bright windows protects against UV radiation and EM waves. Materials with shape memory effects have potential applications in tissue engineering and medical devices like pacemakers. However, current materials have limitations in EMI performance and thickness. Future research aims to improve filler-polymer connections in composite films for better EMI protection.

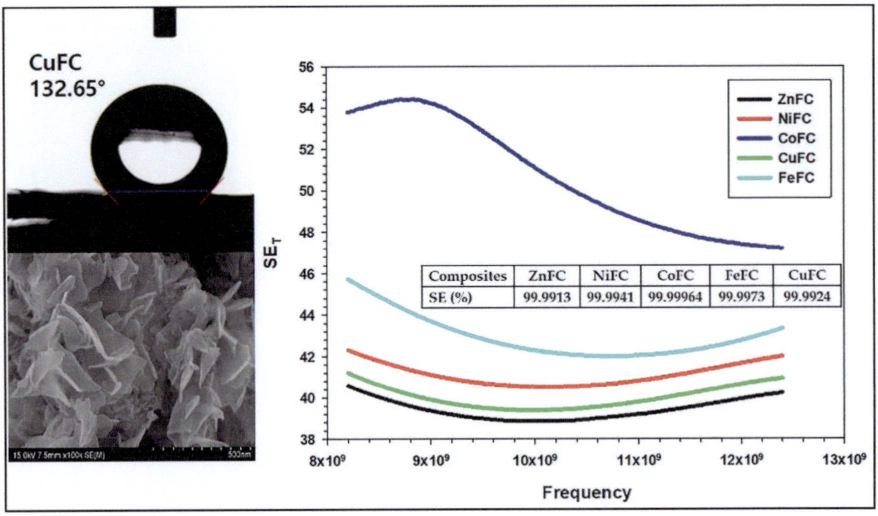

Fig. 9.5 EMI shielding of the hydrophobic, flexible, lightweight carbonless nanoplate composites [28]

The aim is to develop high-power density materials for nanogenerators that not only provide effective EMI shielding for sensitive devices but also support tissue healing and device functionality [27]. Figure 9.5 illustrates the EMI shielding performance of hydrophobic, flexible, and lightweight carbonless nanoplate composites. These materials are applied in a circularly polarized microwave sensor (CPMS) designed for medical applications—specifically, breast cancer detection (BCD) and integration into wireless body area networks (WBANs).

The sensor comprises a circularly polarized textile-based monopole antenna on a cotton substrate, and it functions within the 1.8–8 GHz range with an average gain of 6 dB. A slot design enhances bandwidth and facilitates circular polarization with a 3 dB axial ratio. Tested on and off the body near breast models, the CPMS employs near-field microwave imaging to safely identify tumors by analyzing electric field components. This underscores the potential of textile antennas in wearable tech and medical monitoring systems [14].

Here, microwave technology for post-stroke brain imaging assesses therapy effectiveness with a unique helmet prototype. Through preliminary tests on a brain tissue-mimicking phantom, the device shows promise in imaging stroke targets. Its differential imaging approach aids clinicians in monitoring therapy progress, addressing current monitoring gaps [29]. In Fig. 9.6, the Prototype 3D microwave imaging system includes VNA, switching matrix, and head phantom with helmet connected by coaxial cables.

A rapid caffeine extraction method was developed from roasted coffee beans using microwave-assisted extraction at 2450 MHz. Optimization factors included extraction time (2–6 min), liquid-to-solid ratio (5–15 mL/g), and microwave power (336–595 W). Water was used as the solvent, with optimal conditions determined as

Fig. 9.6 Prototype 3D microwave imaging system includes VNA, switching matrix, and head phantom with helmet connected by coaxial cables [29]

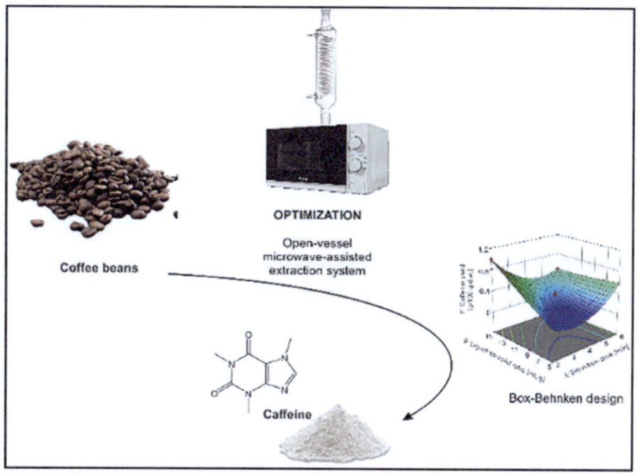

Fig. 9.7 Optimization of the microwave-assisted extraction of caffeine from roasted coffee beans [30]

2 min extraction time, 15 mL/g liquid-to-solid ratio, and 500 W microwave power, resulting in a 1.01/100 g dry weight yield of caffeine. Energy consumption was low, and the isolated caffeine was high purity [30] (Fig. 9.7).

9.5 Future of Electromagnetic Interference Shielding

The future of electromagnetic interference (EMI) shielding is advancing toward lighter, more efficient, and multifunctional solutions, driven by the increasing use of electronics in complex environments, especially in aerospace, medical, automotive, and wearable devices. Here are key areas shaping its future.

9.5.1 Advanced Materials

Nanomaterials like graphene, carbon nanotubes, and MXenes offer ultra-lightweight, flexible, and highly conductive options for EMI shielding. They are beautiful for high-frequency interference in smaller devices. Metamaterials, such as Engineered materials with unique electromagnetic properties, can be tailored to selectively block specific frequencies, providing targeted EMI shielding with minimal weight. Conductive polymers combining polymers with conductive particles like silver, nickel, or copper allows for flexible, corrosion-resistant coatings. Polymers can also be printed or sprayed, enabling easier manufacturing for complex shapes.

9.5.2 Multifunctional Shielding

The Thermal Management for Future EMI shields might integrate thermal management to dissipate heat from high-power electronics. For structural integration in the aerospace and automotive industries, EMI shielding will increasingly be integrated into structural components to save space and weight, making shields part of the design rather than an add-on. The Smart and Adaptive Shielding used as shielding materials may become dynamic, responding to changes in frequency or intensity of EMI to adjust protection levels automatically.

9.5.3 Miniaturization for IoT and Wearables

As devices shrink, there is a greater need for ultra-thin, flexible shielding. This includes printable EMI shields directly applied to circuit boards or integrated into wearable textiles for consumer electronics and medical devices. Flexible coatings and thin films, often created with nanomaterial-based inks, are being developed to apply effective shielding without increasing device thickness or weight.

9.5.4 Cost-Effective Sustainable Solutions

There's growing interest in using recyclable, biodegradable materials for EMI shielding to reduce waste. This could involve carbon-based alternatives to traditional metal-based shielding, which are often costly and hard to recycle. Manufacturing innovations, such as conductive inks or 3D-printed materials, also allow for faster and cheaper production, which is critical as demand increases across consumer electronics and IoT.

9.5.5 Enhanced Shielding for High-Frequency 5G and Beyond

As 5G, 6G, and other high-frequency technologies proliferate, EMI shielding must adapt to handle interference at these higher frequencies. Shielding materials capable of filtering a wide range of frequencies efficiently without adding bulk are a focus for future development, especially for high-performance communication systems and automotive radar.

9.5.6 Biocompatible Shielding for Medical Devices

EMI shielding in implantable and wearable medical devices is increasingly important for patient safety and function. Biocompatible materials that can protect sensitive electronics from EMI while ensuring compatibility with body tissues are an ongoing research focus.

9.5.7 Electromagnetic Shielding as Part of Cyber Security

Besides blocking EMI, shielding may prevent data leaks and electronic eavesdropping. Future designs may integrate EMI shielding to protect sensitive data by isolating devices from potential hacking or interference sources.

Advancements in EMI shielding materials are enabling more compact, efficient, and versatile electronics across industries by meeting the demand for lightweight, resilient, and adaptive solutions. Integrating thermal management into shielding, particularly in high-power applications like electric vehicles, aerospace, and telecommunications is a promising area of innovation. Materials such as graphene and carbon fiber provide EMI protection and heat dissipation, while phase change materials (PCMs) help manage temperature spikes, offering dual-function shielding.

New materials like liquid metals and soft, stretchable composites expand EMI shielding's potential in flexible applications, such as wearables and robotics, by maintaining effectiveness even when deformed. Meanwhile, adaptive and self-healing materials are being developed to respond dynamically to frequency shifts and improve durability in fluctuating environments, like telecommunications and defense. These advancements address modern technological demands by providing sustainable, high-performance EMI shielding that adapts to increasingly complex electronic systems across diverse fields.

9.6 Gaps for Future Research

Research gaps in EMI shielding materials exist in several areas, reflecting the need for enhanced functionality, sustainability, and adaptability to rapidly evolving electronic and environmental demands. Key gaps include the following.

9.6.1 Material Efficiency and Thinness

- **Optimizing Performance at Reduced Thickness**: Many advanced EMI materials like graphene and carbon nanotubes offer high shielding efficiency. However, producing ultra-thin, adequate layers that maintain performance in diverse applications remains challenging. Research is needed for compact electronic devices to achieve optimal conductivity, absorption, and durability at thinner dimensions.
- **High-Frequency Shielding**: With the rise of 5G, 6G, and high-frequency radar systems, developing materials that shield effectively across wide and higher frequency bands is critical. Many current materials work well at lower frequencies but have limited effectiveness in these advanced, high-frequency ranges.

9.6.2 Thermal Management Integration

- **Balancing EMI Shielding and Thermal Conductivity**: Although graphene and carbon composites provide both EMI protection and heat dissipation, maximizing these properties simultaneously without compromising one over the other is an ongoing challenge. Developing materials with balanced, high EMI shielding and thermal management properties would enhance safety and efficiency, especially in high-power electronics.
- **Durable Phase Change Materials (PCMs)**: PCMs hold promise for temperature regulation in EMI shields but have limitations, including a narrow temperature range and phase instability over time. Research into more stable, broad-range PCMs that can endure electronics' dynamic thermal cycles is needed.

9.6.3 Flexible, Stretchable, and Wearable Electronics

- **Long-Term Performance in Flexible Applications:** Flexible and stretchable EMI materials like liquid metals and conductive polymers still face challenges maintaining their shielding properties over repeated deformations. This is critical for wearables, robotics, and medical devices, where longevity and stability are essential.

- **Biocompatibility for Medical Applications:** As EMI shielding becomes necessary for implantable and wearable medical devices, developing biocompatible and non-toxic shielding materials is crucial. Many current materials lack compatibility with human tissues and require extensive testing to ensure safety.

9.6.4 Sustainability and Recycling

- **Eco-friendly and Recyclable Materials:** Traditional EMI materials, especially metal-based options, are often difficult to recycle and contribute to environmental waste. Research into fully recyclable, biodegradable, or plant-based EMI shielding materials, such as nanocellulose composites, is gaining interest but still lacks extensive testing for broad applications.
- **Cost-Effective Manufacturing:** Producing advanced materials like graphene, carbon nanotubes, and MXenes at scale remains expensive, limiting their use in consumer electronics. Developing cost-effective and scalable manufacturing techniques that maintain performance standards is a crucial research gap for widespread adoption.

9.6.5 Self-healing and Adaptive Properties

- **Developing Reliable Self-Healing Mechanisms**: Although some self-healing EMI materials are in development, these are often limited to minor damage and specific conditions. Research into effective self-healing mechanisms in diverse, real-world situations (e.g., temperature and humidity variations) is needed to extend material lifespans.
- **Frequency-Adaptive Materials**: Environments like telecommunications and defense require shielding that adjusts to shifting frequencies, so more advanced materials that can automatically tune their EMI properties in response to electromagnetic changes are necessary but remain underdeveloped.

9.6.6 Complex Multi-layer and Composite Designs

- **Multifunctional Integration**: Current composite designs combining multiple materials for dual or multifunctional shielding (e.g., structural support, thermal management) face issues with interfacial compatibility, weight, and cost. Research into better bonding techniques and lightweight composites is essential for more robust, integrated solutions.
- **Standardization and Testing Methods**: As EMI shielding materials become more complex and multifunctional, developing standardized testing methods for

multi-layered materials that assess EMI shielding effectiveness and additional functionalities is essential for accurate evaluation and industry-wide adoption.

Addressing these research gaps is critical to developing next-generation EMI shielding materials that meet the demands of advanced electronic systems while offering durability, adaptability, and sustainability across applications.

9.7 Conclusions

EMI shielding materials have a bright future due to increasing demand in electronics, 5G, EVs, and aerospace. Innovations are focused on creating lightweight, flexible, and multifunctional materials with high shielding effectiveness and additional properties like thermal management and mechanical strength. Nanomaterials, conductive polymers, and hybrid composites are becoming popular for their unique characteristics. However, challenges remain in improving these materials' durability, scalability, and cost-effectiveness, understanding multi-frequency shielding mechanisms, and exploring eco-friendly options. Overcoming these obstacles will lead to the development of next-generation EMI shielding materials that can meet the needs of modern electronics.

References

1. Chhetri, S., Kuila, T.: Polymer composites with 3D graphene architectures as high-performance EMI shielding materials: a review. RSC Appl. Polym. **2**, 507–533 (2024)
2. Akbar, M.I., Armynah, B., Tahir, D.: Sandwich composite structure (SCS): review of wood-based materials with polymers and metals as a filler for electromagnetic interference (EMI) shielding. Ind. Crops Prod. **215**, 118619 (2024)
3. Maruthi, N., Faisal, M., Raghavendra, N.: Conducting polymer based composites as efficient EMI shielding materials: a comprehensive review and future prospects. Synth. Met. **272**, 116664 (2021)
4. Shelly, D., Jin, F.-L., Lee, S.-Y., Park, S.-J.: Emerging trends in MXene-polymer composites for electromagnetic shielding applications. Compos. Commun. **51**, 102056 (2024)
5. Gao, Q., Wang, X., Schubert, D.W., Liu, X.: Review on polymer/MXene composites for electromagnetic interference shielding applications. Adv. Nanocomposites **1**, 52–76 (2024)
6. Yang, Y., Li, J., Wang, Z., Ju, X., Duan, H., Tang, Y.: Structural design and performance regulation of green electromagnetic interference shielding conductive polymer composites: a review. Adv. Nanocomposites **1**, 290–303 (2024)
7. Liang, C., Qiu, H., Zhang, Y., Liu, Y., Gu, J.: External field-assisted techniques for polymer matrix composites with electromagnetic interference shielding. Sci. Bull. **68**, 1938–1953 (2023)
8. Kruželák, J., Kvasničáková, A., Hložeková, K., Hudec, I.: Progress in polymers and polymer composites used as efficient materials for EMI shielding. Nanoscale Adv. **3**, 123–172 (2021)

9. Zhang, Q., Wang, Q., Cui, J., Zhao, S., Zhang, G., Gao, A., Yan, Y.: Structural design and preparation of $Ti_3C_2T_x$ MXene/polymer composites for absorption-dominated electromagnetic interference shielding. Nanoscale Adv. **5**, 3549–3574 (2023)
10. Liu, J.-T., Zheng, Y.-C., Hou, X., Feng, X.-R., Jiang, K., Wang, M.: Structured carbon for electromagnetic shielding and microwave absorption from carbonization of waste Polymer: a review. Chem. Eng. J. **496**, 154013 (2024)
11. Jiang, X., Xu, W., Chen, Y., Li, J.: Design and microwave absorption performance study of SiC-Fe_3O_4 emulsified asphalt mixture. Materials **17**, 3935 (2024)
12. Thomassin, J.-M., Jérôme, C., Pardoen, T., Bailly, C., Huynen, I., Detrembleur, C.: Polymer/carbon based composites as electromagnetic interference (EMI) shielding materials. Mater. Sci. Eng. R. Rep. **74**, 211–232 (2013)
13. Chen, J., Liu, Y.-L., Sun, D.-X., Qi, X.-D., Yang, J.-H., Wang, Y.: Recent progress in structural design of graphene/polymer porous composites toward electromagnetic interference shielding application. Chem. Eng. J. **495**, 153586 (2024)
14. Elsheakh, D.N., Elgendy, Y.K., Elsayed, M.E., Eldamak, A.R.: Circularly polarized textile sensors for microwave-based smart bra monitoring system. Micromachines **14**, 586 (2023)
15. Protyai, M.I.H., Bin Rashid, A.: A comprehensive overview of recent progress in MXene-based polymer composites: their fabrication processes, advanced applications, and prospects. Heliyon **10**, e37030 (2024)
16. Tian, K., Hu, D., Wei, Q., Fu, Q., Deng, H.: Recent progress on multifunctional electromagnetic interference shielding polymer composites. J. Mater. Sci. Technol. **134**, 106–131 (2023)
17. Cao, Y., Zhu, L., Chen, S., Fang, J., Liang, L., Shen, Y.: Conducting polymers based composite materials towards anticorrosion applications. Synth. Met. **309**, 117769 (2024)
18. Dong, Y., Yu, H., Feng, Y., Feng, W.: Structure, properties and applications of multi-functional thermally conductive polymer composites. J. Mater. Sci. Technol. **200**, 141–161 (2024)
19. Tesei, D., Jewczynko, A., Lynch, A.M., Urbaniak, C.: Understanding the complexities and changes of the astronaut microbiome for successful long-duration space missions. Life **12**, 495 (2022)
20. Salmond, W., Walsh, J., Gorman, A.: Eternity in low earth orbit: icons on the international space station. Religions **11**, 611 (2020)
21. Ali, Z., Yaqoob, S., Yu, J., D'Amore, A., Fakhar-e-Alam, M.: A comparative review of processing methods for graphene-based hybrid filler polymer composites and enhanced mechanical, thermal, and electrical properties. J. King Saud Univ. Sci. **36**, 103457 (2024)
22. Orasugh, J.T., Temane, L.T., Ray, S.S.: Nanocellulose-based conductive composites: a review of systems for electromagnetic interference shielding applications. Int. J. Biol. Macromol. **277**, 133891 (2024)
23. Akazzim, Y., Jofre, M., El Mrabet, O., Romeu, J., Jofre-Roca, L.: UWB-modulated microwave imaging for human brain functional monitoring. Sensors **23**, 4374 (2023)
24. Zheng, S., Wang, Y., Zhu, Y., Zheng, C.: Recent advances in the construction and properties of carbon-based nanofillers/polymer composites with segregated network structures. Mater. Today Commun. **36**, 106773 (2023)
25. Li, Z., Sun, Y., Hu, F., Liu, D., Zhang, X., Ren, J., Guo, H., Shalash, M., He, M., Hou, H., El-Bahy, S.M., Pan, D., El-Bahy, Z.M., Guo, Z.: An overview of polymer-based thermally conductive functional materials. J. Mater. Sci. Technol. **218**, 191–210 (2025)
26. Ali, Z., Yaqoob, S., Yu, J., D'Amore, A.: Critical review on the characterization, preparation, and enhanced mechanical, thermal, and electrical properties of carbon nanotubes and their hybrid filler polymer composites for various applications. Compos. Part C: Open Access **13**, 100434 (2024)
27. Zheng, S., Wang, Y., Wang, X., Lu, H.: Research progress on high-performance electromagnetic interference shielding materials with well-organized multilayered structures. Mater. Today Phys. **40**, 101330 (2024)
28. Raagulan, K., Ghim, J.S., Braveenth, R., Jung, M.J., Lee, S.B., Chai, K.Y., Mi Kim, B., Lee, J.: EMI shielding of the hydrophobic, flexible, lightweight carbonless nano-plate composites. Nanomaterials **10**, 2086 (2020)

29. Tobon Vasquez, J.A., Scapaticci, R., Turvani, G., Bellizzi, G., Rodriguez-Duarte, D.O., Joachimowicz, N., Duchêne, B., Tedeschi, E., Casu, M.R., Crocco, L., Vipiana, F.: A prototype microwave system for 3D brain stroke imaging. Sensors **20**, 2607 (2020)
30. Savić, I.M., Savić Gajić, I.M., Gajić, D.G.: Optimization of the microwave-assisted extraction of caffeine from roasted coffee beans. Foods **13**, 2333 (2024)

Index

S. K. Koul et al., *Microwave Absorbers*, Engineering Materials,
https://doi.org/10.1007/978-981-96-9596-6

MIX
Papier aus verantwortungsvollen Quellen
Paper from responsible sources
FSC® C105338

If you have any concerns about our products,
you can contact us on
ProductSafety@springernature.com

In case Publisher is established outside the EU,
the EU authorized representative is:
Springer Nature Customer Service Center GmbH
Europaplatz 3, 69115 Heidelberg, Germany

Printed by Libri Plureos GmbH
in Hamburg, Germany